# 2014 IEEE SOI-3D-Subthreshold Microelectronics Technology Unified Conference

## (S3S 2014)

Millbrae, California, USA
6-9 October 2014

IEEE Catalog Number: CFP14SOI-POD
ISBN: 978-1-4799-7440-5

**Copyright © 2014 by the Institute of Electrical and Electronic Engineers, Inc**
**All Rights Reserved**

*Copyright and Reprint Permissions*: Abstracting is permitted with credit to the source. Libraries are permitted to photocopy beyond the limit of U.S. copyright law for private use of patrons those articles in this volume that carry a code at the bottom of the first page, provided the per-copy fee indicated in the code is paid through Copyright Clearance Center, 222 Rosewood Drive, Danvers, MA 01923.

For other copying, reprint or republication permission, write to IEEE Copyrights Manager, IEEE Service Center, 445 Hoes Lane, Piscataway, NJ  08854.  All rights reserved.

***\*\*\*This publication is a representation of what appears in the IEEE Digital Libraries.  Some format issues inherent in the e-media version may also appear in this print version.***

IEEE Catalog Number:          CFP14SOI-POD
ISBN 13:                                978-1-4799-7440-5

**Additional Copies of This Publication Are Available From:**

Curran Associates, Inc
57 Morehouse Lane
Red Hook, NY  12571 USA
Phone:          (845) 758-0400
Fax:              (845) 758-2633
E-mail:         curran@proceedings.com
Web:            www.proceedings.com

# SESSION 1:
## Plenary Session
**8:20am**
Sequoia

## SESSION CHAIRS
Olivier Faynot, *CEA-Leti* ● Steven Vitale, *MIT Lincoln Laboratory* ●
Zvi Or-Bach, *MonolithIC 3D*

| | | | |
|---|---|---|---|
| 8:20am | 1.1 | **The Role of the Cloud in Machine to Machine & Internet of Things Computing: Best Practices, Key Insights and Guidance to connecting devices to the cloud**<br>Bruno Terkaly; *Microsoft, San Francisco, CA (invited talk)* | 1 |
| 9:00am | 1.2 | **10 Years of ULV: A Look Back and Where Do We Go From Here?**<br>Alice Wang; *MediTek (invited talk)* | N/A |
| 9:40am | 1.3 | **Key Technology Trends and Their Impact on the Future of the Semiconductor Industry**<br>Mark Edelstone; *Morgan Stanley Investment Banking (invited talk)* | N/A |

# SESSION 2A:
## SOI Platforms
**10:40am**
Sequoia

## SESSION CHAIR
Joao Antonio Martino, *University of São Paulo*

| | | | |
|---|---|---|---|
| 10:40am | 2a.1 | **Advanced High K/Metal SOI technologies for 32nm and Beyond**<br>M. Horstmann; *Global Foundries, Dresden, Wilschdorfer, Landstraße (invited talk)* | 4 |
| 11:20am | 2a.2 | **28nm FDSOI: Platform and Products**<br>P. Flatresse; *STMicroelectronics (invited talk)* | N/A |

# SESSION 2B:
# SubVt – Analog and RF

**10:40am**
Redwood

## SESSION CHAIR
Jing Xie, *Qualcomm Research*

---

**10:40am**    2b.1    **Boltzmann Energetics of Neuromorphic Systems**    N/A
G. Cauwenberghs; *UC San Diego (invited talk)*

**11:20am**    2b.2    **A 361nA Thermal Run-away Immune VBB Generator Using Dynamic**    **7**
**Substrate Controlled Charge Pump for Ultra Low Sleep Current Logic on**
**65nm SOTB**
H. Nagatomi[1], N. Sugii[2], S. Kamohara[2], K. Ishibashi[1]; *[1]The University of Electro-Communications, [2]Low-power Electronics Association & Project*

**11:40am**    2b.3    **A 53µW -82dBm Sensitivity 920MHz OOK Receiver Design Using Bias**    **9**
**Switch Technique on 65nm SOTB CMOS Technolgy**
H.M. Thien[1], N. Sugii[2], K. Ishibashi[1]; *[1]The University of Electro-Communications, [2]Low-power Electronics Association & Project*

---

# SESSION 3A:
# FinFET

**1:00pm**
Sequoia

## SESSION CHAIRS
Maud Vinet, *CEA-Leti* • Ali Khakifirooz, *Spansion*

---

**1:00pm**    3a.1    **SOI FinFET versus Bulk FinFET for 10nm and Below**    **11**
T.B. Hook[1], F. Allibert[3], K. Balakrishnan[4], B. Doris[2], D. Guo[2], N. Mavilla[5], E. Nowak[1], G. Tsutsui[2], R. Southwick[2], J. Strane[2], X. Sun[2]; *[1]IBM SRDC Essex Junction, VT, [2]IBM Albany, NY, [3]SOITEC Albany, NY, [4]IBM Yorktown Heights, NY, [5]IBM Bangalore (invited)*

**1:40pm**    3a.2    **High Mobility Ω-Gate Nanowire P-FET on cSGOI Substrates Obtained by**    **14**
**Ge Enrichment Technique**
P. Nguyen[1, 3], S. Barraud[1], M. Koyama[1], M. Cassé[1], F. Andrieu[1], C. Tabone[1], F. Glowacki[1], J.-M. Hartmann[1], V. Maffini-Alvaro[1], D. Rouchon[1], N. Bernier[1], D. Lafond[1], M.-P. Samson[2], F. Allain[1], C. Vizioz[1], D. Delprat[3], B.-Y. Nguyen[3], C. Mazuré[3], O. Faynot[1], M. Vinet[1]; *[1]CEA-LETI, Minatec campus, Grenoble Cedex, France, [2]STMicroelectronics Crolles, France, [3]SOITEC, Bernin, France*

| 2:00pm | 3a.3 | **Dielectric Isolated FinFETs on Bulk Substrate   16** |

D. Lu[1], K. Cheng[1], P. Morin[2], N. Loubet[2], T. Hook[1], D. Guo[1], A. Khakifirooz[1], P. Oldiges[1], B. Doris[1], K. Rim[1], A. Jacob[3], H. Bu[1], M. Khare[1]; *[1]IBM Research, Albany, NY, [2]STMicroelectronics, [3]GLOBALFOUNDRIES*

| 2:20pm | 3a.4 | **Elastic Relaxation in Intrinsically-Strained Fins: Simulations, Physical   18 and Electrical Characterization** |

F. Allibert[1], P. Morin[2], H. He[3], J. Li[3], W. Schwarzenbach[1], N. Loubet[2], A. Khakifirooz[3,*], B. DeSalvo[4], B. Doris[3]; *[1]SOITEC, Bernin, France, [2]STMicroelectronics, [3]IBM Research, [4]CEA-Leti at Albany Nanotech, Albany, NY, *now with Spansion, Inc.*

# SESSION 3B:
# SubVt – Digital
## 1:00pm
Redwood

## SESSION CHAIR
Adrian Ionescu, *École Polytechnique Fédérale de Lausanne*

| 1:00pm | 3b.1 | **Toward Robust Subthreshold Circuit Design: Variability and Soft Error   21 Perspective** |

M. Hashimoto; *Dept. Information systems Engineering, Osaka University, Japan & JST, CREST (invited talk)*

| 1:20pm | 3b.2 | **Energy Efficiency Benefits of Subthreshold-Optimized Transistors for Digital   23 Logic** |

P.J. Grossmann, S.A. Vitale, P.W. Wyatt; *MIT Lincoln Laboratory, Lexington, MA*

| 2:00pm | 3b.3 | **Ultra-Wide Voltage Range 32-bit RISC CPU with Timing-Error   25 Prevention in 28nm CMOS** |

M. Hiienkari[1], J. Teittinen[1], L. Koskinen[1], M. Turnquist[2], M. Kaltiokallio[2], J. Mäkipää[3], A. Rantala[3], M. Sopanen[3]; *[1]University of Turku, Technology Research Center, Turku, Finland, [2]Aalto University Department of Micro And Nanosciences, [3]VTT Technical Research Centre of Finland*

| 2:20pm | 3b.4 | **A Reduced-Memory FIR Filter Using Approximate Coefficients for   27 Ultra-Low Power SoCs** |

A.Klinefelter, B.H. Calhoun; *University of Virginia, Charlottesville, VA*

# SESSION 4:
## Special Invited 3D Hot Topics
**1:00pm**
Sequoia

### SESSION CHAIR
Zvi Or-Bach, *MonolithIC 3D*

**3:00pm**   **4.1**   **Monolithic 3D Integration: a powerful alternative to classical 2D scaling**   **29**
M. Vinet[1], P. Batude[1], C. Fenouillet-Beranger[1], F. Clermidy[1], L. Brunet[1], O. Rozeau[1], J.M. Hartmann[1], O. Billoint[1], G. Cibrario[1], B. Previtali[1], C. Tabone[1], B. Sklenard[1], O. Turkyilmaz[1], F. Ponthenier[1], N. Rambal[1], M.P. Samson[2], F. Deprat[1], V. Lu[2], L. Pasini[2], S. Thuries[1], H. Sarhan[1], J.-E. Michallet[1], O. Faynot[1]; *[1]CEA-Leti, Minatec, [2]STMicroelectronics, Grenoble, France (invited talk)*

**3:30pm**   **4.2**   **Design Challenges and Solutions for Ultra-High-Density Monolithic 3D ICs**   **32**
S. Panth[1], S. Samal[1], Y.S. Yu[2], and S.K. Lim[1]; *[1]Dept. of Electrical and Computer Engineering, Georgia Institute of Technology, Atlanta, USA, [2]Hankyong National University, Korea (invited talk)*

**4:00pm**   **4.3**   **3D Memory with Shared Lithography Steps: The Memory Industry's Plan to**   **34**
**"Cram More Components onto Integrated Circuits"**
D. C. Sekar, *Rambus Labs (invited talk)*

**4:30pm**   **4.4**   **Monolithic 3D Integration in a CMOS Process Flow**   **37**
E.A. Fitzgerald[1,2], S.F. Yoon[1,3], C.S. Tan[1,3], T. Palacios[1,2], X. Zhou[1,3], L.S. Peh[1,2], C.C. Boon[1,3], D.A. Kohen[3], K.H. Lee[3], Z.H. Liu[3], P. Choi[1,3]; *[1]Low Energy Electronic Systems, SMART Singapore, [2]Massachusetts Institute of Technology Cambridge, MA, [3]Nanyang Technological University, Singapore (invited talk)*

**5:00pm**   **4.5**   **Monolithic 3D Integration Advances and Challenges: From Technology to**   **40**
**System Levels**
M.S. Ebrahimi, G. Hills, M.M. Sabry, M. M. Shulaker, H. Wei, T.F. Wu, S. Mitra, H.-S. Philip Wong; *Department of Electrical Engineering and Department of Computer Science, Stanford University (invited talk)*

# SESSION 5:
# Joint Poster Session and Reception
**5:30 to 7:30pm**
Poplar / Westin Ballroom Foyer

## SESSION CHAIR
Chang-Lee Chen, *MIT Lincoln Laboratory*

5.1 **Performance Assessment of ULP Analog/RF MOSFET Architectures   42**
D. Ghosh, A. Kranti; *Low Power Nanoelectronics Research Group, Electrical Engineering Discipline, Indian Institute of Technology, Indore, India*

5.2 **On the Cryogenic Performance of Ultra-Low-Loss, Wideband SPDT RF Switches Designed   44 in a 180 nm SOI-CMOS Technology**
A.S. Cardoso, P.S. Chakraborty, A.P. Omprakash, N. Karaulac, P. Saha, J.D. Cressler; *School of Electrical and Computer Engineering, Georgia Institute of Technology, Atlanta, GA*

5.3 **Analog Performance of Short-Channel Asymmetric Self-Cascode of Junctionless   46 Nanowire nMOS Transistors**
M. de Souza, R.T. Doria, R.D. Trevisoli, M.A. Pavanello; *Electrical Engineering Department, Centro Universitário da FEI, São Bernardo do Campo, Brazil*

5.4 **Multi-Threshold Design Methodology of Stacked Si-Nanowire FETs   49**
Y.-B. Liao[1], M.-H. Chiang[2]; [1]*Institute of Microelectronics, Department of Electrical Engineering, National Cheng Kung University, Tainan, Taiwan,* [2]*MS Degree Program on Nano-IC Engineering, Department of Electrical Engineering, National Cheng Kung University, Tainan, Taiwan*

5.5 **Analog Building Block Design in 14nm FinFET Using Inversion Coefficient   52**
A. Wang, V. Dhawan, C.-J. R. Shi; *Department of Electrical Engineering, University of Washington, Seattle WA, Supported by Semiconductor Research Corporation (SRC) under Task 1836.095 and the State of Washington Joint Center for Aerospace Technology Innovation (JCATI).*

5.6 **28 nm FD SOI Technology Platform RF FoM   55**
B.K. Esfeh[1], V. Kilchytska[1], V. Barral[2], N. Planes[3], M. Haond[3], D. Flandre[1], J.-P. Raskin[1]; [1]*ICTEAM, Université catholique de Louvain, Louvain-la-Neuve, Belgium,* [2]*CEA-Leti, MINATEC Campus, Grenoble, France,* [3]*ST-Microelectronics, Crolles, France*

5.7 **An SOI Based Integrated Gate-Drivers for Automotive Application   58**
K. Toshiyuki, Y. Hiya, K. Kinoshita, H. Tomita, N. Hoshikawa; *Electronics Development Div. 3 Toyota Motor Corporation, Toyota Aichi, Japan*

5.8 **Effect of Back Gate on Parasitic Bipolar Effect in FD SOI MOSFETs   60**
F. Liu[1], I. Ionica[1], M. Bawedin[2], S. Cristoloveanu[1]; [1]*IMEP-LAHC, MINATEC, Grenoble, France,* [2]*Université Montpellier II, Montpellier, France*

5.9 **High Temperature Performance of Flexible SOI FinFETs with Sub-20 nm Fins   62**
A. Diab, G.A. Torres Sevilla, M.T. Ghoneim, M.M. Hussain; *Integrated Nanotechnology Lab, King Abdullah University of Science and Technology, Saudi Arabia*

**5.10**    **Effects of Back-Gate Bias on Switched-Capacitor DC-DC Converters in UTBB FD-SOI**   **64**

M.J. Turnquist[1], G. de Streel[3], D. Bol[3], M. Hiienkari[2], L. Koskinen[2]; [1]*Dept. of Micro- and Nanosciences, Aalto University, [2]University of Turku, Technology Research Center, [3]ICTEAM Institute, Université catholique de Louvain*

**5.11**    **An Optimal Probing Method of Pre-Bond TSV Fault Identification for 3D Stacked ICs**   **66**

B. Zhang, V.D. Agrawal; *Department of Electrical and Computer Engineering, Auburn University, Auburn AL*

**5.12**    **Power Supply Voltage Detection and Clamping Circuit for 3-D Integrated Circuits**   **69**

D. Pathak, I. Savidis; *Department of Electrical and Computer Engineering, Drexel University, Philadelphia PA*

**5.13**    **Study of Fin-Tunnel FETs with Doped Pocket as Capacitor-less 1T DRAM**   **72**

A. Biswas, A.M. Ionescu; *STI-IEL-NANOLAB, Ecole Polytechnique Fédérale de Lausanne, Switzerland*

# SESSION 6A: FDSOI

**8:00am**
Sequoia

## SESSION CHAIRS

Manfred Horstmann, *Global Foundries* ● Bich-Yen Nguyen, *SOITEC*

8:00am     6a.1    **UTBB/FDSOI: reasons for a success**   **74**

M. Haond, *STMicroelectronics, Crolles, France (invited talk)*

8:40am     6a.2    **Piezoresistivity in Unstrained and Strained SOI MOSFETs**   **76**

R. Berthelon[1], M. Cassé[1], D. Rideau[2], O. Nier[1,2,3], F. Andrieu[1], E. Vincent[2] ,G. Reimbold[1]; [1]*CEA-Leti, MINATEC Campus, Grenoble, France, [2]STMicroelectronics, Crolles, France, [3]IMEP-LAHC, MINATEC, Grenoble, France*

9:00am     6a.3    **nFET FDSOI Activated by Low Temperature Solid Phase Epitaxial**   **78** **Regrowth: Optimization Guidelines**

L. Pasini[1,2,3], P. Batude[1], M. Cassé[1], L. Brunet[1], P. Rivallin[1], B. Mathieu[1], J. Lacord[1], S. Martinie[1], C.Fenouillet-Beranger[1], B. Previtali[1], N. Rambal[1], M. Haond[2], G. Ghibaudo[3], M. Vinet[1]; [1]*CEA-Leti, MINATEC Campus, Grenoble, France, [2]STMicroelectronics, Crolles, France, [3]IMEP-LAHC, MINATEC/INPG, Grenoble, France*

9:20am     6a.4    **In-depth Characterization of Hole Transport in 14nm FD-SOI pMOS**   **80** **Devices**

M. Shin[1,3], M. Shi[1], M. Mouis[1], A. Cros[2], E. Josse[2], G.T. Kim[3], G. Ghibaudo[1]; [1]*IMEP-LAHC, Grenoble INP, Minatec, Grenoble, France, [2]STMicroelectronics, Crolles, France, [3]School of Electrical Engineering, Korea University, Seoul, South Korea*

| 9:40am | 6a.5 | **Influence of Underlap on UTBB SOI MOSFETs in Dynamic Threshold Mode**    82 |
| --- | --- | --- |

K.R.A. Sasaki[1], M. Aoulaiche[2], E. Simoen[2], C. Claeys[2,3], J. A. Martino[1]; *[1]LSI/PSI/USP, University of Sao Paulo, Sao Paulo, Brazil, [2]imec, Leuven, Belgium, [3]E.E. Dept., KU Leuven, Leuven, Belgium*

# SESSION 6B:
# SubVt-Low Voltage Devices 1
## 8:00am
Redwood

## SESSION CHAIR
Dan Radak, *IDA*

| 8:00am | 6b.1 | **Spin-Transfer-Torque Devices for Boolean and Non-Boolean Computing**    N/A |
| --- | --- | --- |

K. Roy, *Purdue University (invited talk)*

| 8:40am | 6b.2 | **Near-0.1V Ultra-low Voltage Operation of SOTB 1M Logic Gates**    85 |
| --- | --- | --- |

Y. Ogasahara, M. Hioki, T. Nakagawa, T. Sekigawa, T. Tsutsumi, H. Koike; *Nanoelectronics Research Institute, AIST, Tsukuba, Japan*

| 9:00am | 6b.3 | **Performance Prediction for Multiple-Threshold 7nm-FinFET-Based Circuits Operating in Multiple Voltage Regimes Using a Cross-Layer Simulation Framework**    88 |
| --- | --- | --- |

S. Chen, Y. Wang, X. Lin, Q. Xie, M. Pedram; *University of Southern California, CA, USA*

| 9:20am | 6b.4 | **A Cross-Layer Design Framework and Comparative Analysis of SRAM Cells and Cache Memories using 7nm FinFET Devices**    90 |
| --- | --- | --- |

A. Shafaei, S. Chen, Y. Wang and M. Pedram; *Department of Electrical Engineering, University of Southern California, Los Angeles, CA, USA*

| 9:40am | 6b.5 | **Efficient Ultra Low Power Rectification at 13.56 MHz for a 10 µA Load Current**    92 |
| --- | --- | --- |

P.-A. Haddad, G. Gosset, J.-P. Raskin and D. Flandre; *Institute of Information and Communication Technologies, Electronics and Applied Mathematics (ICTEAM), Université catholique de Louvain, Louvain-la-Neuve, Belgium*

# SESSION 7A:
# SOI Circuit Design
**10:20am**
Sequoia

## SESSION CHAIRS

Toshiro Hiramoto, University of Tokyo ● Francisco Gamiz, University of Granada

10:20am     7a.1    **FDSOI Circuit Design for a Better Energy Efficiency**    N/A
                              E. Beigné, *CEA-Leti (invited talk)*

11:00am     7a.2    **Experimental Model of Adaptive Body Biasing for Energy Efficiency**    94
                              **in 28nm UTBB FD-SOI**
                              M. Cochet[1,2], B. Pelloux-Prayer[1], M. Saligane[1,2,3], S. Clerc[1], P. Roche[1], J.-L. Autran[2], D. Sylvester[3]; *[1]STMicroelectronics, Crolles, France, [2]IM2NP – Aix-Marseille University, France, [3]University of Michigan, Ann Arbor, MI*

11:20am     7a.3    **UTBB FD-SOI Front- and Back-Gate Coupling Aware Random**    96
                              **Telegraph Signal Impact Analysis on a 6T SRAM**
                              K.C Akyel[1,3], L.Ciampolini[1], O. Thomas[2], D.Turgis[1], G.Ghibaudo[3]; *[1]STMicroelectronics, Crolles, France, [2]CEA-Leti Campus Minatec, Grenoble, France, [3]IMEP-LAHC, Minatec, Grenoble, France*

11:40am     7a.4    **Mixed-Single Well 8T SRAM Bitcell for Wide Voltage Range in 28nm**    98
                              **FDSOI**
                              A. Makosiej[1], N. Planes[2], R. Ranica[2], L. Ciampolini[2] and O. Thomas[1]; *[1]Univ. Grenoble Alpes; CEA-Leti, MINATEC Campus, Grenoble, France, [2]STMicroelectronics, Crolles, France*

# SESSION 7B:
# SubVt – Low Voltage Devices 2
**10:20am**
Redwood

## SESSION CHAIR

Paul Franzon, *North Carolina State University*

10:20am     7b.1    **The Center for Energy Efficient Electronics Science: the Search for**    N/A
                              **Really Low Threshold Voltage**
                              E. Yablonovitch, *UC Berkeley (invited talk)*

| 11:00am | 7b.2 | **A Tunnel-FET SRAM Array for Energy-Efficient Embedded Memory Blocks in Reconfigurable Computing Platforms** | 100 |

M.F. Amir, A.R. Trivedi, S. Mukhopadhyay; *School of Electrical and Computer Engineering, Georgia Institute of Technology, Atlanta, GA, USA*

| 11:20am | 7b.3 | **Impacts of Work Function Variation and Line-Edge Roughness of TFET on FinFET Devices and Logic Circuits** | 102 |

C.-J. Chen, Y.-N. Chen, M.-L. Fan, V. P.-H. Hu, P. Su and C.-T. Chuang; *Department of Electronics Engineering & Institute of Electronics, National Chiao Tung University, Taiwan*

| 11:40am | 7b.4 | **OxRAM-based Pulsed Latch for Non-Volatile Flip-Flop in 28nm FDSOI** | 104 |

A. Levisse[1], N. Jovanović[1], E. Vianello[1], J.-M. Portal[2], O. Thomas[1]; *[1]Univ. Grenoble Alpes, Grenoble, France; CEA-Leti, MINATEC Campus, Grenoble, France, [2]IM2NP, Aix-Marseille Université Marseille, France*

# SESSION 8A:
# Novel SOI Structures

**8:00am**
Sequoia

## SESSION CHAIR
Philippe Flatresse, *STMicroelectronics*

| 8:00am | 8a.1 | **Electron-Hole Bilayer Deep Subthermal Electronic Switch: Physics, Promise and Challenges** | 106 |

A.M. Ionescu[1], C. Alper[1], J.L. Padilla[1], L. Lattanzio[1], P. Palestri[2];*[1]NANOLAB, Ecole Polytechnique Federale de Lausanne, Lausanne, Switzerland, [2]DIEGM, University of Udine, Udine, Italy (invited)*

| 8:40am | 8a.2 | **Beyond TFET: Alternative Mechanisms for SMOS -Compatible Sharp-Switching Devices** | 109 |

S. Cristoloveanu; *IMEP-LAHC (invited)*

| 9:20am | 8a.3 | **A2RAM: Low-power 1T-DRAM Memory Cells Compatible with Planar and 3D SOI Substrates** | 111 |

F.Gamiz[1], N.Rodriguez[1], C.Marquez[1], C.Navarro[2] and S.Cristoloveanu[2]; *[1]Nanoelectronics Laboratory, Department of Electronics, University of Granada, Granada Spain, [2]IMEP, Grenoble INP MINATEC, Grenoble, France (invited)*

# SESSION 8B:
# SubVt – Energy Harvesting for Low Power Circuits
## 8:00am
Redwood

## SESSION CHAIR
Dennis Buss, *Texas Instruments*

8:00am      8b.1      **Ultra Low Voltage Energy Harvesting: Challenges and Design**    **N/A**
**Methodologies**
S. Mukhopadhyay, *Georgia Tech (invited talk)*

8:40am      8b.2      **Bias-Flip Technique for Frequency Tuning of Piezo-Electric Energy**    **113**
**Harvesting Devices: Experimental Verification**
S. Zhao[1], Y. Ramadass[2], J.H. Lang[3], J. Ma[1], D. Buss[2,3]; *[1]Tianjin University, School of Electronic Information Engineering, Tianjin, P.R. China, [2]Texas Instruments, Inc., Dallas, TX, USA, [3]Massachusetts Institute of Technology, EECS Dept. Cambridge, MA, USA*

9:00am      8b.3      **Adaptive Subthreshold Switched Capacitor Voltage Boost for**    **116**
**Thermoelectric Generation**
R. Brito[1], M. Barba[1], P. Palakurthi[1], D. Nemir[2], E. MacDonald[1];
*[1]Electrical and Computer Engineering, University of Texas at El Paso, El Paso, Texas, USA, [2]TXL Group, Inc., El Paso, Texas, USA*

9:20am      8b.4      **Subthreshold RF Powered Digital Circuits**    **N/A**
P.D. Franzon, P. Gadfort, J. Schabel, W. Xu; *North Carolina State University (invited talk)*

# SESSION 9A:
# SOI Substrates
## 10:20am
Sequoia

## SESSION CHAIRS
Sorin Cristoloveanu, *IMEP* • Frederic Allibert, *SOITEC*

10:20am      9a.1      **SOI Substrate Solutions for Recent Advanced Device Applications**    **118**
N. Noto, O. Ishikawa, H. Aga, T. Ishizuka, I. Yokokawa and M. Nakano; *Technology and Development Division, Shin-Etsu Handotai Co. Ltd., Chiyoda, Tokyo, Japan (invited talk)*

| 11:00am | 9a.2 | **The Role of Radiation Effects in SOI Technology Development**   **120** |
| | | L. Palkuti[1], M. Alles[2], H. Hughes[3]; *Defense Threat Reduction Agency, Ft Belvior, VA,* *[2]Vanderbilt University, Nashville, TN, [3]Naval Research Laboratory, Washington, DC (invited talk)* |

| 11:20am | 9a.3 | **A Very Low Power CMOS 28FDSOI Programmable Fractional**   **122** **Frequency Divider for Wifi-WiGig** |
| | | M. Vallet[1], O. Richard[1], Y. Deval[2], D. Belot[1]; *[1]STMicroelectronics, Crolles, France,* *[2]IMS Bordeaux, Talence, France* |

| 11:40am | 9a.4 | **Recent Advances and Future Trends in SOI for RF Applications**   **125** |
| | | A. Joshi, T.-Y. Lee, Y.-Y. Chen and D. Whitefield; *Skyworks Solutions Inc.* *Irvine, CA (invited talk)* |

# SESSION 9B:
# SubVt – Radiation Effects on Low Voltage Circuits and Late News
**10:20am**
Redwood

## SESSION CHAIR
Lew Cohn, NRO

| 10:20am | 9b.1 | **SEU Hardening: Incorporating an Extreme Low Power Bitcell Design**   **128** **(SHIELD)** |
| | | A. Pescovsky[1], O. Chertkow[1], L. Atias[1], A. Fish[2]; *[1]VLSI Systems Center,* *Ben-Gurion University of the Negev, Be'er Sheva, [2]Faculty of Engineering,* *Bar-Ilan University, Ramat Gan* |

| 10:40am | 9b.2 | **Impact of Ultra-Low Voltages on Single-Event Transients and Pulse**   **131** **Quenching** |
| | | J.R. Ahlbin, P. Gadfort; *Information Sciences Institute, University of Southern California,* *Arlington, VA* |

| 11:00am | 9b.3 | **Compensation of Total Ionizing Dose Effects in ULV SoCs through Adaptive**   **134** **Voltage Scaling** |
| | | J. De Vos, V. Kilchytska, D. Flandre, D. Bol; *ICTEAM Institute, Université catholique de* *Louvain (UCL) Louvain-la-Neuve, Belgium* |

| 11:20am | 9b.4 | **Near-Threshold Voltage Operation of Nonvolatile SRAM Cell Based on**   **136** **Pseudo-Spin-FinFET Architecture** |
| | | Y. Shuto, S. Yamamoto, S. Sugahara; *Imaging Science and Engineering Laboratory, Tokyo* *Institute of Technology, Yokohama, Japan* |

11:40am    9b.5    **More than an Order of Magnitude Energy Improvement of FPGA by**    **138**
**Combining 0.4V Operation and Multi-Vt Optimization of 20k Body Bias Domains**
H. Koike[1], C. Ma[1,2], M. Hioki[1], Y. Ogasahara[1], T. Tsutsumi[2], T. Nakagawa[1], T. Sekigawa[1]; *[1]National Institute of AIST, [2]Meiji University, Japan*

# SESSION 10:
# Special Invited MEMS Hot Topics
**1:00pm**
Sequoia

## SESSION CHAIR
Jeremy Muldavin, *MIT Lincoln Laboratory*

1:00pm    10.1    **Inertial MEMS using SOI**    **N/A**
B. Taheri, *Freescale (invited talk)*

1:20pm    10.2    **Si Resonators and MEMS using SOI**    **N/A**
T. Kenney, *Stanford (invited talk)*

1:40pm    10.3    **SOI Wafer Manufacturing for MEMS**    **N/A**
A. Cornell, *Okmetic (invited talk)*

2:00pm    10.4    **SOI Based Inertial Sensors**    **N/A**
A. Shkel, *UCI (invited talk)*

# SESSION 11:
# 3D New Developments
**3:00pm**
Sequoia

## SESSION CHAIR
Olivier Faynot, *CEA-Leti*

3:00pm    11.1    **Monolithic IC Integration Key Alignment Aspects for High Process Yield**    **140**
T. Uhrmann, T. Wagenleitner, T. Glisner, M. Wimplinger, P. Linder; *EV Group, St. Florian am Inn, Austria*

| 3:20pm | 11.2 | **New Precision Alignment Methodology for CMOS Wafer Bonding** 142 |

**New Precision Alignment Methodology for CMOS Wafer Bonding** 142

I. Sugaya[1], H. Mitsuishi[1], H. Maeda[1], M. Okada[1], K. Okamoto[1,2]; *[1]Nikon Corporation, Yokohama, Japan, [2]Osaka University, Osaka, Japan*

3:40pm  11.3  **Precision Bonders - A Game Changer for Monolithic 3D** 145

Z. Or-Bach, B. Cronquist, Z. Wurman, I. Beinglass, A. Henning; *MonolithIC 3D, Inc., San Jose, CA*

4:00pm  11.4  **Fully Functional Fine-grain Vertically Integrated 3D Focal Plane** 148 **Neuromorphic Processor**

M. Di Federico[1,2], P. Julián[1,2], A. G. Andreou[3], P.S. Madolesi[1,4]; *[1]Departamento de Ingeniería Eléctrica y de Computadoras, Universidad Nacional del Sur, Bahía Blanca, Argentina, [2]CONICET, Argentina, [3]Electrical and Computer Engineering Department, Johns Hopkins University, USA, [4]CIC, Pcia. Bs. As., Argentina*

# SESSION 12:
# Late News Session
**4:20pm**
Sequoia

## SESSION CHAIR
Bruce Doris – *IBM*

4:20pm  12.1  **Smart Co-Integration of Light Sensitive Layers with FDSOI Transistors for** 150 **More than Moore Applications**

L. Grenouillet, B. De Salvo, L. Brunet, J. Coignus, C. Tabone, J. Mazurier, C. Le Royer, P. Grosse, M.A. Jaud, P. Rivallin, Z. Chalupa, O. Rozeau, O. Faynot, M. Vinet; *CEA-Leti, Minatec Campus, Grenoble, France*

4:40pm  12.2  **Prototype of Multi-Stacked Memory Wafers Using Low-Temperature** 152 **Oxide Bonding and Ultra-Fine-Dimension Copper Through-Silicon Via Interconnects**

W. Lin[1], J. Faltermeier[1], K. Winstel[1], S. Skordas[1], T. Graves-Abe[2], P. Batra[2], K. Herman[2], J. Golz[2], T. Kirihata[2], J. Garant[2], A. Hubbard[1], K. Cauffman[1], T. Levine[1], J. Kelly[1], D. Priyadarshini[1], B. Peethala[1], R. Patlolla[1], M. Shoudy[1], J.J. Demarest[1], J. Wynne[1], D. Canaperi[1], D. McHerron[1], D. Berger[2], S. Iyer[2]; IBM Corporation Systems and Technology Group, *[1]Albany, NY, [2]Hopewell Junction, NY*

5:00pm  12.3  **A 262nW Analog Front End with a Digitally-Assisted Low Noise Amplifier** 155 **for Batteryless EEG Acquisition**

P. Bhargava[1], W. D. Hairston[2], R. M. Proie[1]; *[1]U.S. Army Research Laboratory Sensors and Electron Devices Directorate, [2]Human Research and Engineering Directorate*

# 2014 SOI-3D-Subthreshold Microelectronics Technology Unified Conference

40th ANNUAL CONFERENCE and 20th ANNUAL SHORT COURSE PROGRAM
Sponsored by the Institute of Electrical and Electronics Engineers, Inc. and the Electron Devices Society

Westin San Francisco Airport
Millbrae, California
October 6th thru 9th 2014

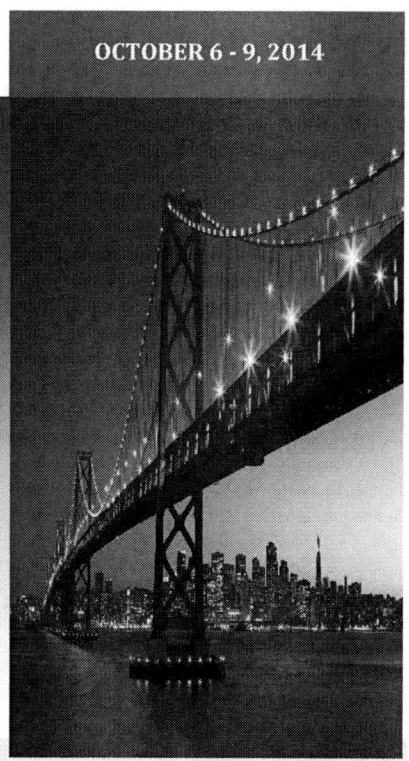

**OCTOBER 6 - 9, 2014**

# 2014 SOI-3D-Subthreshold Microelectronics Technology Unified Conference (S3S)

Westin San Francisco Airport
Millbrae, CA

## Contents

General Chairs Message

Committee List

Introduction and History

## Technical Sessions

Session 1: Plenary
Session 2a: SOI Platforms
Session 2b: SubVt - Analog and RF
Session 3a: FinFET
Session 3b: SubVt - Digital
Session 4: Special Invited 3D Hot Topics
Session 5: Joint Poster Session
Session 6a: FDSOI
Session 6b: SubVt - Low Voltage Devices 1
Session 7a: SOI Circiut Design

Session 7b: SubVt - Low Voltage Devices 2
Session 8a: Novel SOI Structures
Session 8b: SubVt - Energy Harvesting for Low Power Circuits
Session 9a: SOI Substrates
Session 9b: SubVt - Radiation Effects on Low Voltage Circuits and Late News
Session 10: Special Invited MEMS Hot Topics
Session 11: 3D New Developments
Session 12: Late News

## Author Index

# Welcome to the 2014 IEEE SOI-3DI–Subthreshold Microelectronics Technology Unified Conference (S3S)

Last year we entered into a new era as the IEEE S3S Conference. The transition from the IEEE International SOI Conference to the IEEE S3S conference was successful by any measurement. The first year of the new conference leading-edge experts from 3D Integration, Sub-threshold Microelectronics and SOI fields gathered and we established a world class international venue to present, learn and debate about these exciting topics. The overall participation at the first year of the new conference grew by over 50%, and the overall quality and quantity of the technical content grew even more. This year we are looking forward to continuing to enhance the content of the 2014 S3S Conference. On Monday, Oct. 6 we will feature two Short Courses that will run in parallel. Short courses are an educational venue where new comers can gain overview and generalists can learn more details about new and timely topics. The short course on Monolithic 3D will be a full day deep dive into the topic of three-dimensional integration without the use of stacking patterned wafers by using through vias. Already there are extremely successful examples of monolithic 3D Flash Memory. Looking beyond this initial application, we will explore the application of monolithic 3D to alternate memories like RRAM, CMOS systems with silicon and other channel materials like III V. In addition, a significant portion of the short course will be dedicated to the exciting opportunity of Monolithic 3D in the context of CMOS Logic. The other short course we will offer this year is entitled Power Efficient Chip Technology. This short course will address several key aspects of power-efficiency including low power transistors and circuits. The course will also review in detail the impact of design and architecture on the energy-efficiency of systems. The short course chairs as well as the instructors are world class leading experts from the most prestigious industry and academic institutions.

The regular conference sessions will start on Tuesday Oct 7 with the plenary session which will feature presentations from Wall Street (Morgan Stanley Investment Banking), Microsoft and Media Tek. After the plenary session we will hear invited talks and this year's selection of outstanding papers from international researchers from top companies and universities. The most up to date results will be shared. Audience questions and one on one interaction with presenters is encouraged. Back by popular demand we will have 2 Hot Topics Sessions this year. The first Hot Topic Session is scheduled for Tuesday Oct. 7th and will feature exciting 3DI topics. The other Hot Topics session is scheduled for Thursday Oct 9 and will showcase new and exciting work in the area of MEMS. Our unique poster session and reception format will have a short presentation by the authors followed by one on one interaction to review details of the poster with the audience, in a friendly atmosphere, around a drink. Last year we had regular posters as well as several invited posters with very high quality content and we anticipate this year's poster session to be even better than last years. We are offering a choice of two different fundamentals classes on Wednesday afternoon. One of the Fundamentals classes will focus on Robust Design of Subthreshold Digital and Mixed Circuits, with tutorials be the world's leading experts in this field. The SOI fundamentals course is focused on RF SOI Technology Fundamentals and Applications.

Keeping in line with tradition, on Wednesday night we will have a hearty cook out with delicious food and drink followed by the Panel Session entitled *Cost and Benefit of Scaling Beyond 14nm*. Panel speakers from financial, semiconductor equipment, technology, and academic research institutions will gather along with the audience to debate this timely topic. Although Thursday is the last day of the conference we will have stimulating presentations on novel devices, energy harvesting, radiation effects along with the MEMS Hot Topic Session and Late News Session. As always we will finish the conference with the award ceremony for the best papers.

Our conference has a long tradition of attracting presenters and audience members from the most prestigious research, technology and academic institutions from around the world. There are many social events at the S3S Conference as well as quiet time where ideas are discussed and challenged off line and people from various fields can learn more about other fields of interest from leading experts. The conference also offers many opportunities for networking with people inside and also outside ones area. The venue this year is San Francisco. We chose this location to attract the region's leading experts from Academia and Industry. If you have free time we encourage you to explore San Francisco which is famous for a multitude of cultural and culinary opportunities.

I would like to express my personal thanks for the hard work and dedication of the executive, senior and technical committees, as well as the conference manager Joyce Lloyd, for putting together the outstanding program. I would also like to thank the invited speakers for presenting their exciting research, and to the authors who are the heart and soul of our conference for sharing the exciting new results that they generated for the purpose of furthering our understanding of SOI 3DI and Sub Vt.

*Bruce Doris, General Chair*

# 2014 CONFERENCE COMMITTEE

## Executive Committee

**GENERAL CHAIR**
Bruce Doris
IBM Research
bdoris@us.ibm.com

**TECHNICAL PROGRAM CHAIR**
Olivier Faynot
CEA-LETI
olivier.faynot@cea.fr

**LOCAL ARRANGEMENTS CHAIR**
Bich-Yen Nguyen
SOITEC
bich-yen.nguyen@soitec.com

**SOI TECHNICAL CHAIR**
Olivier Faynot
CEA-LETI
olivier.faynot@cea.fr

**SUB VT TECHNICAL CHAIR**
Steven Vitale
MIT
steven.vitale@ll.mit.edu

**3D TECHNICAL CHAIR**
Zvi Or-Bach
MonolithIC 3D
zvi@monolithic3d.com

**TREASURER & REGISTRATION CHAIR**
Frederic Allibert
Soitec
Frederic.Allibert@soitec.com

**PUBLICITY & DEVELOPMENT CHAIR**
Meishoku Masahara
National Institute of AIST
m.masahara@aist.go.jp

**PUBLICITY & DEVELOPMENT CHAIR**
Mike Fritze
USC-ISI
mfritze@isi.edu

## Senior Commitee

**POWER EFFICIENCY SHORT COURSE CHAIR**
Frederic Allibert
Soitec
Frederic.Allibert@soitec.com

**MONOLITHIC 3D SHORT COURSE CHAIR**
Paul Franzon
North Carolina State
paulf@ncsu.edu

**MONOLITHIC 3D SHORT COURSE CHAIR**
Zvi Or-Bach
MonolithIC 3D
zvi@monolithic3d.com

**SUB VT FUNDAMENTALS CLASS CHAIR**
Mostafa Emam
Incize
mostafa.emam@incize.com

**SOI FUNDAMENTALS CLASS CHAIR**
Nobuyuki Sugii
LEAP
n-sugii@leap.or.jp

**RUMP & POSTER CHAIR**
Chang-Lee Chen
MIT Lincoln Laboratory
clchen@ll.mit.edu

**SOI MULTIMEDIA CHAIR**
Joao Antonio Martino
University of Sao Paulo
martino@lsi.usp.br

**SUBVT/3D MULTIMEDIA CHAIR**
Alex Fish
Bar-Ilan University
alexander.fish@gmail.com

## Advisory Committee

**Yang Du**
Qualcomm
ydu@qti.qualcomm.com

**Toshiro Hiramoto**
University of Tokyo
hiramoto@nano.iis.u-tokyo.ac.jp

**Malgorzata Jurczak**
imec
jurczakm@imec.be

**Carlos Mazure**
Soitec
carlos.mazure@soitec.com

**Les Palkuti**
DTRA/NTS
Les.palkuti@dtra.mil

**Mario Pelella**
On Semiconductor
mario.pelella@onsemi.com

**Jean Luc Pelloie**
ARM
jean-luc.pelloie@arm.com

# 3D Technical Committee

**Mukta Farooq**
IBM
farooqm@us.ibm.com

**Paul Franzon**
North Carolina State University
paulf@ncsu.edu

**Subu Iyer**
IBM
ssiyer@us.ibm.com

**Mitsumasa Koyanagi**
Tohoku University
koyanagi@bmi.niche.tohoku.ac.jp

**Patrick Leduc**
CEA-Leti
patrick.leduc@cea.fr

**Zvi Or-Bach**
MonolithIC 3D
zvi@monolithic3d.com

# SOI Technical Committee

**Mike Alles**
Vanderbilt University
mike.alles@vanderbilt.edu

**Frederic Allibert**
Soitec
Frederic.Allibert@soitec.com

**Uygar Avci**
Intel
uygar.e.avci@intel.com

**Ibrahim Ban**
Intel
ibrahim.ban@intel.com

**Gary Bronner**
Rambus
gbronner@rambus.com

**Chang-Lee Chen**
MIT Lincoln Laboratory
clchen@ll.mit.edu

**Yuh-Yue Chen**
Skyworks
yuhyue.chen@skyworksinc.com

**Kangguo Cheng**
IBM
cheng@us.ibm.com

**Bruce Doris**
IBM
bdoris@us.ibm.com

**Olivier Faynot**
CEA-Leti
olivier.faynot@cea.fr

**Samuel Fung**
Taiwan Semiconductor
Manufacturing Company
khfung@tsmc.com

**Michel Haond**
STMicroelectronics
michel.haond@st.com

**Toshiro Hiramoto**
University of Tokyo
hiramoto@nano.iis.u-
tokyo.ac.jp

**Ru Huang**
Peking University
ruhuang@pku.edu.cn

**Keiji Ikeda**
AIST-GNC / Toshiba
keiji.ikeda@aist.go.jp

**Ali Khakifirooz**
Spansion
khaki@us.ibm.com

**Jong-Ho Lee**
Seoul National University
jhl@snu.ac.kr

**Gosia Malgorzata**
imec
jurczakm@imec.be

**Joao Antonio Martino**
University of Sao Paulo
martino@lsi.usp.br

**Meishoku Masahara**
National Institute of AIST
m.masahara@aist.go.jp

**Carlos Mazure**
Soitec
carlos.mazure@soitec.com

**Bich-Yen Nguyen**
Soitec
bich-yen.nguyen@soitecusa.com

**Les Palkuti**
DTRA/NTS
Les.palkuti@dtra.mil

**Mario Pelella**
On Semiconductor
mario.pelella@onsemi.com

**Jean-Luc Pelloie**
ARM
jean-luc.pelloie@arm.com

**Changhwan Shin**
University of Seoul
shinch02@gmail.com

**Nobuyuki Sugii**
LEAP
n-sugii@leap.or.jp

**Maud Vinet**
CEA-Leti
vinet@us.ibm.com

**Steven Vitale**
MIT Lincoln Laboratory
steven.vitale@ll.mit.edu

**Geng Wang**
IBM
wanggeng@us.ibm.com

# SubVt Technical Committee

**John Ahlbin**
ISI
jahlbin@isi.edu

**Amara Amara**
ISEP
amara.amara@isep.fr

**David Bol**
Catholic University of
Louvain
david.bol@uclouvain.be

**Dennis Buss**
Texas Instruments
buss@ti.com

**Ben Calhoun**
UVA
bcalhoun@virginia.edu

**Lew Cohn**
NRO
lewis.cohn@nrl.navy.mil

**Yang Du**
Qualcomm
ydu@qti.qualcomm.com

**Mostafa Emam**
Incize
mostafa.emam@incize.com

**Alex Fish**
Bar Ilan University
alexander.fish@biu.ac.il

**Paul Franzon**
North Carolina State
paulf@ncsu.edu

**Mike Fritze**
ISI
mfritze@isi.edu

**Pascale Gouker**
MIT Lincoln Laboratory
pgouker@ll.mit.edu

**Sumeet Gupta**
Penn State
skgupta@psu.edu

**David Hansquine**
Qualcomm
dhansqui@qti.qualcomm.com

**Adrian Ionescu**
Ecole Polytechnique
Federale de Lausanne
adrian.ionescu@epfl.ch

**Les Palkuti**
DTRA/NTS
Les.palkuti@dtra.mil

**Dan Radack**
IDA
dradack@ida.org

**Mingoo Seok**
Columbia University
ms4415@columbia.edu

**Makoto Takamiya**
University of Tokyo
mtaka@iis.u-tokyo.ac.jp

**Steven Vitale**
MIT Lincoln Laboratory
steven.vitale@ll.mit.edu

**David Wentzloff**
University of Michigan
wentzlof@umich.edu

**Joyce Lloyd**
IMF Conference
Management
Joyce@imf.la

# INTRODUCTION and HISTORY

The IEEE S3S Conference is a technical forum for the presentation and discussion of research results relating to semiconductor-on-insulator technology. The technical papers deal with a range of topics including materials preparation and characterization, device physics, modeling, circuit fabrication and applications. The Conference is sponsored by the Electron Devices Society of IEEE. In 2013 the IEEE International SOI Conference and the IEEE Subthreshold Microelectronics Conference co-located to create the IEEE S3S Conference. Below is a brief history of the Conference.

**1975 IEEE SOS Technology Workshop**
September 8-10, 1975
Waystation Hotel, South Lake Tahoe, CA
Conference Chairman: Frank Micheletti, Rockwell
Technical Program: Dave Dumin, Allied Chemical
Local Arrangements: Tom Toumbs
Treasurer: Robert Nielsen
42 papers submitted, 150 attendees

**1976 IEEE SOS Technology Workshop**
September 11-12, 1976
Stanford University Palo Alto, CA
Conference Chairman: Dave Dumin, Allied Chemical
Technical Program: Robert Nielson
Local Arrangements: R. Mueller
Treasurer: C.S. Kim
38 papers submitted, 136 attendees

**1977 IEEE SOS Technology Workshop**
September 28-30, 1977
Lodge at Vail, Vail, Colorado
Conference Chairman: Robert Stewart, RCA
Technical Program: D. Howard Phillips, Rockwell
Local Arrangements: Roger Stewart
Treasurer: R.M. Bergman
Banquet Speaker: Larry Lopp, Hewlett Packard
47 papers submitted, 150 attendees

**1978 IEEE SOS Technology Workshop**
October 4-6, 1978
Lodge at Vail, Vail, CO
Conference Chairman: D. Howard Phillips, Rockwell
Technical Program: Alfred Ipri, RCA
Local Arrangements: Roger Steward
Treasurer: Robert Nielsen
Banquet Speaker: Bobby Buchanan, Air Force Cambridge
27 paper submitted, 120 attendees

**1979 IEEE SOS Technology Workshop**
October 4-6, 1979
Carefree Inn, Carefree, Arizona
Conference Chairman: Alfred Ipri, RCA
Technical Program: R.J. Hollingsworth
Local Arrangements: C. Grenato
Treasurer: R.M. Bergman
33 papers submitted, 150 attendees

**1980 IEEE SOS Technology Workshop**
September 30-October 2, 1980
Stone Bridge Inn, Snowmass, CO
Conference Chairman: Pete Hudson
Technical Program: Steve Ramer
Treasurer: Sidney Marshall
27 papers submitted, 120 attendees

**1981 IEEE SOS Technology Workshop**
October 6-8, 1981
Sun River Resort, Sunriver, Oregon
Conference Chairman: Ranjeet Pancholy, Hughes
Technical Program: Mike Splinter
Local Arrangements: Quent Cassen
Treasurer: J.A. Mouten
Banquet Speaker: Peter Hudson, ERADCOM
37 papers submitted, 110 attendees

**1982 IEEE SOS Technology Workshop***
October 5-7, 1982
Provincetown Inn, Provincetown, MA
Conference Chairman: Achilles G. Kokkas, RCA
Technical Program: Quent Cassen
Local Arrangements: Jim Ford
Treasurer: M. Czarnecki
Banquet Speaker: Frank Micheletti, Rockwell
38 papers submitted, 150 attendees
*First Workshop with SOI papers accepted

**1983 IEEE SOS Technology Workshop**
October 4-6, 1983
Snow King Resort, Jackson Hole, Wyoming
Conference Chairman: Quent Cassen, Aerospace
Technical Program: Jal K. Hakhur, Rockwell
Local Arrangements: Eliezer D. Richmont, NRL
Treasurer: Marian G. Peebles, Rockwell
Banquet Speaker: D. Howard Phillips, Rockwell
44 papers submitted, 140 attendees

**1984 IEEE SOS Technology Workshop**
October 2-4, 1984
Hilton Head Island & Beach Resort, Hilton Head Island, SC
Conference Chairman: Jal K. Hakhu, Rockwell
Technical Program; Jim Ford, CDC
Local Arrangements: George Celler, AT&T Bells Labs
Treasurer: Jo Ann Landers, Rockwell
Banquet Speaker: Bob Conklin, REC Electronics
39 papers submitted, 140 attendees

## 1985 IEEE SOS/SOI Technology Conference
October 1-3, 1985
Prospector Square Resort, Park City UT
Conference Chairman: Jim Ford, CDC
Technical Program: George Celler, AT&T
Local Arrangements: Anil Gupta, Hughes
Treasurer: Jack Mee, Rockwell
Banquet Speaker: Jim Mendel, Stanford University
42 papers submitted, 140 attendees

## 1986 IEEE SOS/SOI Technology Conference
October 3-5, 1986
South Seas Plantation, Captiva Island, FL
Conference Chairman: George Celler, AT&T
Technical Program: Anil Gupta, Hughes
Local Arrangements: Marie Burham, Motorola
Treasurer: Brooke Jones, Rockwell
Banquet Speaker: William Gregory, Sandia National Labs
70 papers submitted-160 attendees

## 1987 IEEE SOS/SOI Technology Conference**
October 6-8, 1987
Tamarron Resort, Durango, CO
Conference Chair: Anil Gupta, Hughes
Technical Program: Marie Burham, Motorola
Local Arrangements: Jack Mee, Rockwell
Treasurer: Gracie Davis, Northrup Electronics
Banquet Speaker: Al Tasch, University of Texas at Austin
89 papers submitted 190 attendees
** The first time abstracts were published

## 1988 IEEE SOS/SOI Technology Conference
October 3-5, 1988
Sea Palms Resort, St. Simons Island, GA
Conference Chair: Marie Burnham, Motorola
Technical Program: Jean-Pierre Colinge, Hewlett-Packard
Local Arrangements: Gracie Davis, DNA
Treasurer: Michael Duffy, D. Sarnoff Research
Rump and Poster: Dimitris Ioannou, University of Maryland
Banquet Speaker: David Webb, University of North Dakota
101 papers submitted, 210 attendees

## 1989 IEEE SOI Technology Conference
October 3-5, 1989
High Sierra Hotel, State, S. Lake Tahoe, NV
Conference Chairman: Jean-Pierre Colinge, Hewlett Packard
Technical Program: Gracie Davis, DNA
Local Arrangements: Michael Duffy, D. Sarnoff Research
Treasurer: Jerry Brandewie, Rockwell
Rump and Poster: Dimitris Ioannou, University of Maryland
Banquet Speaker: John Moll, Hewlett-Packard
108 papers submitted, 250 attendees

## 1990 IEEE/SOI Technology Conference
October 2-4, 1990
Marriott's Casa Marina Resort, Key West, FL
Conference Chairman: Gracie Davis, DNA
Technical Program: Michael Duffy, D. Sarnoff Research
Local Arrangements: Jerry Brandewie, SEMATECH/Rockwell
Treasurer: John Schott, USAF RADC/ESR
Rump and Poster: Wade Krull, Harris Semiconductor
Banquet Speaker: Yuji Okuto, NEC Corp. Japan
86 papers submitted, 225 attendees
Best Paper Award: *"Dual-MOSFET structure for suppression of kink in SOI MOSFETs at room and liquid helium temperatures"*, M. H. Gao, J.P. Colinge, L. Aauwers, S.H. Wu, and Claeys

## 1991 IEEE International SOI Conference
October 1-3, 1991
Hyatt Regency Beaver Creek Resort, Vail CO
General Chair: Michael Duffy, D. Sarnoff Research
Technical Program: Jerry Brandewie, SEMATECH/Rockwell
Local Arrangements: John Schott, USA Rome Lab
Treasurer: Witek Maszara, Allied-Signal
Rump and Poster: Peter Roitman, NIST
Banquet Speaker: Jeff Miller, JPL
97 papers submitted, 180 attendees
Best Paper Award: *"Correlation of the leakage and charge pumping in SIMOX gated diodes"*, H. Seghir, S. Cristoloveanu, R. Jerisian, J. Qualid, A. Auberton-Herve

## 1992 IEE International SOI Conference
October 6-8, 1992
Marriott at Sawgrass Resort, Ponte Vedra Beach, FL
General Chair: Jerry Brandewie, SEMATCH/Rockwell
Technical Program: John Schott, USAF Rome Lab.
Local Arrangements: Witek Maszara, Allied-Signal
Treasurer: Jay Schrankler, Honeywell
Rump and Poster: Don Mayer, Aerospace Corp.
Banquet Speaker: Will Stackhouse, JPL
94 papers submitted, 188 attendees
Best Paper Award: *"Performance limitations of current deep submicron fully depleted SOI MOSFETs"*, J.G. Fossum, S. Krishnan, and P. C. Yeh

## 1993 IEEE International SOI Conference
October 1-3. 1993
The Autry Resort, Palm Springs, CA
General Chair: John Schott, , USAF Rome Lab
Technical Program: Witek Maszara, Allied-Signal
Local Arrangements: Jay Schrankler, Honeywell
Treasurer: Don Mayer, The Aerospace Corp.
Rump and Poster: Scott Tyson, Mission Research Corp.
Banquet Speaker: R. Ervin Taylor, University of CA
75 papers submitted, 177 attendees
Best Paper Award: *"A Unified Model of Threshold Voltage, Subthreshold Slope and Interface Coupling in Thin Film SOI MOSFETs"*, A.M. Ionescu, S. Cristoloveanu, A. Chovet, A. Hassein-Bey, A. Rusu

**1994 IEEE SOI International Conference**
October 6-8, 1994
White Elephant Resort, Nantucket Island, MA
General Chair: Witek Maszara, Allied Signal
Technical Program: Jay Schrankler, Honeywell
Local Arrangements: Don Mayer, The Aerospace Corp.
Treasurer: Jerry G. Fossum, University of Florida
Rump and Poster: Lisa Allen, Ibis Technology Corp.
Banquet Speaker: George Hazarigg, Science Foundation
120 Papers submitted, 203 attendees
Best Paper Award: *"The Applications of Silicon-on-Insulator (SOI) Technology for the Fabrications of Fully Scanned Active Matrix Flat Panel Displays"*, A.C. Ipri, G. Dolny, F.L. Hsueh, R.G. Stewart, D. Jose, M. Spitzer, D.P. Vu, M. Batty, R. Khormaei, S. Thayer, T. Keyser, G. Becker, M. Tilton, R. Rhoades

**1995 IEEE International SOI Conference\*\*\***
October 2-5, 1995
Westward Look Resort, Tucson, AZ
General Chair: Jay Schrankler, Honeywell
Technical Program: Don Mayer, The Aerospace Corp.
Local Arrangements: Jerry G. Fossum, University of FL.
Treasurer: Lisa Allen, Ibis Technology Corp.
Rump and Poster: Ted W. Houston, Texas Instruments
Short Course: James E. Chung, MIT
Banquet Speaker: Dr. Young-Soo Kim, Samsung
147 papers submitted, 242 attendees
Best Paper Award: *"Comparison of Plasma-Induced Charging Damage in Bulk and SOI MOSFETs"* M.J. Sherony, A.H. Chen, K.R. Mistry, D.A. Antoniadis, and B.S. Doyle
\*\*\* The first Short Course presented

**1996 IEEE SOI International Conference**
September 30-October 3, 1996
Sanibel Harbour Resort & Spa, Sanibel Island, FL
Conference Chair: Don Mayer, The Aerospace Corp.
Technical Program: Jerry G. Fossum, University of Florida
Local Arrangements: Lisa Allen, Ibis Technology Corp.
Treasurer: Akira Yoshino, NEC, Japan
Rump and Poster: Stephen Krause, Arizona State University
Short Course: Jean Pierre Colinge
Banquet Speaker: Dr. Leon Alkalai, JPL
Best Paper Award: *"Mechanistic Studies of Hydrophilic Wafer Bonding and Si Exfoliation for SOI Fabrication"*, M.K. Weldon, V. Marsico, Y.J. Chabal, S.B. Christman, E.E. Chaban, D.C. Jacobson, J.B.Sapjeta, A. Pinczuk, B.S. Dennis, A.P. Mills, C.A. Goodwin, D. –M. Hsieh

**1997 IEEE SOI International Conference**
October 6-9, 1997
Tenaya Lodge at Yosemite, Fish Camp, California
Conference Chair: Jerry Fossum, University of FL
Technical Program: Lisa P. Allen, Ibis Technology Corp.
Treasurer: Ted Houston, Texas Instruments
Rump and Poster: Subramanian S. Iyer, IBM Corp.
Short Course: Jason Woo, UCLA
Banquet Speaker: Dr. Leon Alkalai, JPL
123 papers submitted, 240 attendees
Best Paper Award: *"Mechanism of Silicon Exfoliation by Hydrogen Implantation and He, Li and Si Co-implantation"*, M.K. Weldon, V. Marsico, Y.J. Chabal, M. Collot, Y. Caudano, S.B. Christman, E.E. Chaban, D.C. Jacobson W.L. Brown, J.B. Sapjeta, C.A. Goodwin, D. M. Hsieh, Bell Laboratories, Lucent Technologies, A. Agarwal, V.C. Venezia, and T.E. Haynes, Oak Ridge National Laboratory, and W.B. Jackson, Xerox Palo Alto Research Center, USA

**1998 IEEE International SOI Conference**
October 5-8, 1998
Indian River Plantation, Stuart, FL
General Chair: Lisa P. Allen, Ibis Technology Corp.
Technical Program: Ted Houston, Texas Instruments
Local Arrangements: Jason Woo, UCLA
Treasurer/Registration: Harold Hovel, IBM Corporation
Rump and Poster: John McKitterick, Allied-Signal
Short Course: Dimitris Ioannou, Geo. Mason Univ.
Banquet Speaker: James D. Meindl, Georga Tech.
173 papers submitted, 240 attendees
Best Paper Award: *"A0.5-V Data-Storage Circuit for Triple-Threshold MTCMOS/SIMOX LSIs"* by T. Douseki and M. Harada

**1999 IEEE International SOI Conference**
October 4-7, 1999
Doubletree Hotel Sonoma County, Rohnert Park, California
General Chair: Ted Houston, Texas Instruments
Technical Program: Jason Woo, UCLA
Local Arrangements: Harold Hovel, IBM Corporation
Treasurer/Registration: Dimitris Ioannou, Geo. Mason Univ.
Rump and Poster: Brian Doyle, Intel Corp.
Short Course: Mike Liu, Honeywell
Publicity: John Conley, Dynamics Research Corp.
Banquet Speaker: Jean Pierre Colinge, UC Davis
98 papers submitted, 235 attendees
Best Paper Award: *"Ultimately Thin SOI MOSFETs: Special Characteristics and Mechanisms"* T. Ernst, D. Munteanu, S. Cristoloveanu, T. Ouisse, N. Hefyene, S. Horiguchi, Y. Ono, Y. Takahashi & , K. Murase

**2000 IEEE International SOI Conference**
October 2-5, 2000
Sheraton Colonial Hotel & Golf Club, Wakefield, MA
General Chair: Jason Woo, UCLA
Technical Program: Harry Hovel, IBM Corporation
Local Arrangements: Dimitris Ioannou, Geo. Mason Univ.
Treasurer/Registration: Mike Liu, Honeywell
Rump and Poster: Mike Mendicino, Motorola
Short Course: Brian Doyle, Intel Corp.
Publicity: John Conley, JPL
103 papers submitted, 258 attendees
Best Paper Award: *"A High-Performance Body-Charge-Modulated SOI Sense Amplifier"*, J.B. Kuang, D.H. Allen, C.T. Chuang

**2001 IEEE International SOI Conference**
October 1-4, 2001
Sheraton Tamarron Resort, Durango, CO
General Chair: Harry Hovel, IBM Corporation
Technical Program: Dimitris Ioannou, Geo. Mason Univ.
Local Arrangements: Mike Liu, Honeywell
Treasurer/Registration: Mike Mendicino, Motorola
Rump and Poster: John Conley, JPL
Short Course: James Burns, MIT/Lincoln Lab
Publicity: Christophe Tretz, Advanced Micro Devices
111 papers submitted, 138 attendees
Best Paper Award: *"Characterization of Fully Depleted SOI Transistors after Removal of the Silicon Substrate"*, J. Burns, K. Warner, P. Gouker

## 2002 IEEE International SOI Conference
October 7 - 10, 2002
Williamsburg Lodge, Williamsburg, VA
General Chair: Dimitris Ioannou, George Mason Univ.
Technical Program: Mike Liu, Honeywell
Local Arrangements: Mike Mendicino, Motorola
Treasurer/Registration: James Burns, MIT/Lincoln Lab
Rump and Poster: Olivier Faynot, CEA-LETI
Short Course & Publicity: Christophe Tretz, AMD
123 papers submitted, 197 attendees
Best Paper Award:*"Partial SOI/SON formation by He+
Implantation & Annealing"* by A. Ogura

## 2003 IEEE International SOI Conference
September 29 - October 2, 2003
Newport Beach Marriott, Newport Beach, CA
General Chair: Mike Liu, Honeywell
Technical Program: Mike Mendicino, Motorola
Local Arrangements: James Burns, MIT/Lincoln Lab
Treasurer/Registration: Christophe Tretz, IBM
Rump and Poster: Pierre Fazan, EPFL
Short Course: Toshiro Hiramoto, University of Tokyo
Publicity: Rajiv Joshi, IBM, Research Division
118 papers submitted, 181 attendees
Best Paper Award: *"A New Block Refresh Concept for SOI
Floating Body Memories"* P. Fazan, S. Okhonin, M. Nagoga

## 2004 IEEE International SOI Conference
October 4 - 7, 2004
Francis Marion Hotel, Charleston, SC
General Chair: Mike Mendicino, Freescale Semiconductor
Technical Program: James Burns, MIT/Lincoln Lab
Local Arrangements: Christophe Tretz, IB M
Treasurer/Registration: Toshiro Hiramoto, Univ. of Tokyo
Rump and Poster: Mario Pelella, AMD
Short Course: Pierre Fazan, EPFL
Publicity: Christophe Tretz, IBM
127 papers submitted, 172 attendees
Best Paper Award: *"Experimental Gate Misalignment Analysis
on Double Gate SOI MOSFETs"* J. Widiez, F. Daugé, M. Vinet,
T. Poiroux, B. Previtali, M. Mouis, S. Deleonibus

## 2005 IEEE International SOI Conference
October 3 - 6, 2005
Hyatt Regency Waikiki, Honolulu, HI
General Chair: James A. Burns, MIT/Lincoln Lab
Technical Program: Christophe Tretz, IBM
Local Arrangements: Toshiro Hiramoto, University of Tokyo
Treasurer/Registration: Pierre Fazan, EPFL
Rump and Poster: Hector Sanchez, Freescale Semiconductor
Short Course: Mario Pelella, AMD
145 papers submitted, 164 attendees
Best Paper Award: *"A Novel Self-Aligned Substrate Diode and
Resistor Structure for SOI Technologies"* by M.M. Pelella, G.
Burbach, A. Salman, S. Beebe, D. Chan, J. Buller

## 2006 IEEE International SOI Conference
October 2 - 5, 2006
Holiday Inn Select, Niagara Falls, NY
General Chair: Christophe Tretz, IBM
Technical Program: Toshiro Hiramoto, University of Tokyo
Local Arrangements: Pierre Fazan, EPFL
Treasurer/Registration: Mario Pelella, AMD
Rump and Poster: Carlos Mazure, SOITEC
Short Course: Hector Sanchez, Freescale Semiconductor
100 papers submitted, 139 attendees
Best Paper Award: *"State of the art 200 GHz passive
components and circuits integrated in advanced thin SOI
CMOS technology on High Resistivity substrate"* by F.
Gianesello, D. Gloria, C. Raynaud, S. Montusclat, S. Boret, C.
Clément, Ph. Benech, J.M. Fournier, G. Dambrine

## 2007 IEEE International SOI Conference
October 1 - 4, 2007
Miramonte Resort & Spa, Indian Wells, CA
General Chair: Toshiro Hiramoto, University of Tokyo
Technical Program: Pierre Fazan, Innovative Silicon, Inc.
Local Arrangements: Mario Pelella, AMD
Treasurer/Registration: Carlos Mazure, SOITEC
Publicity & Development: Christophe Tretz, IBM
Rump and Poster: Wade Xiong, Texas Instruments
Short Course: Les Palkuti, DTRA/TDNR
Fundamentals Class: Mario Pelella, AMD
Multimedia: Christophe Tretz, IBM
85 papers submitted, 136 attendees
Best Paper Award: *"High performance, highly reliable FD/SOI
I/O MOSFETs in contemporary high-performance PD/SOI
CMOS"* by V.P. Trivedi, B. Winstead, P. Choi, L. Kang, T Lou,
M. Khazhinsky, A. Haggag, S. Parsons, H. Sanchez, M.
Moosa, V. Kolagunta, and J. Cheek
Best Poster Award: *"Geometric Magnetoresistance and
Mobility Behavior in Single-Gate and Double-Gate SOI
Devices"* by N. Rodriguez, L. Donetti, F. Gamiz, and S.
Cristoloveanu

## 2008 IEEE International SOI Conference
October 4 – 7, 2008
Mohonk Mountain House, New Paltz, NY
General Chair: Pierre Fazan, Innovative Silicon, Inc.
Technical Program: Mario Pelella, tauMetrix
Local Arrangements: Carlos Mazure, SOITEC
Treasurer/Registration: Wade Xiong, Texas Instruments
Publicity & Development: Toshiro Hiramoto, University of
Tokyo
Rump and Poster: Jean-luc Pelloie, ARM
Short Course: Vyshnavi Suntharalingam, MIT Lincoln Lab
Fundamentals Class: Malgorata Jurczak, IMEC
Multimedia: Harry Liu, Seagate
82 papers submitted, 132 attendees
Best Paper Award: *"Capping-Metal Gate Integration
Technology for multiple-VT CMOS in MuGFETs"* by A. Veloso,
L. Witters, M. Demand, I. Ferain, N. J. Son, B. Kaczer, Ph. J.
Roussel, C. Adelmann, S. Brus, O. Richard, H. Bender, T.
Conard, R. Vos, R. Rooyackers, S. Van Elshocht, N. Collaert,
K. De Meyer, S. Biesemans, M. Jurczak
Best Poster Award: *"Sub-45nm Fully-Depleted SOI CMOS
Subthreshold Logic for Ultra-Low-Power Applications"* by D.
Bol, R. Ambroise, D. Flandre and J.-D. Legat; Université
catholique de Louvain

**2009 IEEE International SOI Conference**
October 5 – 8, 2009
Crowne Plaza Hotel, Foster City, CA
General Chair: Mario Pelella, tauMetrix
Technical Program: Carlos Mazure, SOITEC
Local Arrangements: Wade Xiong, TI
Treasurer/Registration: Malgorata Jurczak, IMEC
Publicity & Development: Pierre Fazan, Innovative Silicon, Inc.
Rump and Poster: Bruce Doris, IBM
Short Course: Jean-Luc Pelloie, ARM
Fundamentals Class: Sameul Fung, TSMC
Multimedia: Vishal Trivedi, Freescale
81 papers submitted, 114 attendees
Best Paper and Best Student Paper Award: *"SRAM Yield Enhancement with Thin-Box FD-SOI"* by C. Shin, M.H. Cho, Y Tsukamoto, B.Y. Nguyen, B. Nikolic, and T. J. King Liu

**2010 IEEE International SOI Conference**
October 11 – 14, 2010
Catamaran Resort and SPA, San Diego, CA
General Chair: Carlos Mazure, SOITEC
Technical Program: Wade Xiong, Sematech
Local Arrangements: Malgorata Jurczak, IMEC
Treasurer/Registration: Jean-Luc Pelloie, ARM
Publicity & Development:. Mario M. Pelella, tau-Metrix, inc.
Rump and Poster: Makoto Fujiwara, Toshiba
Short Course: Bruce Doris, IBM
Fundamentals Class: Vishal Trivedi, Freescale
Multimedia: Samuel Fung, TSMC
82 papers submitted, 133 attendees
Best Paper: *"Field Effect Resistor, A Single-Device-At-Pad Solution for ESD Protection In Deeply Scaled SOI Technology"* by S. Cao, A.A. Salman, J.H. Chun, S.G. Beebe, M.M. Pelella, R.W. Dutton

**2011 IEEE International SOI Conference**
October 3 – 6, 2011
Tempe Mission Palms Hotel and Conference Center, Tempe, AZ
General Chair: Wade Xiong, AMD GmbH
Technical Program: Malgorzata Jurczak, imec
Local Arrangements: Jean-Luc Pelloie, ARM
Treasurer/Registration: Bruce Doris, IBM
Publicity & Development: Carlos Mazure, SOITEC.
Rump and Poster: Vishal Trivedi, Freescale
Short Course: Bich Yen Nguyen, SOITEC
Fundamentals Class: Makoto Fujiwara, Toshiba
Multimedia: Olivier Faynot, CEA-LETI
76 papers submitted, 108 attendees
Best Paper: *"Implant Approaches and Challenges for 20nm Node and Beyond ETSOI Devices"* by S. Ponoth, M. Vinet, L. Grenouillet, A. Kumar, P. Kulkarni, Q. Liu, K. Cheng, B. Haran, N. Possémé, A. Khakifirooz, N. Loubet, S. Mehta, J. Kuss, V. Destefanis, N. Berliner, R. Sreenivasan, Y. Le Tiec, S. Kanakasabapathy, S. Schmitz, T. Levin, S. Luning, T. Hook, M. Khare, G. Shahidi, B. Doris

**2011 Subthreshold Microelectronics Conference**
September 26-27, 2011
MIT Lincoln Laboratory, Lexington, MA
General Chair: Steven Vitale, MIT Lincoln Laboratory
Technical Chair: Pascale Gouker, MIT Lincoln Laboratory
Local Arrangements: Anne Cappucci, MIT Lincoln Laboratory
Treasurer: Jeff Knecht, MIT Lincoln Laboratory
Banquet Speaker: Dr. Susan Heilman, Museum of Science
52 papers submitted, 94 attendees

**2012 IEEE International SOI Conference**
October 1-4, 2012
Meritage Resort and Spa, Napa, CA
General Chair: Malgorzata Jurczak, imec
Technical Program: Jean-Luc Pelloie, ARM
Local Arrangements: Bruce Doris, IBM
Treasurer/Registration: Olivier Faynot, CEA-Leti
Publicity & Development: Carlos Mazure, Soitec
Rump and Poster: Bich-Yen Nguyen, Soitec
Short Course: Denis Flandre, UCL &
Olivier Faynot, CEA-Leti
Fundamentals Class: Uygar Avci, Intel
Multimedia: Chang-Lee Chen, MIT Lincoln Lab
66 papers submitted, 104 attendees
Best Paper: *"6T SRAM Design for Wide Voltage Range in 28nm FDSOI"* by O. Thomas, B. Zimmer, B. Pelloux-Prayer, N. Planes, K-C. Akyel, L. Ciampolini, P. Flatresse and B. Nikolic

**2012 IEEE Subthreshold Microelectronics Conference**
October 9-10, 2012
Westin Waltham Boston Hotel, Waltham, MA
General Chair: Steven Vitale, MIT Lincoln Laboratory
Co-Chair: Pascale Gouker, MIT Lincoln Laboratory
Technical Program Chair: Yang Du, Qualcomm
Poster Session Chair: Dennis Buss, Texas Instruments
Publicity & Publications: David Bol, Catholic University of Louvain
Finance & Local Arrangements: Bob Alongi, IEEE Boston
35 papers submitted, 107 attendees

**2013 IEEE IEEE SOI-3D-Subthreshold Microelectronics Technology Unified Conference**
October 7-10, 2013
Hyatt Regency Hotel and Spa, Monterey, California
General Chair: Jean-Luc Pelloie, ARM
Technical Program: Bruce Doris, IBM Research
Local Arrangements: Olivier Faynot, CEA-Leti
Treasurer/Registration and Publicity & Development Chair: Bich-Yen Nguyen, SOITEC
SOI Technical Chair: Bruce Doris, IBM Research
SubVt Technical Chair: Steven Vitale, MIT
3D Technical Chair: Subramanian Iyer, IBM
Rump and Poster: Michel Haond, STMicroelectronics
SOI Short Course Chair: Uygar Avci, Intel
3D Short Course Chair: Mukta Farooq, IBM
SOI Fundamentals Class Chair: Chang-Lee Chen, MIT Lincoln Laboratory
SubVt Fundamentals Chair: Alex Fish, Bar-Ilan University
SOI Multimedia Chair: Meishoku Masahara, AIST
SubVt Fundamentals Chair: David Bol, Université catholique de Louvain
78 papers submitted, 138 attendees
Best Paper: *"SRAM Row Decoder Design for Wide Voltage Range in 28nm UTBB-FDSOI"* by O. Thomas, B. Zimmer, B. Pelloux-Prayer, N. Planes, K.-C. Akyel, L. Ciampolini, P. Flatresse and B. Nikolić

# The Role of the Cloud in Machine To Machine & Internet of Things Computing

## Best Practices, Key Insights and Guidance to connecting devices to the cloud

Bruno Terkaly
Principal Software Engineer, Mobile and Cloud
Microsoft
San Francisco, CA
bterkaly@microsoft.com

*Abstract*— **There is an unmistakable proliferation of small devices that are performing a wide range of tasks, such as using sensors to measure and record the environment around them, power robots and actuators, check for intruders, and beyond.**

*Keywords—raspberry pi, Arduino, azure, cloud, mobile, microsoft (key words)*

### I. INTRODUCTION

By 2020 nearly every human on earth will be connected to about six devices. It has never been more economical or easy for the average person to prototype and produce a consumer or commercial device manufacture and sale. For less than $100, devices like the Arduino or Raspberry Pi provides you a microcontroller fitted with a small CPU, memory, and storage. Moreover, there is a plethora of expansion boards that plug into these devices, such as motor controls, GPS, Ethernet, LCD displays, sensors, actuators and more. These tiny devices come equipped with either a proprietary operating system or GNU/Linux with up to 512MB or RAM.

But these devices rarely have value unless they are connected to a cloud service on the backend. The cloud provides many useful services, including data analysis, issuing commands to control the device, security, provisioning, and management.

This talk will focus on the key considerations that companies, entrepreneurs, and individuals will face when creating and supporting these devices at scale.

| Discussion Point | Description | Questions Answered |
|---|---|---|
| Device Spectrum | At the low end of the spectrum is the Arduino. The higher-end devices include the raspberry pi. A myriad of expansion boards can be attached to the small microcontrollers. | What other capabilities with respect to CPU, memory, and storage?<br><br>What expansion boards are available and what purpose do they serve? |
| Device Software | A variety of operating systems are supported on these devices. Programming tools and compilers make it possible to create custom software that run on these devices. | What are the capabilities of these operating systems in terms of storage, memory, processing speed, and networking?<br><br>What type of programming tools are available?<br><br>What programming languages are supported in the device itself? |

978-1-4799-7440-5/14 $31.00 © 2014 IEEE

| Cloud Backend | The cloud provides many useful services for these devices in a connected world. | What role does the cloud play with respect to security?<br><br>Which networking connectivity solutions that should be considered for device-to-cloud communication?<br><br>How does the cloud help alleviate bandwidth challenges and minimize power consumption on the device?<br><br>What are the four patterns for device to cloud service message communication? |
|---|---|---|

## II. CHOOSING A DEVICE

### A. Selecting CPU, memory, and storage

At the low end of the spectrum is the Arduino with only 4K of memory. The raspberry pie represents the higher end of the spectrum, running a full-blown version of Linux.

### B. Expanding device capabilities

The ability to expand your device capabilities is astounding. You can connect everything from motor controls to infrared remote control, GPS, wireless networking, home automation to sensor and beyond.

Even if you limit yourself to look in all sensors, there is remarkable variety: Acoustic, sound, vibration, Automotive, transportation, Chemical, Electric current, electric potential, magnetic, radio, Environment, weather, moisture, humidity, Flow, fluid velocity, Ionizing radiation, subatomic particles, Navigation instruments, Position, angle, displacement, distance, speed, acceleration, Optical, light, imaging, photon, Pressure, Force, density, level, Thermal, heat, temperature, Proximity, presence

## III. WRITING SOFTWARE FOR YOUR DEVICE

Most of these devices come pre-equipped with some type of operating system, and support for programming tools. Many of these devices are even capable of running a Web server or supporting an external display that can provide a user interface for interaction.

### A. Programming tools, compilers and languages

Software developers work with code editors and compilers to create executables that run on the device. This presentation will provide an overview about what's available and what makes sense given specific use case scenarios. Practically all popular languages are supported on these devices. It is even possible to write code with C# on top of Mono running on Linux, for example.

### B. Client connectivity to the cloud

How a client device connects it to the cloud is important. Choosing an efficient communication protocol, including both networking and messaging is critical, as it affects performance, power consumption, and security.

## IV. MICROSOFT AUZRE – AN IDEAL CLOUD BACK-END FOR IoT DEVICES

Special consideration must be given to the cloud backend. Microsoft Azure is ideally suited because of its support for the Service Bus, which provides a hosted, secure, and widely available infrastructure for widespread communication, large-scale event distribution, naming, and service publishing.

### A. Addressability and network connectivity

Virtual private networks fail miserably in providing a simple, efficient, and secure way to connect a client device to the cloud. Many of these devices are not under direct physical control, so making them part of a virtual network dramatically compromises security.

Typical HTTP request response architectures drains battery life on devices. Moreover, the payload associated with the request/response payload is unnecessarily huge and can be ridiculously wasteful.

Finally, most, if not all of these devices sit behind a network address translation layer, posing a big problem for addressability and connectivity. The Microsoft Azure Service Bus greatly simplifies connectivity acting as a relay service, and serving as a proxy for the cloud backend.

### B. Scalabilty, Storage, Big Data, Web Services

One of the most important aspects of the cloud is its ability to scale on demand. The amount of devices put into production is generally unknown in the beginning and can change dramatically throughout the lifecycle of the devices in the field. Having the capability to scale as needed, whether scaling up or scaling down, is critical.

The amount of data that devices can generate staggering. Having a storage infrastructure that provides the necessary security, scale, and support for various data formats is mandatory. Microsoft Azure supports traditional blobs, key-value, column family, graph, relational, and document stores.

978-1-4799-7440-5/14 $31.00 © 2014 IEEE

The ability to analyze this data is also critically important. Support for Hadoop, reporting and data analytics is an essential part of the cloud backend. I signing probability to predictions, as well as finding patterns is an important part of the infrastructure needed to support these data producing devices.

The cloud backend also needs to support a variety of operating systems, languages and open-source software. Microsoft Azure supports everything from Python to C#, Node.js, C/C++, as well as a variety of functional languages typically used in machine learning scenarios.

## V. CONCLUSION

This paper reflects the content that will be presented during my talk. It provides a comprehensive deep dive into the challenges and solutions for individuals and companies looking to enter the brave new world of Machine to Machine computing or the Internet of Things. The discussion will address the needs of the executive decision-makers, architects, software developers, and hardware enthusiasts. Real-life examples will be brought into the discussion as well as concrete recommendations and examples.

### REFERENCES

[1] Bruno Terkaly, Ricardo Villalobos, "Azure Insider: Microsoft Azure Service Bus and the Internet of Things", MSDN Magazine, February 2014, http://msdn.microsoft.com/en-us/magazine/200112c8-d2db-4299-9307-e589572994af

[2] Bruno Terkaly, Ricardo Villalobos, "Azure Insider: Microsoft Azure Service Bus and the Internet of Things, Part 2", MSDN Magazine, March 2014, http://msdn.microsoft.com/en-us/magazine/9da09295-1cce-4f0a-a497-3a1a992cc92b

[3] Bruno Terkaly, Ricardo Villalobos, Thomas Conte,"Azure Insider: Telemetry Ingestion and Analysis Using Microsoft Azure Services", June 2014 , http://msdn.microsoft.com/en-us/magazine/b739b829-da9f-42eb-b396-3c00d2fb5360

# Advanced High K/Metal SOI technologies for 32nm and beyond

Manfred Horstmann

*GLOBALFOUNDRIES Dresden, Wilschdorfer Landstraße 101, D-01109 Dresden*
*Email: manfred.horstmann@globalfoundries.com*

## INTRODUCTION

Within the semiconductor industry, pure leading edge foundries serve a special mission by delivering state-of-the-art competitive logic performance with a strong focus on system-on-chip (SoC) solutions. Therefore they have to support a broad portfolio of different technology options on each node. GLOBALFOUNDRIES is an industry leader by representing this particular business model and has a long experience in leading-edge semiconductor manufacturing and technology capabilities, particular on silicon on insulator (SOI) based high performance microprocessors (table 1). To achieve a "high performance per watt" figure of merit, technology elements like partial depleted (PD) -SOI, strained-Si, ultra low K BEOL and HKMG were needed together with an efficient multiple core- and power-efficient design.

Those technology elements were developed and optimized for multiple generations beginning from an 180nm down to the 28nm technology node which runs currently in high volume production.

In particular the 28nm node should remain for a long time at the sweet spot in the Foundry Industry for yield, performance and cost reasons.

This node will be the foundation to add technology features like Flash, HV, MEMS etc. but also will enable a quick productization of new innovations like fully depleted extreme thin (ET) planer SOI devices with back bias options (BB), reducing power consumption even further.

**TABLE 1.** GLOBALFOUNDRIES' path to 28nm. Key innovations implemented in the past.

| Technology Node | Year of introduction | Innovation |
|---|---|---|
| 180nm Bulk | 2000 | Cu interconnect |
| 130nm Bulk | 2002 | Low-k dielectric |
| 130nm PD SOI | 2003 | SOI-substrates |
| 90nm PD SOI | 2004 | Strained Silicon |
| 65nm PD SOI | 2006 | Multi strain FETs |
| 45nm PD SOI | 2008 | Immersion Litho |
| 32nm PD SOI | 2010 | Gate first HKMG & Ultra low k |
| 28nm Bulk | 2011 | Poly SiON, TFMH |
| 28nm Bulk/ ET SOI (opt.) | 2012/13 | Gate first HKMG, TFMH |

## CURRENT 32NM PD SOI/28NM BULK IN HIGH VOLUME PRODUCTION

GLOBALFOUNDRIES is the first foundry that ramped up a 32nm HKMG technology based on SOI into a high volume production. To overcome the scaling limitations of conventional Poly/SiON based gate stacks like increased tunneling and leakage currents a High-k/Metal Gate (HKMG) technology was needed based on HfO2 materials. The 32nm solution is using a HKMG-Gate First approach that shares the conventional process flow, design flexibility, design elements and benefits of all previous nodes based upon Poly/SiON gate stack technology. This technology is compatible to commonly used strained-Si techniques. The 32nm-PD-SOI-HKMG has been running in very high volume production with at least >750K wafer shipped up to date, meeting D0 as well as all performance targets. Figure 1 shows the 32nm PD SOI D0 reduction over time in comparison with 28nm Bulk from Q2 2011 until Q2 2014.

978-1-4799-7440-5/14 $31.00 © 2014 IEEE

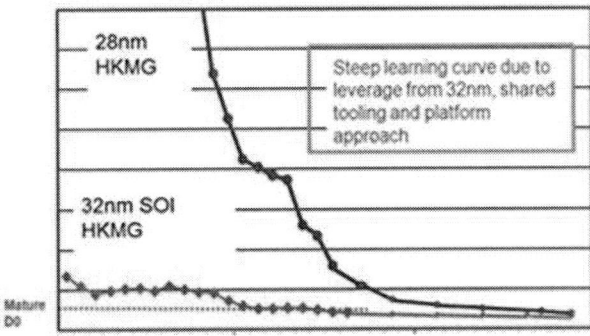

**FIG. 1** 28nm vs 32nm HKMG PD SOI D0 reduction over time, from Q2 2011 to Q2 2014

## 28NM TECHNOLOGY

GLOBALFOUNDRIES has developed several 28nm technologies in Poly/SiON and HKMG technology. The technologies are developed as a platform and share MOL and BEOL modules as much as possible. The 28nm FEOL is tailored to the customer's needs and uses gate-first HKMG architecture leveraging 32nm HKMG PD SOI or a conventional Poly/SiON gate stack. This concept allowed a steep learning curve for defect reduction (Fig. 1)

The 28nm HKMG technology portfolio contains super low-power (SLP) and high performance (HPP) technologies. These technologies are applied to wide variety of applications like high performance micro- or graphic-processors and wired networking chips as well as mobile computing and digital consumer products in the low power/mobile area. Within the 28nm base technologies multiple supply and threshold voltages as well as gate dielectric thicknesses are supported. In addition, several passive devices and various SRAM cells are provided.

## THE ROAD AHEAD IN 28NM CMOS

As already mentioned the 28nm node is currently at the sweet spot in Foundry Industry for yield, performance and cost. It requires no double patterning and conventional planar design styles can be used.

This node will be the basis to add technology features like Flash, HV, MEMS etc. but also will enable new innovations like fully depleted devices, reducing power consumption even further. For the later the ITRS roadmap [1] shows a continuing path of stressor and HKMG scaling. Evolving

transistor trends like fully depleted SOI and Multi-Gate MOSFET are described. Non-CMOS device architectures such as tunnel FETs (TFETs) are currently in research and could perhaps serve a CMOS replacement in future. Fig. 2 shows the evolution of fully depleted SOI and FinFETs. Both approaches emerge from the state of the art concepts based on super-steep-retrograde well profiles (SSRW). The first approach leads from hybrid tri-gate devices to FinFETs on bulk and ultimately to FinFETs based on SOI [2,3]. Excellent gate control was demo'd for Lgate down to 25nm. IDS of about 1000µA/µm for pfet and 1200µA/µm for nfet are reported at IOFF 100nA/µm, at Vdd=1.0V.

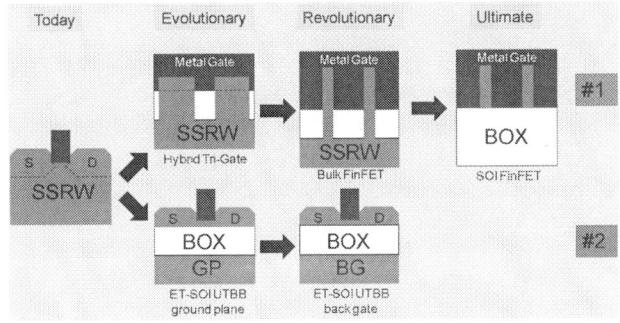

**FIG. 2** Possible device evolution paths [2].

The other approach starts with fully depleted devices on extremely thin silicon layers (ET-SOI) moving towards structures with a scaled box thickness and back bias gates (ET-SOI-UTBB). Here, innovations are driven to prevent drain field penetration thru the box oxide into the device channel. For such devices a similar device performance was demonstrated at a gate length of 25nm [4, 5].

In addition SOI ET-UTBB FETs show similar electrostatics compare to FinFETs for a contacted back gate. Another advantage is the $V_{TH}$ tunability by back biasing applications. $V_{TH}$ tunability and low GIDL currents are favorable advantages of ET-SOI-UTBB devices especially for SoC applications reducing the power consumption even further. Those planar device architectures allow the continuation of previous technology manufacturing and design experience. Random doping fluctuations (RDF) do not play a role for any of those advanced devices.

This solution allows the leveraging of the 28nm platform foundation using the same BEOL and

978-1-4799-7440-5/14 $31.00 © 2014 IEEE

MOL modules. The FEOL can be adjusted and tailored to fit the needs of SOI and back biasing. This enables a quick ramp up meeting 28nm D0 in a short time.

## CONCLUSION

A foundry's mission is to deliver competitive device performance and flexibility to support a variety of SoC offerings. In particular the 28nm node should remain for a long time at the sweet spot in Foundry Industry for yield, performance and cost. This node will be the basis to add technology features like Flash, HV, MEMS etc. but also will enable new innovations like fully depleted devices.

## REFERENCES

1. 2011-edition of the ITRS roadmap,
2. A. Wei et al, "Advanced Foundry CMOS" SSDM Conf. 2011, Nagoya Japan.
3. T. Yamashita et al., "Sub-25nm FinFET with Advanced Fin Formation and Short Channel Effect Engineering", VLSI 2011 Tech. Dig., p. 14-15.
4. .A. Keshavarzi et al., "Architecting Advanced Technologies for 14nm and Beyond with 3D FinFET Transistors for the Future SoC Applications", 2011 IEDM Tech. Dig., p 67-70.
5. K. Chang et al., „ETSOI CMOS for System-on-Chip Applications ", 2011 VLSI Tech. Dig., p. 128-130..
6. Q. Liu et al., „Impact of Back Bias on Ultra-Thin Body and BOX (UTBB) Devices", 2011 VLSI Tech. Dig. p 160-162.

# A 361nA Thermal Run-away Immune VBB Generator using Dynamic Substrate Controlled Charge Pump for Ultra Low Sleep Current Logic on 65nm SOTB

Hiroki Nagatomi,

Nobuyuki Sugii*, Shiro Kamohara*, and Koichiro Ishibashi

The University of Electro-Communications,*Low-power Electronics Association & Project,

Email: ishibashi@edu.cc.uec.ac.jp, Tel: +81-42-443-5188

**Abstract** -This paper proposed an on-chip low power Body Bias Generator (VBBGEN) for ultra low leakage at 65nm SOTB (Silicon on Thin Buried Oxide) logic circuits at sleep mode. In the results of post layout simulation, the VBBGEN can generate and apply up to -2V body bias at a supply voltage of 0.5V with a current consumption of less than 361nA. By using the VBBGEN, it is expected that sleep current of CPU on SOTB is decreased by more than two orders of magnitude. In addition, the VBBGEN also has a function that prevents thermal run away of SOTB logic circuits.

(Keywords: SOTB, Body bias, Low leakage, Energy harvesting, Charge pump)

## 1. Introduction

SOTB (Silicon on Thin Buried Oxide) device is an FD-SOI in which body bias $V_b$ is applied from back body through a thin BOX layer. Fig.1 shows the structure of the SOTB device. The threshold voltage ($V_{th}$) of the SOTB device can be controlled by applying body bias from P-well (VBN) and N-well (VBP). It is reported that 32bit CPU on SOTB operates at ultra low energy of 3.4PJ/cycle [1], [2]. Besides, a large back bias voltage could be applied to the SOTB device that is effective to reduce the subthreshold leakage current at standby mode [1]~[4]. This paper proposes a fully-integrated on-chip body bias generator (VBBGEN) that makes possible two orders magnitude smaller sleep current for energy harvesting applications such as sensor network.

## 2. Proposed charge pump circuit

The proposed charge pump circuit (CP) generates large RBB from low supply voltages (VDD). In this paper, the body bias $V_b$ is expressed by the (1).

$$V_b = VBN = -(VBP-VDD) \qquad (1)$$

Generally, the power efficiency of dickson's CP [5] is low due to loss of MOSFET's $V_{th}$ predominantly. Therefore, many organizations proposed methods to reduce loss of $V_{th}$ using forward bias [6], charge transfer switch (CTS) [7], cross coupled inverter [8], etc. However, in case of the SOTB device, the reverse current flows from the next stage of the charge pump in the switching term and leakage flows from previous stage of the charge pump in the off term. These currents lower power efficiency and output voltage (Vout) because $V_{th}$ of the SOTB device is low. Fig.2 is the proposed DSCCP (dynamic substrate controlled charge pump). Boost inverters (INV1~INV5) generate and apply reverse bias to the diode connected MOSFETs in the off term to increase $V_{th}$. By using SOTB device, the performance of the DSCCP becomes higher because the body bias coefficient shown by (2) of the SOTB device is high [2].

$$\gamma = \Delta V_{th}/\Delta V_b = C_{BOX}/C_{OX} \approx 0.15 \qquad (2)$$

Fig.3 shows simulated Vout of the dickson's CP, the CTS CP [6] and the proposed DSCCP using SPICE model of SOTB device. In this figure, Vout of the CTS CP is much higher than that of dickson's CP, however, Vout of the CTS CP is decreased as temperature increases. Due to these characteristics, logic circuits with CTS CP have possibility of thermal run away at high temperatures. Vout of the DSCCP is lower than that of the CTS CP at temperatures less than 70˚C. At temperatures larger than 70˚C, it is larger than that of the CTS CP. The power efficiency of the DSCCP is higher than that of the CTS CP at temperature more than 20˚C due to large reverse current and leakage of the CTS CP (Fig.4).

## 3. The proposed circuit (VBBGEN)

We have designed VBB generator (VBBGEN) using the DSCCP. The VBBGEN is composed of 4 blocks: ring oscillators, clock boosters, 5 stages DSCCPs and switch circuits (Fig.5). It is not necessary to compose a comparator and a voltage reference circuit for VBBGEN because VBBGEN at sleep mode allows a small amount of ripple. VBP (VBN) is connected to body terminals of PMOS (NMOS) of the ring oscillators to make feedback. The switch circuits (SW) changes state of body bias of the SOTB logic circuit. Fig.6 shows a layout of the VBBGEN. DSCCPs are consisted of a 2pF capacitor for each stage. The VBBGEN occupied only $245 \times 128\mu m^2$.

## 4. The results of post layout simulations

Fig.7 shows the characteristic of VDD vs VBP, VBN at T=25˚C. The output resistance (Ro) is 260Mohm for VBP and 210Mohm for VBN respectively because the substrate impedance of the SOTB is very high [8]. The VBBGEN can generate VBP=2.54V, VBN=-2.06V at VDD of 0.5V. In the range of sub-threshold voltage, the VBBGEN can generate VBP=0.59V, VBN=-0.45V at VDD of 0.15V. Fig.8 shows the temperature characteristic. When temperature changes from 25˚C to 70˚C, both VBP and |VBN| rise (about $\Delta VB$=-0.15V). This means that the VBBGEN suppress lowering $V_{th}$ in logic circuits at high temperatures, thereby preventing thermal run away. Fig.9 shows the timing diagram of VBBGEN at output capacitance (Co) of 10pF. The switch circuits can change body bias VBP (VBN) to 0.5V (0V) correctly only when STBB signal rise up to VDD. Table 1 shows the current consumption of the VBBGEN with different conditions of VDD and temperature. In the condition of VDD=0.5V, T=25˚C, a current consumption is only 361nA. This means that the VBBGEN does not occupy a large fraction of the total current consumption of the whole IC. Therefore, it is expected that sleep current of SOTB CPU [1] can be decreased by more than two orders of magnitude on-chip completely by using the proposed VBBGEN.

## Acknowledgement

This work was performed as "Ultra-Low Voltage Device Project" funded and supported by the Ministry of Economy, Trade and Industry (METI) and the New Energy and Industrial Technology Development Organization (NEDO)

H. Nagatomi is currently with Renesas Electronics Corporation

## Reference

[1] K. Ishibashi, et.al., Cool Chips, XVII, April 2014.
[2] Y. Yamamoto, et.al., VLSIC, 2013, p.p.T212.
[3] S. Morohashi, et.al., S3S, 2013, p.p.165
[4] H. Nagatomi, et.al., ICDV, 2013, p.p.42.
[5] J.F. Dickson, JSSC, 1976, p.p.374.
[6] Po-Hung Chen, et.al., CICC, 2010, p.p.1.
[7] Xueqiang Wang, et.al., ASICON'09, p.p.320.
[8] M. D. Ker, et.al., JSSC, 2006, p.p.1100.

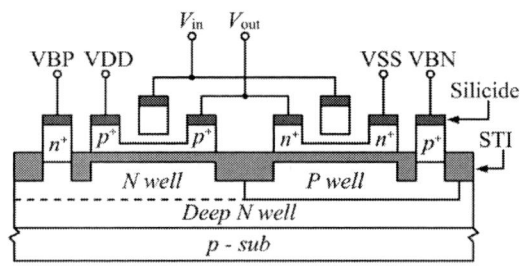

Fig.1. Structure of active region for SOTB inverter

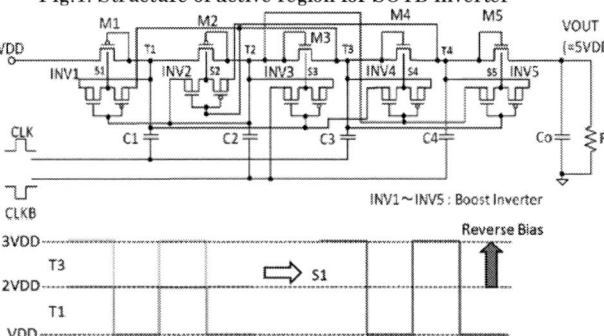

Fig.2 Schematic and operating waveform of DSCCP

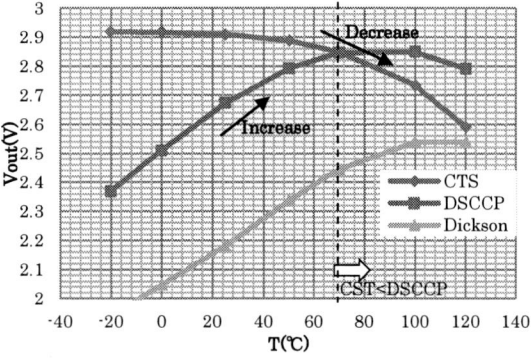

Fig.3 Temperature characteristic of CPs

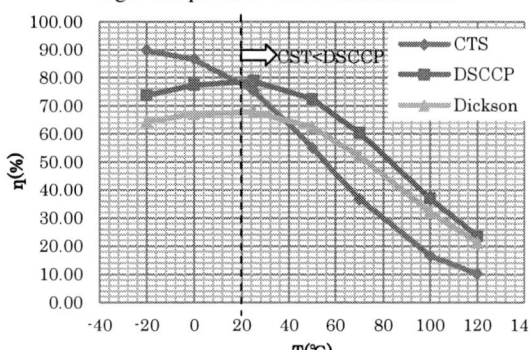

Fig.4 Power efficiency of CPs

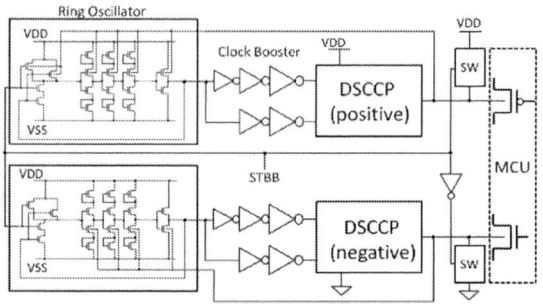

Fig.5 Schematic of VBBGEN

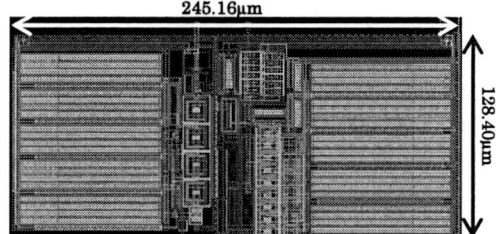

Fig.6 Layout picture of VBBGEN

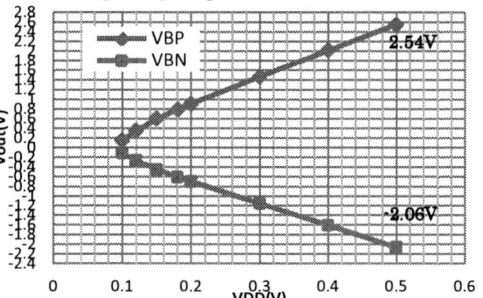

Fig.7 Output voltage of VBBGEN

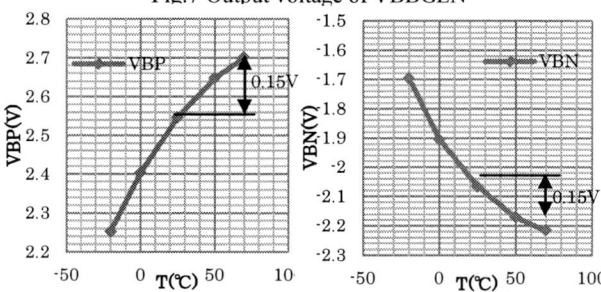

Fig.8 Temperature characteristic of VBBGEN
(right: VBP    left: VBN)

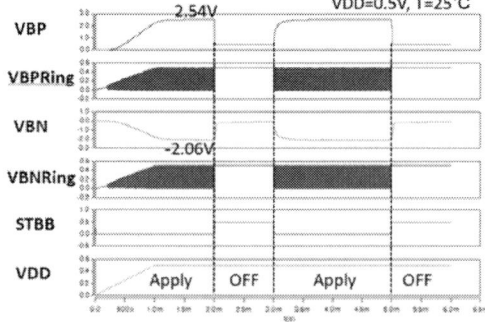

Fig.9 Timing diagram of switching body bias

Table 1. current consumption of VBBGEN

| VDD/T | -20°C | 0°C | 25°C | 70°C | 100°C |
|-------|-------|-----|------|------|-------|
| 0.5V | 180.53nA | 237.19nA | 360.77nA | 1.08μA | 2.21μA |
| 0.55V | 228.84nA | 307.67nA | 475.01nA | 1.37μA | 2.71μA |
| 0.6V | 306.19nA | 421.50nA | 634.95nA | 1.77μA | 3.08μA |

978-1-4799-7440-5/14 $31.00 © 2014 IEEE

# A 53µW -82dBm Sensitivity 920MHz OOK Receiver Design Using Bias Switch Technique on 65nm SOTB CMOS Technology

Hoang Minh Thien, Nobuyuki Sugii*, Koichiro Ishibashi

The University of Electro-Communications and *Low-power Electronics Association & Project

Email: minhthienhoang@uec.ac.jp, Tel: +81-80-4946-0406

*Abstract*-This paper presents an ultra-low power receiver design at 920MHz. We proposed a receiver architecture, in which bias switch technique is applied to reduce power consumption significantly. The receiver was simulated and laid out on 65nm SOTB CMOS technology, consuming only 53uW at 0.6V supply voltage. It achieves a sensitivity of -82dBm with a data rate of 10 - 100 kbps.

*Keywords—Ultra-low power receiver; wireless sensor; OOK transceiver; SOTB, CMOS technology*

## I. INTRODUCTION

Recently, SOTB (Silicon on Thin Buried Oxide) CMOS device, one of the FDSOI device, has been developed, on which 13.4pJ/cycle 32 bit CPU was realized [1][4]. This paper presents a receiver design that uses SOTB devices, allowing the receiver to operate at low supply voltage of 0.6V, reducing power consumption. In addition, RF front-end blocks, which consume majority of power in the receiver, are biased by a 50% duty cycle clock to reduce more power consumption to 53µW.

## II. ARCHITECHTURE

Block diagram of the proposed receiver is shown in fig. 1. The RF stages are common source cascaded amplifiers. A clock signal is applied to the bias terminal of the cascaded transistor (fig. 2a) instead of a constant voltage (fig. 2b). The cascaded transistor in fig.2a plays the role of a lock. It opens only in a half of clock cycle. Therefore, in comparison with the constant bias topology in fig. 2b, power consumption of the circuit with the clock bias is reduced by 50%. Fig. 3 shows operation of the receiver with the bias switch technique. After the RF amplifiers, signal is a RF pulse chain at the duration of high bits (fig.3c). At the output of the envelope detector, signal has the same frequency as the bias clock (fig.3d). In this design, we used a 2.5MHz clock for switching bias of the cascaded transistors. The final bit sequence is recovered using a sample & hold followed by baseband blocks (fig. 3e, 3f), similar to the double sampling method presented in [2].

## III. CIRCUIT IMPLEMENTATON

Schematic of the RF front-end is shown in fig. 4. The first state was optimized for gain instead of noise figure.

C1, C3, L1 form input matching network. C2, C4 and C6 are AC couple capacitors; C5, C7 are total capacitances at the output nodes. They are chosen to resonate with L2, L3, respectively. L2, C5 and L3, C7 form output loads of each state at the interest frequency and concurrently play the role of filters to attenuate low frequency signal caused by the bias clock. Total gain of the RF font-end stages is 46dB with power consumption of 45.6uW. The envelope detector is shown in fig.6. M1 and M1' are biased to operate in the weak inversion regime.

The IF amplifier is a fully differential op-am (fig. 5), including two stages. The tail current source of the first stage is split into two halves connected by a capacitor. The transfer function of this amplifier has a zero at DC [3], suppressing DC offset caused by asymmetrical factors at the inputs. Gain of the IF amplifier can be set between 20dB to 40dB. Fig.5 shows AC characteristic of the IF amplifier with a zero at DC. Following the IF amplifier is simple sample&holds including CMOS switches and small capacitors.

The baseband amplifier (fig. 7) is also a two-stage op-amp with differential inputs, single-end output. In the first stage, loads are diode-connected transistors. With this kind of load, a common mode feedback circuit is not necessary. The baseband amplifier has a gain that can be set in the range from 15dB to 30dB.

## IV. POST-LAYOUT SIMULATION RESULT

Core layout of the receiver is shown in fig. 8. It occupies an area of 985µmx600µm. Fig. 9 and table I depict sensitivity at different data rate and performance comparison with other receivers, respectively. It can be seen that the proposed receiver can reach -82dBm sensitivity and consumes only 53µW at 0.6V supply.

## REFERENCES

[1] K. Ishibashi, et al., Cool Chips XVII, Apr. 2014
[2] X. Huang, et al., IEEE ISSCC, pp.222-223, Feb. 2010.
[3] T. Toifl, et al., IEEE JSSCC, vol. 41, no. 4, pp.954–965, Apr. 2006.
[4] R.Tsuchiya, et al., IEDM, pp. 631-634, Dec. 2004.

This work was performed as "Ultra-Low Voltage Device Project" funded and supported by the Ministry of Economy, Trade and Industry (METI) and the New Energy and Industrial Technology Development Organization (NEDO).

978-1-4799-7440-5/14 $31.00 © 2014 IEEE

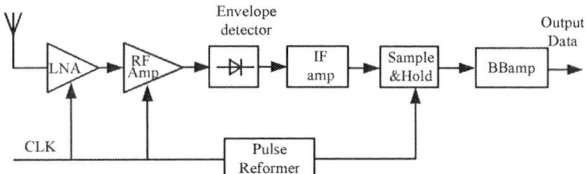

Fig. 1. Simple block diagram of the receiver

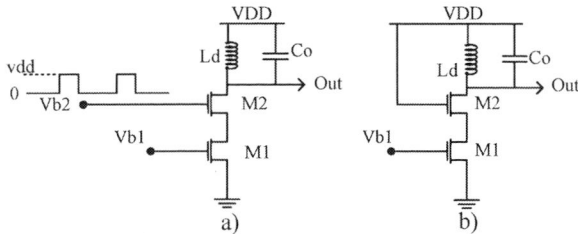

Fig. 2. Bias switch

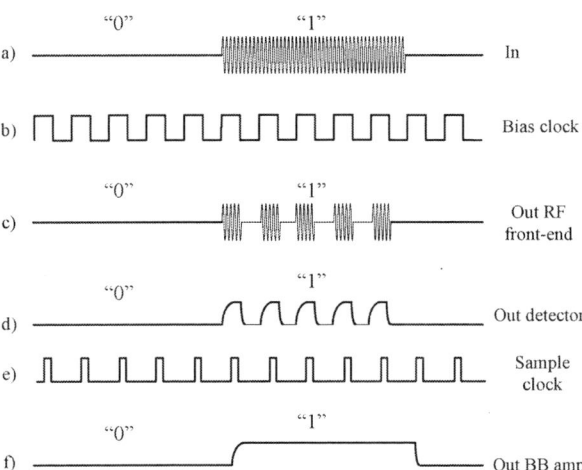

Fig. 3. Timing diagram

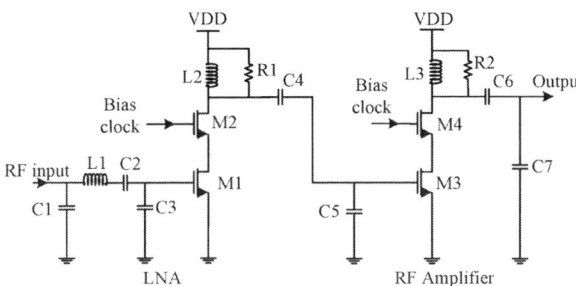

Fig. 4. RF front-end

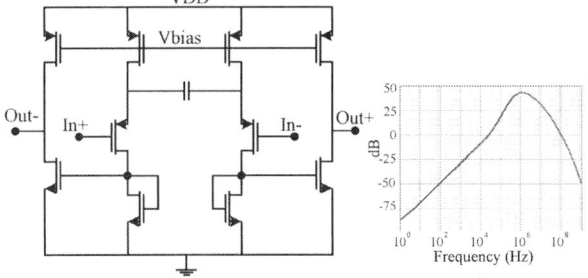

Fig. 5. IF amplifier and AC characteristic

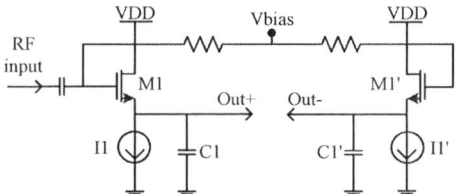

Fig. 6. Envelope detector

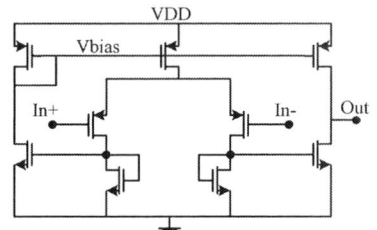

Fig. 7. Baseband amplifier

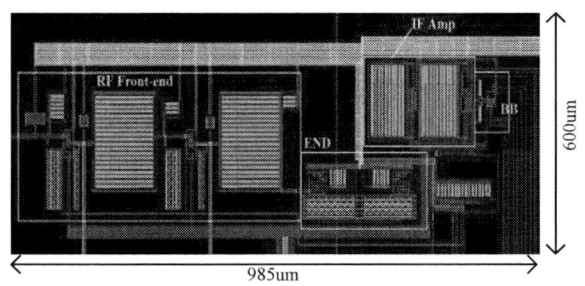

Fig. 8. Core layout

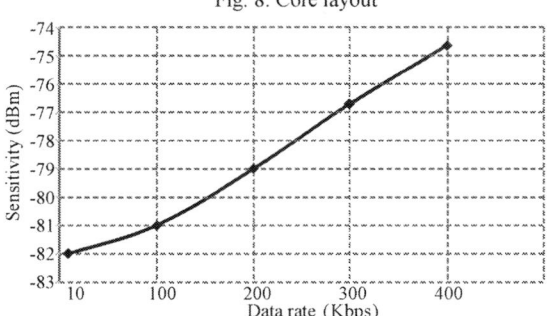

Fig. 9. Sensitivity vs Data rate

TABLE I
PERFORMANCE COMPARISION

| Circuit | (1) | (2) | (3) | This work |
|---|---|---|---|---|
| Technology(nm) | 180 | 90 | 180 | 65 SOTB |
| Modulation Scheme | FSK | OOK | OOK | OOK |
| Operation Freq. (MHz) | 920 | 915 | 920 | 920 |
| Data Rate(Kbps) | 5000 | 10 | 500 | 100 |
| Sensitivity(dBm) | -73 | -83 | -65 | -82 |
| DC power (μW) | 420 | 121 | 4500 | 53 |
| Energy/bit | 84pJ | 12.1nJ | 9nJ | 0.53nJ |

(1) J. Bae, et al., JSSCC, vol. 46, no.4, pp 928–937, Apr. 2011
(2) X. Huang, et al., JSSCC, vol. 47, no. 12, pp 3197–3207, Dec. 2012
(3) W. Gao, et al., JSSCC, vol. 48, no. 11, pp 2717-2733, No. 2013

# SOI FinFET versus Bulk FinFET for 10nm and below

Terence B. Hook*, F. Allibert[o,] K. Balakrishnan[@], Bruce Doris[+], Dechao Guo[+], Narasimha Mavilla[++], E. Nowak*, G. Tsutsui[+], R. Southwick[+] ,J. Strane,[+] Xin Sun[+],

* IBM SRDC, Essex Junction, VT, [+] IBM Albany, NY [o]SOITEC, Albany, NY [@] IBM Yorktown Heights, NY, [++]IBM Bangalore

## Introduction

FinFETs may in principle be built on either bulk [1-3] or SOI [4-5] substrates. In this paper we will review some of the technical issues associated with choice of substrate, directly comparing empirical results on 10nm hardware for which all the other processes are as much the same as possible. Furthermore, we will discuss the challenges beyond the 10nm generation, where fundamental changes in materials may render the debate moot. Our conclusion and prognosis is that SOI was, is, and will continue to be the technically superior choice.

## Process Sequence – cost and variability at the 10nm node

In Fig. 1 is pictured the basic sequence for creating fins in both bulk and SOI substrates. The obvious conclusion is the utter simplicity of the SOI approach relative to the bulk. Once a wafer with the desired thickness is available, formation of the fin is blindingly simple. In contrast, forming a fin of a desired dimension in a bulk substrate requires at least three additional steps of fill, polish, and etch. Cost and complexity aside,

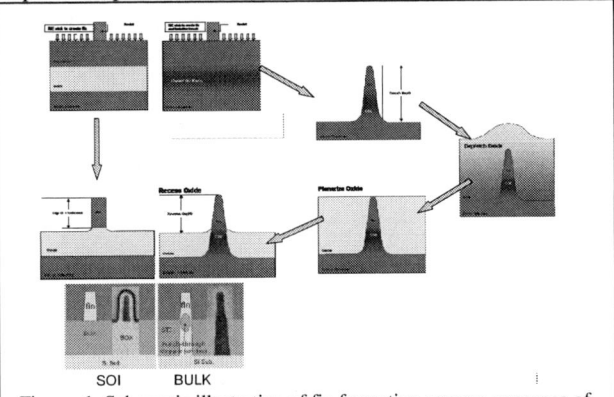

SOI        BULK

Figure. 1. Schematic illustration of fin formation process sequence of bulk and SOI and resultant fin profile. The bulk process requires additional steps and results in a less rectangular fin profile.

an inescapable consequence of such a sequence is a loss of fin height control. In Fig. 2 we show inline

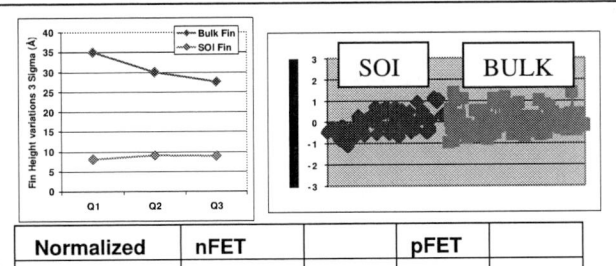

| Normalized | nFET | | pFET | |
|---|---|---|---|---|
| Capacitance | SOI | Bulk | SOI | Bulk |
| Median | 1 | 1.1 | 0.98 | 1.02 |
| Sigma | 0.012 | 0.045 | 0.019 | 0.04 |
| Sigma/Median | 12 | 41 | 19 | 39 |

Figure. 2. Inline data on 10nm bulk and SOI flows showing fin height and fin width control. While the ultimate control of bulk fin height may not have yet been realized, with no effort at all the SOI tolerance is 1/3 that of bulk. Fin width control is similar between the two.

process control data of the fin height for a stream of SOI and bulk hardware across several quarters. Although significant progress has been made in improving the control of the bulk process, the fundamental control capability in bulk is still three times worse than SOI. Electrical data of gate capacitance bears out the above conclusion. Average fin width control is similar for both sequences, as would be expected. However, the shape of the fin associated with the bulk process is typically more tapered than in SOI, as seen in Fig. 1.

## Electrical Differences for Bulk and SOI at the 10nm node

Bulk FinFETs necessarily contain dopants in the body to isolate drain from the source. The degree of doping may perhaps be optimized by adjusting the gate to the source depth (Fig. 3), but the fundamental conclusion is inescapable: bulk fins are doped. The presence of dopants in FinFETs is detrimental to variability, reliability, and performance. To reach the same off-current in the presence of fin doping, the gate work function must be driven closer to band-edge to compensate. In Fig. 4 we show calculations of the electrical impact of the required work-function shift, based on detailed empirical acceleration factors measured on 10nm FinFET hardware. The impact is significant – an SOI design point can support a power supply 80mV larger than that of bulk for the same level of reliability. In terms of performance, carrier scattering by the dopants degrades mobility (Fig. 5) and although some ambiguity remains in the hole

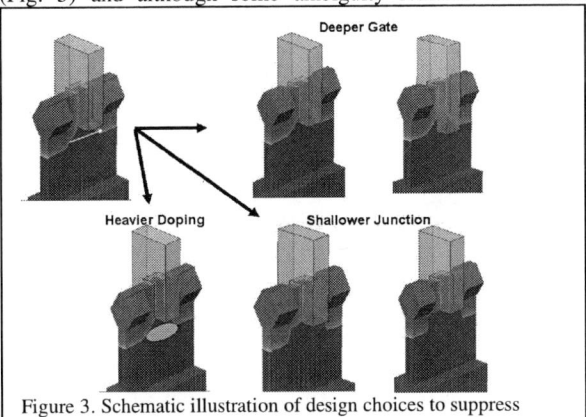

Figure 3. Schematic illustration of design choices to suppress subfin punchththrough

mobilities between bulk and SOI, the adverse impact of the doping in electron transport is clear. One of the most troubling aspects of adding doping is the compromise to the threshold voltage tolerance, particularly in adjacent transistor matching. This has an unfortunate effect on the minimum voltage at which a 6T SRAM will function. Figure 6 shows some data which may be difficult to accept, but illustrates the very real degradation of matching for bulk FinFETs relative to SOI. The curve is from [6]; Pelgrom plots for 10nm SOI and 10nm bulk technologies are shown, and the approximate location of those points is mapped to the doping curve. For those two data points the only explicit difference in processing is related to the fin formation; e.g., the gate formation is nominally identical. The results are nearly as

978-1-4799-7440-5/14 $31.00 © 2014 IEEE

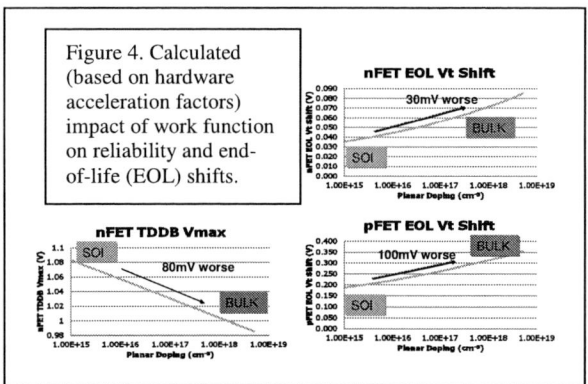

Figure 4. Calculated (based on hardware acceleration factors) impact of work function on reliability and end-of-life (EOL) shifts.

expected from the doping difference, although one might argue that the bulk result is actually slightly poorer than might have been expected from doping alone; such a result may not be surprising in light of what is known about the difference in the

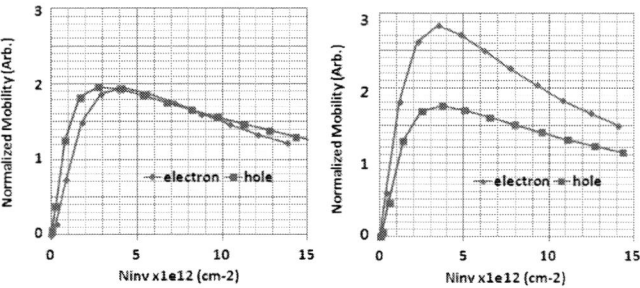

Figure 5. Measured mobility for bulk and SOI FinFETs. The electron mobility benefit is obvious. The potential benefit in the hole transport may possibly be obviated by a reduced strain level in the SOI hardware. From [5].

fin profiles. Another implication of fin taper and nonuniform doping appears in the analog characteristics: the measured self-gain data of the SOI FiNFET is superior to bulk planar devices and also to published data on bulk fins (Fig. 7) [7].

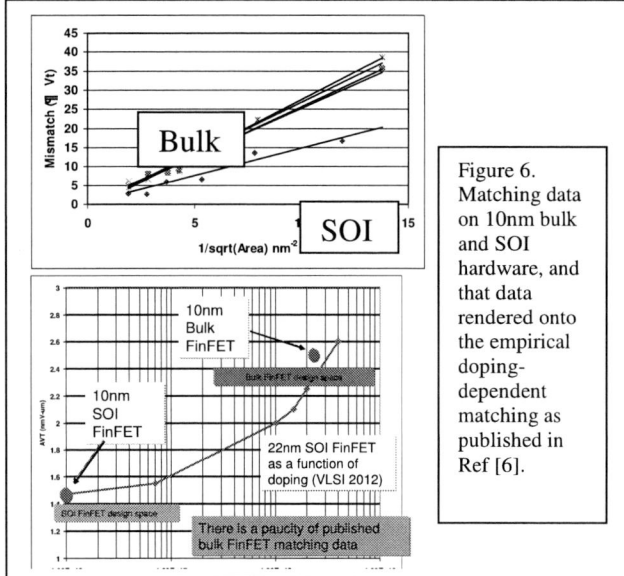

Figure 6. Matching data on 10nm bulk and SOI hardware, and that data rendered onto the empirical doping-dependent matching as published in Ref [6].

## Looking ahead beyond 10nm

Two potential concerns for SOI-based FinFETs have been the effect of self-heating and a shortfall in the ability to exert strain on the channel from the source/drain region (which may be responsible for the equivalence of hole mobility in Fig. 5.) While

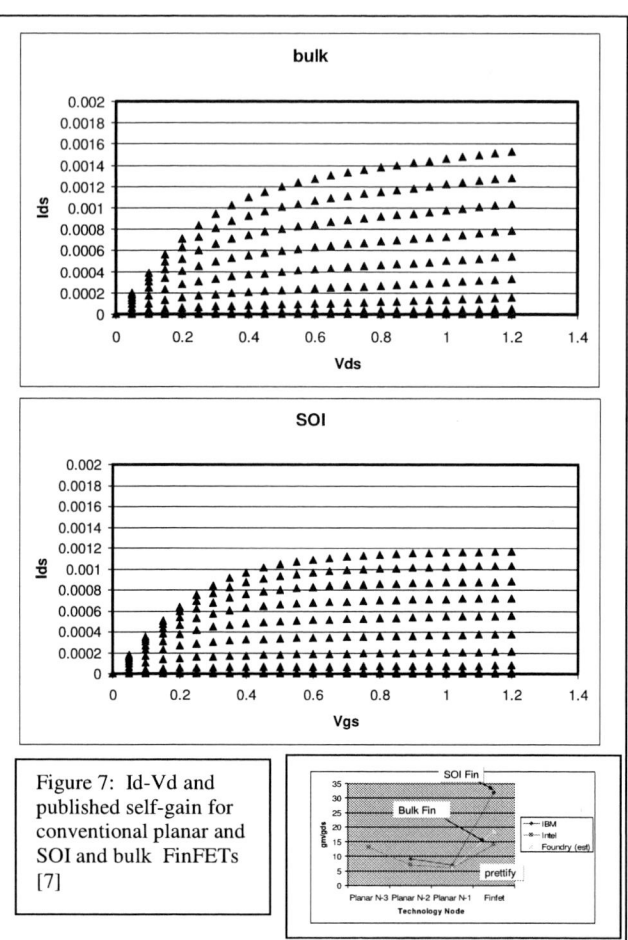

Figure 7: Id-Vd and published self-gain for conventional planar and SOI and bulk FinFETs [7]

both of these aspects may be of some interest at the 10nm node, the future direction of technology renders these issues moot – not so much because SOI FinFETs will be less susceptible to these concerns, but rather because bulk devices will become more so. The difference in self-heating between bulk and SOI depends on

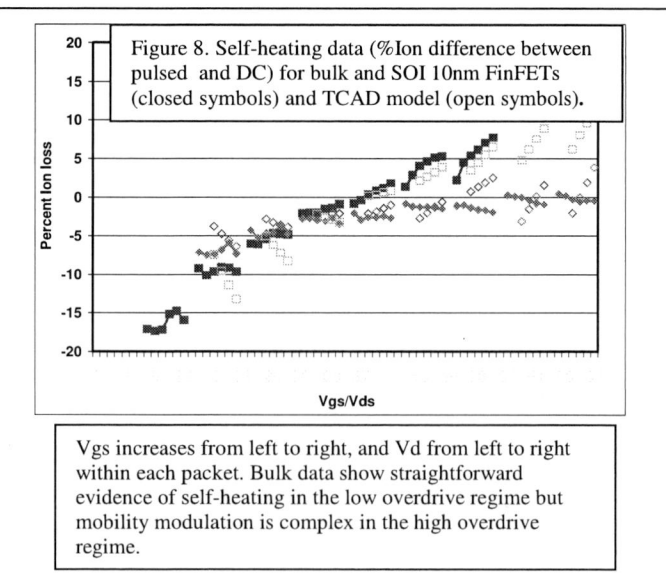

Figure 8. Self-heating data (%Ion difference between pulsed and DC) for bulk and SOI 10nm FinFETs (closed symbols) and TCAD model (open symbols).

Vgs increases from left to right, and Vd from left to right within each packet. Bulk data show straightforward evidence of self-heating in the low overdrive regime but mobility modulation is complex in the high overdrive regime.

many detailed factors such as the thickness of the BOX, the width of the fin through the thickness of the bulk isolation oxide (the thermal conductivity of silicon is degraded in confinement), the relative heat flow up through the metal interconnects and

978-1-4799-7440-5/14 $31.00 © 2014 IEEE

downward to the substrate. Self-heating data for 10nm devices are shown in Fig. 8, up to an extreme of 1.2V, well above the normal operating conditions. Several competing effects may be observed in these data. Following from left to right across the abscissa the gate voltage increases, and within each gate voltage there are several drain voltages. For low overdrive the drain current increases with self-heating, as the threshold voltage is reduced. Both bulk and SOI bulk data show this phenomenon, although bulk to a somewhat lesser degree, as expected. On the right side of this plot – high overdrive – the change in mobility with temperature dominates the behavior. For the SOI case the net result is a reduction of current with self-heating; for the bulk case the doping in the channel changes the mobility response and the result is a flatter dependence. A TCAD-based representation incorporating doping and mobility and thermal resistance is shown for both the SOI and bulk case, with reasonable fidelity to the overall features. The magnitude of the effect in SOI FinFETs is very similar to that of thick-BOX PDSOI, for which a plethora of publications offer circuit-level mitigation techniques [8]; additionally a thinner BOX (perhaps as thin as 20nm, as in FDSOI [9]) may be readily employed and all the advantages of SOI in fin formation retained while decreasing the thermal resistance. Regardless of the tempestuous teapot of anxiety about self-heating, it seems not unlikely that future so-called 'bulk' manifestations will be as much or more susceptible than classic SOI technology. Recent publications suggest that SiGe is a strong candidate for future technology nodes for improved channel mobility [10]. There are at least four profound implications for this in the context of the bulk versus SOI discussion. Firstly, it is well known that thermal conductivity of SiGe is much smaller than that of silicon and InGaAs is not much better [11], We therefore anticipate that self-heating in bulk FiNFETs will be virtually indistinguishable from SOI FinFETs in future generations and that self-heating will need to be dealt with regardless. Finite-element calculations of thermal resistance for

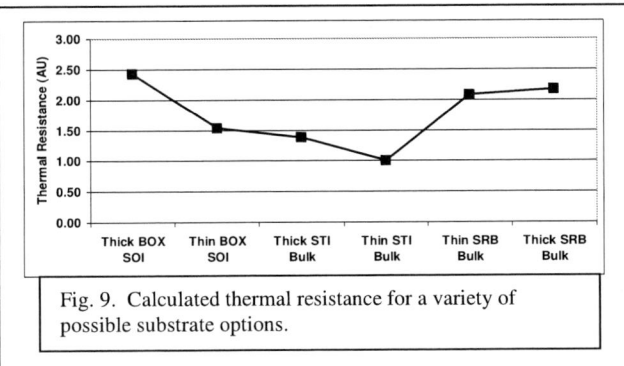

Fig. 9. Calculated thermal resistance for a variety of possible substrate options.

various substrates are shown in Fig. 9. As mentioned above, fortunately the phenomenon has almost no real implications for high-speed circuits and does not limit the scaling or performance. The transition to SiGe (or other high-mobility material) channel recognizes that embedded stressors are ineffective at small gate pitches [12], a strained epitaxy is increasingly irrelevant (Fig. 10). We therefore anticipate that for both SOI and bulk configurations the channel strain must originate in the channel and not be applied externally, and while those approaches may not be identical between bulk and SOI, source/drain epitaxy is not an important difference. Another consequence is that the bandgap is narrower in high-mobility channel materials. Figure 11 shows a calculation of doping required to suppress subfin punchthrough for silicon and Ge fins – various levels of Ge content will be intermediate. One of the defining upper bounds on an acceptable level of doping – band-to-band current – is further exacerbated by the

reduced bandgap as well. Bandgap engineering may be applied to mitigate this impact, but with accompanying cost and complexity. Finally, bulk fins must, as always, be concerned with vertical diffusion of dopants, including Ge itself. For example, it is known that Arsenic diffusion is more rapid in SiGe than in Si, which suggests that preventing the intrusion of punchthrough-stopper doping into the active fin will be more difficult. Therefore we anticipate that with the introduction of narrow bandgap, high mobility channels bulk FinFETs will be even more challenged to suppress punchthrough.

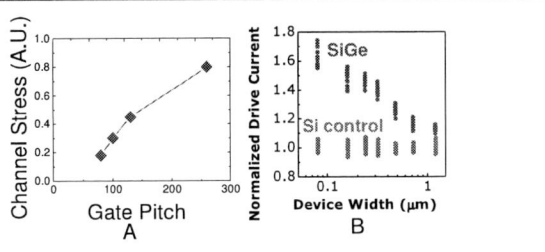

Fig. 10 (A) Simulation for channel stress from embedded SiGe showing diminishing channel strain for smaller gate pitches. (B) Planar fully depleted device data showing significant strain enhancement for SiGe channel pFETs . From Ref [12]

## Summary

We have shown data from bulk and SOI versions of a 10nm technology.

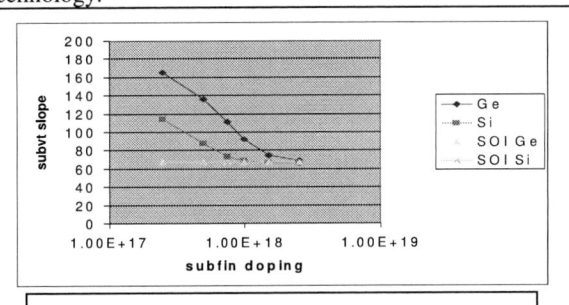

Figure 11. Calculated doping level required to suppress punchthrough (as reflected in the effective subthreshold slope) for silicon and high Ge-content semiconductor.

As predicted, bulk technology has poorer fin height control and transistor matching, mobility, and reliability than the corresponding SOI version. While 10nm-generation SOI fins show more limited embedded stressor potential, and slightly larger self-heating effects, dimensional scaling and migration to a narrow bandgap material such as SiGe will obviate those differences, while further exacerbating bulk subfin punchthrough. Consideration of self-heating will become unavoidable in any case, but fortunately a considerable body of knowledge is available to render the concern secondary. Complex and expensive super-steep profiles and bandgap engineering may serve to reduce the doping in bulk devices but the wisdom of adding costly process features to obtain the dubious advantages of a semiconducting intimacy with the substrate is questionable. SOI FinFETs - shown to be successful at 14nm, shown to be superior at 10nm - look to be even better beyond that.

## Acknowledgement and References

This work was performed by the Research Alliance Teams at various IBM Research and Development Facilities. [1] Auth IEDM 2011, [2] Wu IEDM 2013 [3] Kim EDL 2013, [4 ] Hook ASN 2013 [5] Seo VLSI 2014 [6] Lin VLSI 2012 [7] Jan IEDM 2012 [8] Bernstein and Rohrer 2007 [9] Weber VLSI 2014 [10] Hashemi IEDM 2013. [11] Wang APL 2010 [12] Doris VLSI-T 2013.

978-1-4799-7440-5/14 $31.00 © 2014 IEEE

# High Mobility Ω-Gate Nanowire P-FET on cSGOI Substrates Obtained by Ge Enrichment Technique

P. Nguyen[*ⁿ], S. Barraud[*], M. Koyama[*], M. Cassé[*], F. Andrieu[*], C. Tabone[*], F. Glowacki[*], J.-M. Hartmann[*], V. Maffini-Alvaro[*], D. Rouchon[*], N. Bernier[*], D. Lafond[*], M.-P. Samson[†], F. Allain[*], C. Vizioz[*], D. Delprat[ⁿ], B.-Y. Nguyen[ⁿ], C. Mazuré[ⁿ], O. Faynot[*], M. Vinet[*]

[*]CEA-LETI, Minatec campus, 17 rue des Martyrs, 38054 Grenoble Cedex 9, France ; E-mail: nguyet-phuong.nguyen@cea.fr
[†]STMicroelectronics, 850, rue J. Monnet, 38926 Crolles, France
[ⁿ]SOITEC, Parc Technologiques des Fontaines, 38926 Bernin, France

*Abstract* - Ω-gate nanowires (NW) P-FETs on compressively-strained–SiGe-on-insulator (cSGOI) substrate obtained by the Ge enrichment technique are presented. Effectiveness of cSGOI channel is demonstrated for ultra-scaled P-FET NW ($L_G$=15nm and $W_{NW}$=25nm) with an outstanding $I_{ON}$ current ($I_{ON}$=860µA/µm at $I_{OFF}$=140nA/µm) and a good electrostatics immunity (DIBL=110mV/V). For the first time, $Si_{0.8}Ge_{0.2}$–channel transistors highlight a mobility improvement for narrow NWs down to short gate length compared to Si one (92% for $L_G$=30nm). The hole mobility improvement provided by the strong uniaxial compressive strain coming from cSiGe and cCESL leads to an $I_{ON}$ current improvement of 95% at $L_G$=15nm.

## I. Introduction

NW transistors are today widely recognized as one of the most promising solutions to further continue the Moore's law beyond conventional planar bulk and Fully-Depleted SOI CMOS technologies. Aggressively scaled NW transistors have already been processed [1-4]. Nevertheless, strain-induced performance enhancement in short-channel NWs still needs to be proven and understood. In a previous work, we demonstrated high-performance uniaxially-strained Ω-gate sSOI N-FET NWs down to 10nm gate length with an excellent electrostatic control [1]. For high-performance P-FET NWs, strained-Ge [5] and strained-SiGe materials [6-11] are promising options to improve hole transport thank to high-mobility channels. In this work, Ω-gate SGOI P-FET NWs are processed on cSGOI substrates fabricated by Ge enrichment technique with Ge content of 20%. The substrate structural properties are characterized using SIMS, STEM and Raman spectrometry. Finally, ultra-scaled devices with $L_G$ and channel width $W_{NW}$ down to 15nm are demonstrated. We systematically examined their electrical properties and demonstrated superior device performance in comparison to NWs processed on SOI substrates.

## II. Ge Enrichment Process and Device Fabrication

The cSGOI process (Fig. 1) starts with the epitaxy of 12nm thick $Si_{1-x}Ge_x$ film on 300nm blanket SOI wafer (with 6nm thick Si layer and 145nm buried oxide). cSGOI substrates with Ge content of 20% have been fabricated following the process described in [12]. Cross-sectional TEM image (Fig. 2a) shows the final enriched cSiGe/BOX structure. The SiGe film is 9.5nm thick, and exhibits a perfect single-crystal lattice structure. The ToF SIMS is in excellent agreement with TEM analysis showing the homogeneous Ge content throughout the SiGe layer (Fig. 2b). The strain in the SGOI sample was investigated by Raman spectroscopy (Fig. 3). The negative $\varepsilon_{//}$ value is the signature of compressive strain corresponding to a stress value of $\sigma$~-1.2GPa for this Ge concentration (x~0.2). Then the NW transistors were processed on top of the SGOI substrates using the flow presented in Fig. 4 with channel oriented along the [110] direction. $W_{NW}$ and $L_G$ scaled down to 15nm were achieved. The Si or SiGe thickness (NW height: $H_{NW}$) under the HfSiON/TiN gate is 11.5nm. In-situ boron-doped $Si_{0.7}Ge_{0.3}$ raised S/D were used to thicken by 18nm the access regions of both SOI and SGOI devices (Fig. 5). Fig. 6 shows the TEM image with a perfect interface between $Si_{0.7}Ge_{0.3}$:B RSD and $Si_{0.8}Ge_{0.2}$ channel of Ω-shaped-gate NW.

## III. Electrical Results and Discussion

In order to understand and quantify the effectiveness of SiGe channels, we extracted the hole effective mobility as a function of $N_{inv}$ on long P-FET transistors. A wide range of NW width is studied (25nm $W_{NW}$ 240nm) to evaluate the mobility enhancement when the compressive strain changes from biaxial to uniaxial. Fig. 7a shows that $\mu_{hole}$ enhancement is maintained in wide devices due to biaxial compressive strain and for the narrowest NWs due to a strong uniaxial strain, in agreement with previous measurements of piezoresistive coefficients in P-FET NWs [13-14]. As expected, $\mu_{hole}$ significantly increases with SiGe channel NW devices. We thus measured mobility gain as high as 100% for $x_{Ge}$=0.2 at $N_{inv}$=$10^{13}cm^{-2}$ compared to Si (Fig. 7b). The extraction of the threshold voltage versus the active layer width at long $L_G$ is shown in Fig. 8. A $V_T$ shift of 200mV on $Si_{0.8}Ge_{0.2}$ compared to Si is achieved over the whole range of W. The $V_T$ reduction is coming from the energy band gap difference between Si and $Si_{0.8}Ge_{0.2}$ (almost exclusively in the valence band). The same $V_T$ extraction versus $L_G$ for wide and narrow transistors shows a similar reduction of threshold voltage for SiGe channels (Figs. 9-10). The low-field mobility $\mu_0$ *vs.* $L_G$ (Fig. 11) extracted on NWs shows the beneficial impact of an additional compressive strain induced by cCESL, especially for $L_G$ below 500nm. Indeed, it is the effect of an additional compressive strain induced by a cCESL. In Fig. 12, $\mu_0$ *vs.* $W_{NW}$ is shown: the hole mobility increases as the channel is narrowed for both Si and SiGe material, due to a better hole mobility on the (110) surface orientation [15], *i.e.* on the sidewall of the NWs. However this improvement is noticeably higher for SiGe NWs compared to Si NWs, and maintained down to short $L_G$ (+92% for $L_G$=30nm), which highlights the higher benefits of the SiGe material for these 3D architectures. The ($I_{OFF}$-$I_{ON}$) performance of our P-FET NWs is reported in Fig. 13a. An $I_{ON}$ improvement up to 95% compared to Si is evidenced in $Si_{0.8}Ge_{0.2}$ P-FET NWs down to 15nm gate length at $V_D$=-0.9V, as shown by the $I_{DS}(V_{GS})$ curve (Fig. 13b). With similar low values of $R_{SD}$ (Fig. 14), this $I_{ON}$ enhancement is likely due to the higher hole mobility presented in Fig. 11. A benchmark of $I_{ON}$-$I_{OFF}$ tradeoff demonstrates that our Ω-Gate SGOI NW P-FET is at the state of the art [6] but with lower Ge content ($x_{Ge}$~0.2 *vs* $x_{Ge}$~0.3) (Fig. 15).

## IV. Conclusion

We fabricated and characterized SOI and SGOI P-FET NW transistors with $W_{NW}$ and $L_G$ scaled down to 15nm. SGOI nanowires based on Ge enrichment technique show high-performance ($I_{ON}$=860µA/µm at $I_{OFF}$=14nA/µm) with excellent electrostatics (DIBL=110mV/V). We demonstrated high effectiveness of cSiGe channels and cCESL to improve $I_{ON}$ performance of NW transistors, through mobility improvement down to short gate length.

## Acknowledgements

This work was carried out in the frame of the ST/IBM/LETI joint program.

978-1-4799-7440-5/14 $31.00 © 2014 IEEE

## References

[1] S. Barraud *et al.*, *VLSI*, p.230 (2013)
[2] M. Saitoh *et al.*, *VLSI*, p.11 (2012)
[3] S. Bangsaruntip, *VLSI*, p.21 (2010)
[4] T. Tezuka *et al.*, *Jpn. J.Appl. Phys. 40*, 2866 (2001)
[5] J. Suh *et al.*, *Appl. Phys. Let.* 99, 142108 (2011)
[6] P. Hashemi *et al.*, *VLSI*, p.18 (2013)
[7] T. Irisawa *et al.*, *IEDM Tech. Dig.*, p.709 (2005)
[8] C.E. Smith *et al.*, *IEDM Tech. Dig.*, p.309 (2009)
[9] I. Ok *et al.*, *IEDM Tech. Dig.*, p.777 (2010)
[10] K. Ikeda *et al.*, *VLSI Tech. Dig.*, p.165 (2012)
[11] M.G.H. van Dal *et al.*, *IEDM Tech. Dig.*, p.521 (2012)
[12] F. Glowacki *et al.*, *EuroSOI Conf'* (2013)
[13] M. Cassé *et al.*, *ECS Trans.53*, p.125 (2013)
[14] Y.-M. Niquet *et al.*, *Nanolett.* (2012)
[15] Liu *et al.*, *VLSI* (2006).

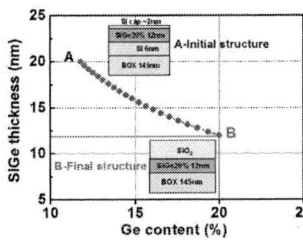

Fig.1. Ge enrichment schematics: the SiGe thickness is plotted as function of the Ge content. Once the enriched SiGe film is 12nm thick, the oxidation is stopped.

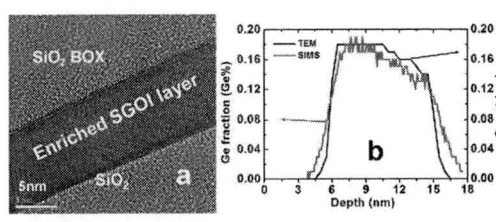

Fig.2. (a) The HR-TEM of final structure (B of Fig. 1) after Ge enrichment. A good crystalline quality is achieved; (b) Ge content profile obtained by ToF-SIMS and TEM are in good agreement.

|  | Si-Si mode (cm$^{-1}$) | $\Delta\omega$(cm$^{-1}$) | %Ge | $\varepsilon_{//}$ | $\sigma$ (Gpa) |
|---|---|---|---|---|---|
| Edge | 516.7 | -5.2 | 17.3 | -0.0066 | -1.19 |
| Middle | 516.5 | -5.4 | 18 | -0.0069 | -1.24 |
| Center | 516.7 | -5.2 | 17.3 | -0.0066 | -1.19 |

Fig.3. Raman spectra measurements on 9.5nm SGOI substrate. From the peak shift $\Delta\omega=\omega_{SiGe}-\omega_{Si}$, the Ge content, the in-plane strain ($\varepsilon_{//}$) was extracted and the corresponding stress ($\sigma$) values were calculated.

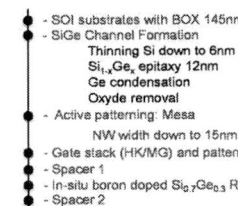

- SOI substrates with BOX 145nm
- SiGe Channel Formation
  Thinning Si down to 6nm
  Si$_{1-x}$Ge$_x$ epitaxy 12nm
  Ge condensation
  Oxyde removal
- Active patterning: Mesa
  NW width down to 15nm
- Gate stack (HK/MG) and patterning
- Spacer 1
- In-situ boron doped Si$_{0.7}$Ge$_{0.3}$ Raised S/D
- Spacer 2
- HDD implant and anneal
- CESL (compressive versus neutral)
- Back-End

Fig.4. Key process steps of Ω-Gate P-FET nanowire transistors.

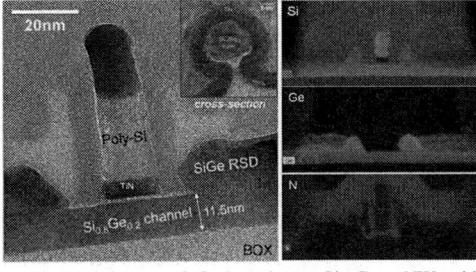

Fig.5. TEM image of Ω-shaped-gate Si$_{0.8}$Ge$_{0.2}$ NW with Si$_{0.7}$Ge$_{0.3}$:B RSD (H$_{NW}$=11.5nm and L$_G$=15nm).

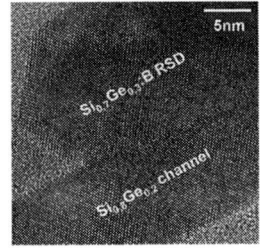

Fig.6. TEM image of perfect interface between Si$_{0.7}$Ge$_{0.3}$:B RSD and Si$_{0.8}$Ge$_{0.2}$ channel of Ω-shaped-gate NW.

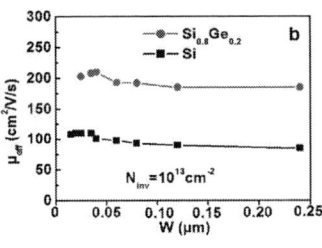

Fig.7. (a) Hole effective mobility values (extracted at N$_{inv}$=10$^{13}$cm$^{-2}$) vs. NW width (W$_{NW}$) in Si and Si$_{0.8}$Ge$_{0.2}$ P-FET NWs; (b) Hole effective mobility of Si and Si$_{0.8}$Ge$_{0.2}$ P-FET NWs vs. N$_{inv}$ (W$_{NW}$=25nm); L$_G$=10µm for both graphs.

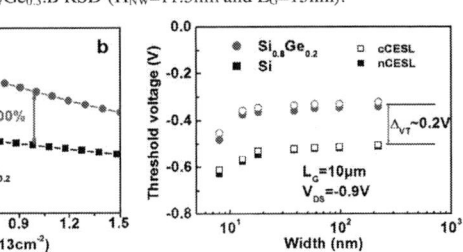

Fig.8. V$_T$ versus active layer width (W) for long gate length (L$_G$=10µm and V$_{DS}$=-0.9V).

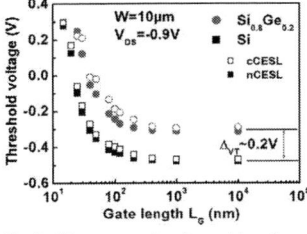

Fig.9. V$_T$ versus L$_G$ for wide planar FET (W=10µm) with Si and Si$_{0.8}$Ge$_{0.2}$ channels (V$_{DS}$=-0.9V).

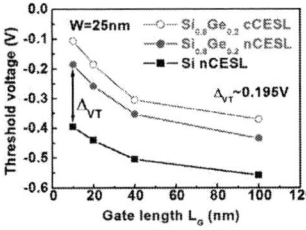

Fig.10. V$_T$ versus L$_G$ for NW FETs with Si and Si$_{0.8}$Ge$_{0.2}$ channels (V$_{DS}$=-0.9V and W$_{NW}$=25nm). A V$_T$ shift of 195mV is achieved at short L$_G$.

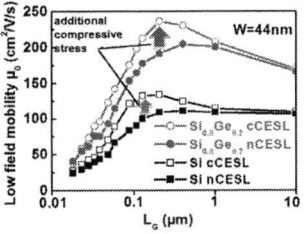

Fig.11. $\mu_0$ vs L$_G$ at W$_{NW}$=44nm extracted on P-FET NWs with Si and Si$_{0.8}$Ge$_{0.2}$ channels (neutral and compressive CESL are shown).

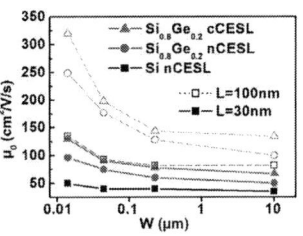

Fig.12. $\mu_0$ vs W extracted on P-FET NWs with Si and Si$_{0.8}$Ge$_{0.2}$ channels for L$_G$=30 and 100nm (neutral and compressive CESL are shown).

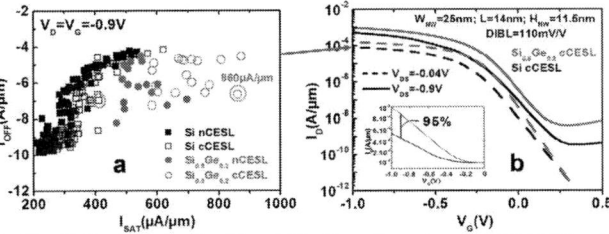

Fig.13. (a) I$_{OFF}$-I$_{ON}$ plot for Si and Si$_{0.8}$Ge$_{0.2}$ channels on P-FET NWs at V$_{DD}$=-0.9V (neutral and compressive CESL are shown); (b) I$_{DS}$(V$_{GS}$) of Ω-Gate Si and Si$_{0.8}$Ge$_{0.2}$ NW transistor with a L$_G$=15nm gate length.

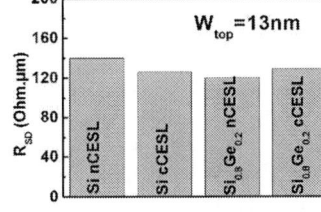

Fig.14. Si$_{0.7}$Ge$_{0.3}$ R$_{S/D}$ extracted on Ω-gate with Si and Si$_{0.8}$Ge$_{0.2}$ channels (neutral and compressive CESL are shown).

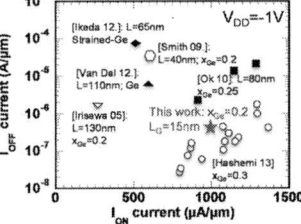

Fig.15. I$_{ON}$-I$_{OFF}$ data of this work compared to published Si$_{1-x}$Ge$_x$ channel FinFETs/Trigates at V$_{DD}$=1V.

978-1-4799-7440-5/14 $31.00 © 2014 IEEE

# Dielectric Isolated FinFETs on Bulk Substrate

Darsen Lu, Kangguo Cheng, Pierre Morin[1], Nicolas Loubet[1], Terence Hook, Dechao Guo, Ali Khakifirooz,
Phil Oldiges, Bruce Doris, Ken Rim, Ajey Jacob[2], Huiming Bu and Mukesh Khare
IBM Research, 257 Fuller Road, Albany, NY, [1]STMicroelectronics, [2]GLOBALFOUNDRIES
Email: ddlu@us.ibm.com , Tel: (914) 945-1774

## INTRODUCTION

Dielectric Isolated (DI) FinFETs exhibit superior electrostatic control compared to bulk FinFET without needing heavy sub-fin punchthrough stop doping, which increases device variability. Bottom oxidation through STI (BOTS) [1] and silicon-on-nothing (SON) are viable techniques to fabricate DI FinFETs on inexpensive bulk substates, as alternative to SOI substrate. In this paper we analyze DI FinFETs in terms of mechanical stress, transport, electrostatics and parasitic capacitances.

## BOTS AND SON PROCESSES

Fig. 1 illustrates the BOTS process as simulated in TCAD [3]. A silicon fin is first formed on bulk substrate with local oxide isolation partly filling the gaps between fins. Subsequently a thin capping nitride is deposited conformally on the sides of the fin and above the hard mask nitride left on top of the fin. After opening the thin nitride layer on horizontal surfaces with an anisotropic etch, local oxide isolation is removed and the exposed silicon is pulled-back laterally with an isotropic etch. Gaps between fins are filled with oxide – an important step for mechanical stability [1]. The bottom of the fins are oxidized from two sides until the active fin is isolated from the substrate. Capping nitride and part of the oxide are removed, leaving the fin standing on oxide.

Fig. 2 illustrates the alternative SON FinFET process. The fin is formed by etching a tri-layer Si/SiGe/Si stack instead of plain silicon. Subsequently the gate and the spacers are formed, mechanically anchoring the fin. Selective etching with HCl removes only the SiGe but not the silicon, creating a tunnel between the active fin and the substrate [2]. Finally, the tunnel is filled with dielectric material such as oxide. The final fin is electrically isolated from the substrate.

## S/D STRESSOR EFFICIENCY

Compressive strain from source/drain stressor is effective to enhance hole transport in pMOSFETs. We assess the effect of source/drain stressor in BOTS, SON, and bulk FinFETs with mechanical simulations. The source/drain stressor is SiGe with 35% germanium, lattice matched to silicon. The SiGe volume for all cases are the same for fair comparison. Since silicon dioxide has lower stiffness (Young's modulus) than silicon, the presence of oxide underlying the channel causes more force from source/drain to be exerted onto the silicon channel. With thickest oxide in BOTS FinFET compared to SON and bulk FinFETs, coupling of source/drain stress into the channel is most effective (Fig. 3).

## ELECTROSTATICS

A unique feature of FinFETs formed with the BOTS process is the fin "tail" – a triangular tip near the bottom of the fin (Fig. 4). Since the tail region is farther away from the gate compared to other parts of the channel, short channel leakage through this region is larger. The fin tail can be reduced by increasing the amount of lateral pullback (Fig. 1b) prior to thermal oxidation (Fig. 4). However, saturation occurs near 5nm. Gate recess (oxide recess after dummy gate removal), which brings the bottom of the gate further into the isolation oxide, mitigates leakage through the fin tail region. Simulation shows that with 5nm gate recess SS degradation due to the presence of fin tail is fully recovered (Fig. 5).

## THERMAL OXIDATION STRESS

Thermal oxidation generates a significant amount of mechanical stress, as in the "birds beak" generated with the LOCOS process. In the BOTS process as the bottom of the fin is oxidized, volume expansion results in a tendency for the fin to move up, causing vertical tensile stress. This stress is greater than 4GPa near the bottom of the fin prior to fin separation from substrate according to process simulations (Fig. 4a). After separation, a vertical movement of about 12nm occurs and stress relaxes (Fig. 4b). Although the process simulator does not capture the fracturing of silicon under high strain, the final structure obtained from the simulation matches the hardware well. Any errors that might be introduced by neglecting the impact of yield stress appear to be minimal.

## PARASITIC CAPACITANCE

In bulk FinFET the control of punchthrough current is required. Gate recess helps with puncthrough control but parasitic capacitance ($Cgs$) is larger (Fig. 7). Although heavier doping improves $Cgs$-$SS$ trade-off, it degrades device transport, increases band-to-band tunneling leakage, and random dopant fluctuation (RDF). Dielectric-isolated FinFETs such as the SOI FinFET shows the lowest $SS$ and $Cgs$ (Fig. 7). For the alternative partial dielectric isolated (PDI) FinFET (Fig. 7b), where the source/drain is connected to the substrate, punchthrough stop dopants are placed underneath the isolation oxide to minimize mobility degradation and RDF. However, gate recess of 30nm or more is required, causing even larger parasitic capacitance.

### REFERENCES

[1] K. Cheng *et al.*, *Symp. VLSI Tech.*, 2014
[2] M. Jurczak *et al.*, *IEEE Tran. ED*, vol. 47, Nov 2000
[3] *Sentaurus Process User Guide*, I-2013-12, Dec 2013

### ACKNOLEDGMENTS

This work was performed by the Research Alliance Teams at various IBM Research and Development Facilities

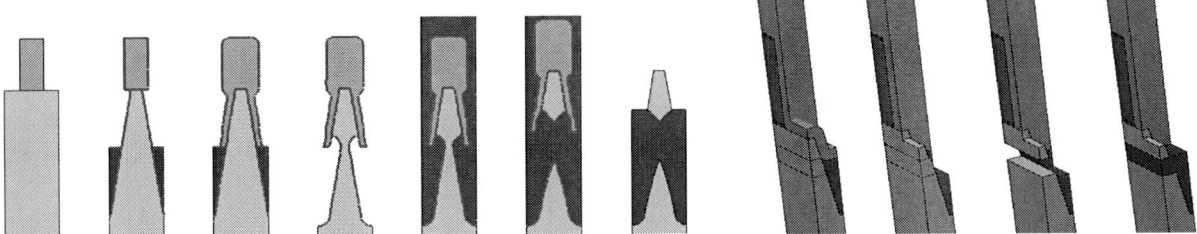

Fig. 1 BOTS process steps to form dielectric-isolated FinFET.

Fig. 2 SON process steps to form dielectric-isolated FinFET.

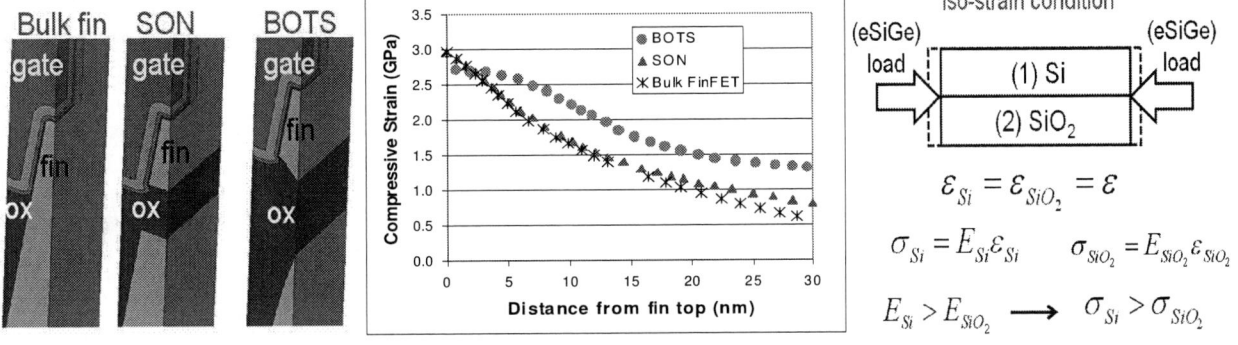

Fig. 3 (a) Bulk, SON and BOTS FinFETs (b) Simulated pFET stress profile along middle-of-fin, where e-SiGe S/D volume are kept identical (c) Explanation of superior stress coupling in the FinFET formed with BOTS process.

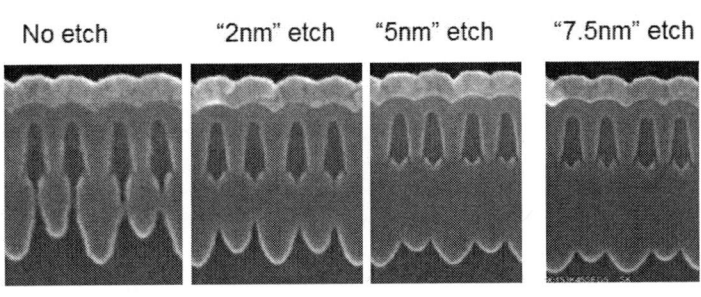

Fig. 4. SEM cross sections showing the fin profiles of dielectric isolated FinFETs fabricated with the BOTS process with varying amounts of lateral pullback etch.

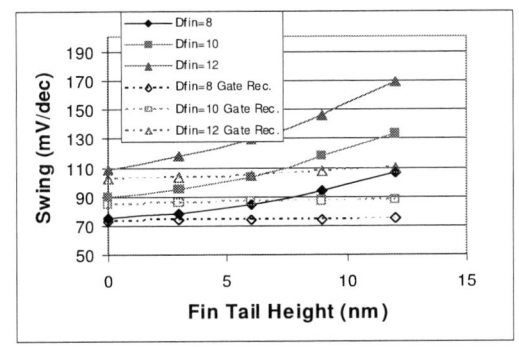

Fig. 5. Sub-threshold swing versus height of the triangular "tail" region at fin bottom, with and without 5nm gate recess.

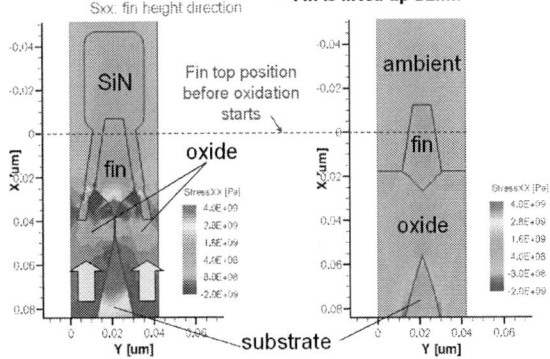

Fig. 6. Stress relaxation and vertical movement of the fin after oxide reaches through fin bottom.

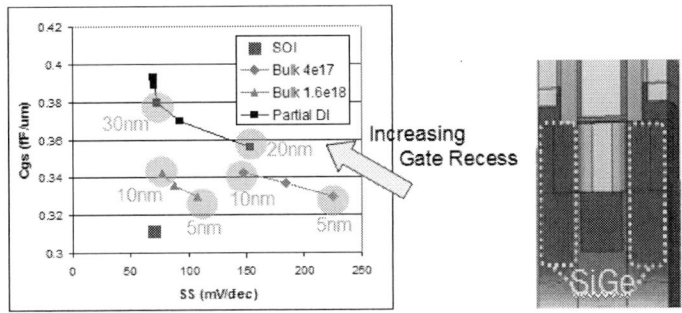

Fig. 7. (a) Parasitic gate-to-source capacitance in depletion model ($V_{gs}=0$), $C_{gs}$ versus sub-threshold swing ($SS$) trade-off. (b) Illustration of partial dielectric isolated FinFET.

978-1-4799-7440-5/14 $31.00 © 2014 IEEE

# Elastic relaxation in intrinsically-strained Fins: Simulations, physical and electrical characterization

F. Allibert[1], P. Morin[2], H. He[3], J. Li[3], W. Schwarzenbach[1], N. Loubet[2], A. Khakifirooz[3,*], B. DeSalvo[4], and B. Doris[3]

[1] Soitec, Parc Technologique des Fontaines, 38190 Bernin, France, [2] STMicroelectronics, [3] IBM Research & [4] CEA, LETI at Albany Nanotech, Albany, NY 12203, *now with Spansion Inc. - Email: frederic.allibert@soitec.com

## INTRODUCTION

Channel strain engineering has been a major mobility booster since the 90nm node [1]. The most common techniques consist in transferring stress into the channel from external stressing elements (Embedded S/D, Stress liners...) [2, 3]. As technology scales, the size of those external stressors is reduced, and they cannot further improve the device performance. New boosting techniques need to be implemented. An attractive solution is to use intrinsically strained channel materials. One specific example is strained-silicon on insulator (sSOI) which is pitch-independent [4]. Strained materials are rather well known and documented on planar structures [5, 6]. With the advent of 3D structures, i.e. FinFETs on bulk Si [7, 8] or SOI [9], it becomes essential to understand the impact of Fin geometry on elastic relaxation. In this paper, we present a thorough study of elastic relaxation in state-of-the art Fins through calibrated 3D simulations supplemented by Nano-Beam Diffraction (NBD) and electrical measurements.

## SIMULATIONS

Mechanical simulations were performed on Fin structures using a finite element method to solve the balance force equations. We considered the anisotropic properties of the strained materials [10] with stiffness parameters $C_{11}$=157GPa; $C_{12}$=60GPa and $C_{44}$=77GPa.

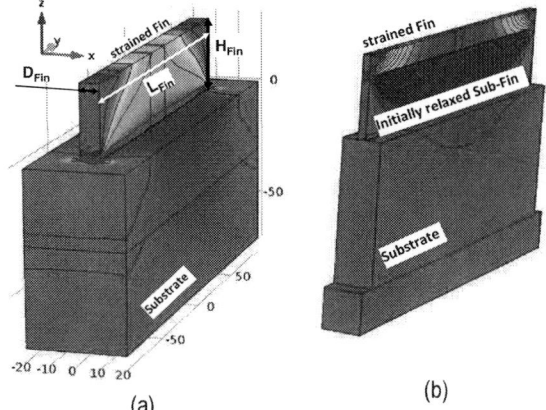

(a)                    (b)

Fig 1: Typical single layer (a) and dual layer (b) simulated structures.

Two intrinsically strained materials were used: (i) with compressive stress in the order of 1.5GPa, and (ii) with tensile stress of 1.3GPa.. Mechanical behavior of (ii) is similar to sSOI. Before patterning, the material is fully biaxially strained.

We studied the evolution of the strain in the structures after Fin formation (i.e. etch and cut), only accounting for elastic relaxation. Simulated structures included a single layer structure (Fig. 1-a) composed only of a strained Fin on an uncut relaxed substrate, and a dual layer structure (Fig. 1-b) composed of a strained Fin on top of an initially relaxed sub-Fin resting on an uncut relaxed substrate.

**Structure definition (Fig. 1-a):**
The Fin's Length, along the y axis is denoted $L_{Fin}$. The Fin is centered on y=0. $L_{Fin}$ was varied between 100 and 1000nm. The Fin's width, $D_{Fin}$, is along the x axis (centered on x=0). $D_{Fin}$=8nm. $H_{Fin}$ denotes the height of the **strained Fin** (i.e. the electrically active part), along the z axis with z=0 at the bottom of the Fin. $H_{Fin}$ was varied between 10 and 60nm. Unless specified, the results are reported for $H_{Fin}$=30nm.
In the dual layer structure (Fig. 1-b), the strained Fin rests on top of a **sub-Fin** ($H_{Sub-Fin}$=60nm). The total height of the dual layer structure is thus $H_{Fin}$ + $H_{Sub-Fin}$.

**Reported stress values:**
We only report here stress values in the y direction (along the Fin). The starting layer is biaxially stressed along (x,y) with no stress in the z (vertical) direction. After Fin cut, the stress is essentially relaxed along the x direction.
Some of this paper's results are reported as the stress in the "channel region", that is, the part of the Fin that would be covered by the gate in a finished FinFET. We define it as the average stress in the 20nm-long portion of the strained layer centered on y=0 (See Fig. 4-b).

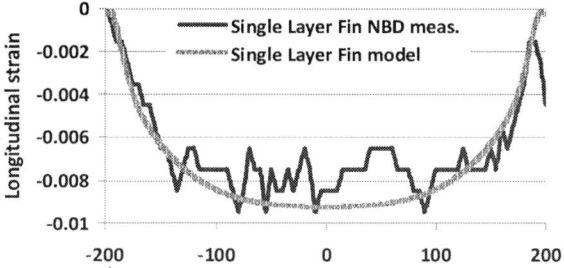

Figure 2: NBD measurement (solid line) and Simulation (dashed line) of the longitudinal strain along the Fin Length for a Strained Fin. Profile drawn along the black arrow on the sketch in Fig. 4-b.

## NANO-BEAM DIFFRACTION

We used the nano-beam electron diffraction TEM technique [11] to determine the lattice deformation along the strained Fins formed in the the single layer configuration (Fig. 2-a).

978-1-4799-7440-5/14 $31.00 © 2014 IEEE

Longitudinal and vertical strains along the length of a 400nm Fin, at mid $H_{Fin}$, were then inferred from the lattice deformation knowing the initial lattice parameter and strain. The resulting profiles were compared to mechanical simulations on a similar structure. An excellent matching is obtained. Both measured and simulated profiles are shown on Fig. 2.

## DEVICE CHARACTERIZATION

NMOS FinFET transistors were fabricated on sSOI wafers. The devices used in this experiment have a fin pitch of 42nm, $H_{Fin}$ and $D_{Fin}$ close to 30nm and 10nm respectively. Contacted Gate Pitch is 70nm. Gate length is in the order of 20nm. Four different Fin Lengths were studied: 145nm, 280nm, 560nm and 1100nm. FinFETs built on SOI wafers with similar features were used as reference.

## RESULTS AND DISCUSSIONS

Fig 3 shows the longitudinal stress in a vertical cross section along a strained Fin with single layer (a) and dual layer structures (b). The elastic relaxation is clearly visible at the edges of the Fin and is more pronounced at its top. When the Fin is cut, the strain is not maintained along the edges: a stress gradient appears.

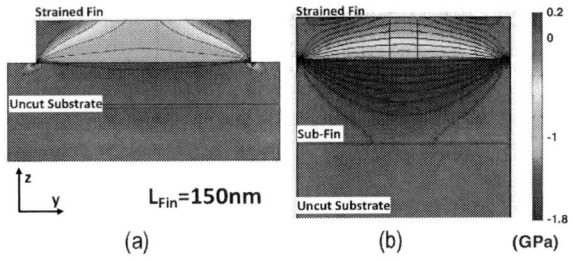

Figure 3: Longitudinal cross section of a strained Fin in single (a) and dual layer configuration. (b) showing stress in the y direction.

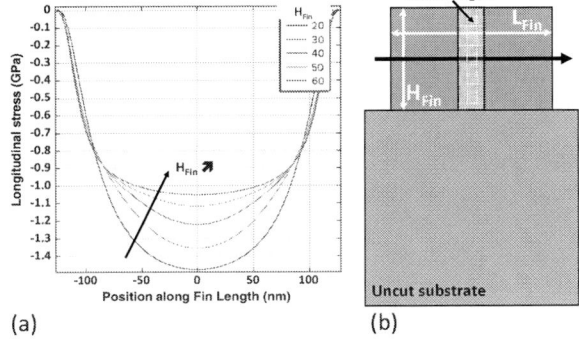

Figure 4: (a) Longitudinal stress profile in a 250nm long strained Fin single layer structure for various $H_{Fin}$. The profile is drawn along the black arrow on the schematics in (b).

The elastic relaxation, and thus the residual stress in the Fin depend on $H_{FIN}$ and $L_{Fin}$.

**Impact of $H_{Fin}$:** Fig. 4-a shows mid-height stress profiles (as defined in Fig. 4-b) along 250nm long Fins with different $H_{Fin}$. In a 20nm tall Fin, little relaxation is observed at the center (y=0) of the Fin. In taller Fins, significant relaxation is

observed. The residual stress in the middle of a 60nm high strained Fin barely exceeds 1GPa. Fig 5 shows the residual stress in the channel section of a 250nm long Fin as a function of Fin height. The taller the Fin, the lower the average stress.

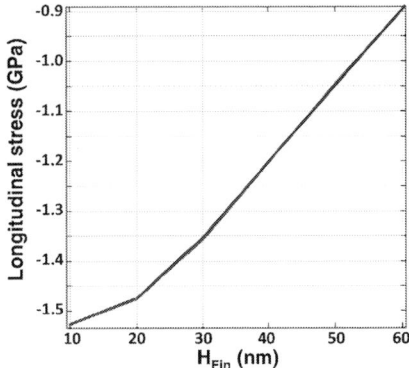

Figure 5: Calculated average longitudinal stress in the channel section of a strained Fin as a function of $H_{Fin}$ (single layer).

**Impact of $L_{Fin}$:** Fig. 6 shows mid-height stress profiles (as defined in Fig. 4-b) along Fins with various $L_{Fin}$. Stress relaxation occurs at the edges of the Fins. In short Fins, the relaxation extends throughout the Fin length and the average strain in the central section decreases. Thus, the shorter the Fin, the lower the stress (Fig. 7).

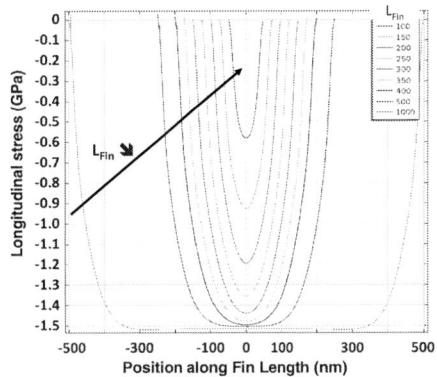

Figure 6: Longitudinal stress profile in a single layer strained Fin. for $H_{Fin}$=30nm and various $L_{Fin}$. The profile is drawn along the black arrow on the schematics in Fig. 4-b.

**Elastic relaxation mechanisms in cut Fins:**
Two main contributions are observed.
1/ Elastic relaxation directly in the Fin:
As can be intuitively understood, when the Fin is cut, the relaxation propagates from the edges of the Fin, where the stress is no-longer maintained. The stress is only maintained by the underlying structure (the uncut substrate for the single layer structure and Sub-Fin for the dual layer structure) at the bottom interface of the Fin. The further from this interface (taller Fins), the larger the elastic relaxation.

978-1-4799-7440-5/14 $31.00 © 2014 IEEE    19

2/ Stress relaxation by transfer of mechanical energy from the Fin to the underlying structure:

The compressive stress in the Fin partially relaxes through elastic deformation of the uncut substrate or sub-Fin. The results are very different between those two structures.

The elastic energy stored in the Fin is in equilibrium with the energy stored in the underlying structure. The stored elastic energy depends on the stress and the volume.

On the single layer structure substrate is not cut. Because the volume of the substrate is much larger than the Fin's, the average stress in the substrate is small. The substrate is not deformed Fig. 3-a shows only pockets of stress close to the edges of the Fin. This implies little relaxation in the Fin.

On the dual layer structure, the sub-Fin region, being cut, has a volume comparable to the Fin's. Compressive stress in the strained-Fin relaxes while tensile stress appears in the sub-Fin region (Fig. 3-b). Because of the structural difference the dual layer structure is much more prone to elastic relaxation than equivalent Fins built using a single layer structure.

**Layout considerations:**

In the studied single layer cases, strain relaxation starts for $L_{Fin}$ shorter than 400nm and becomes significant below 300nm. For the dual layer structure, even at $L_{Fin}$=1000nm some relaxation is seen in the center of the Fin, and it becomes more severe in Fins shorter than 750nm (Fig. 7). For layout purposes, considering a 10% stress loss as acceptable, a 30nm tall Fin on SOI requires the gate to be at least 125nm from the Fin's edge which corresponds to the addition of one dummy device at each end of the Fin, at 70nm pitch. For much taller Fins or Fins on bulk, a larger number of dummy devices is required, which may not be practical. Transistor Folding [13] or techniques to prevent stress relaxation, could be used alternatively to dummy devices.

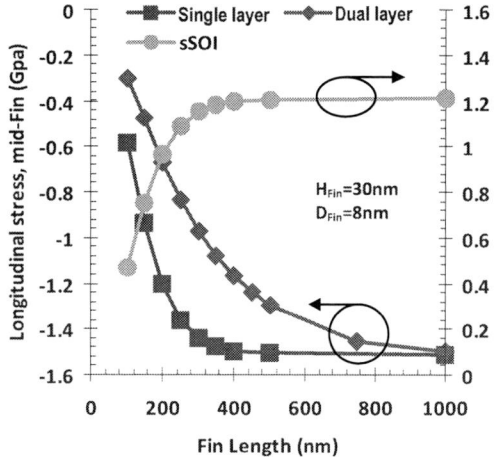

Figure 7: Average longitudinal stress in the middle (y=0) of a sSOI Fin (circles) as well as compressively strained single-layer (squares) and dual layer (diamonds) Finsas a function of $L_{Fin}$.

Since the improvement of NFET performance in sSOI compared to SOI FinFETs is linked to the longitudinal strain, it is expected to observe a decrease of the on-current gain in Fins shorter than 300nm. Fig. 8 shows the measured ratio of sSOI over SOI on-current at $V_D$=50mV and 300mV overdrive. The expected [4, 12] gain of over 20% is confirmed on longer Fins, but performance degrades below 300nm. In 145nm long Fins, sSOI and SOI performances are equivalent. Overlayed on Fig. 8 is the stress in the channel region inferred from the simulations. The reduction in average stress remarkably correlates to the change in performance.

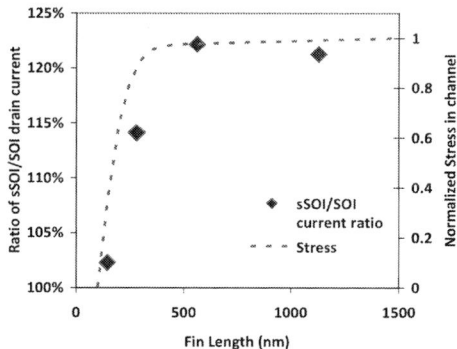

Figure 8: Ratio of measured drain currents of sSOI over SOI (diamonds) and predicted stress in the channel region of sSOI FinFETs (line). The performance correlates well with the simulated stress in the Fin.

## CONCLUSIONS

Elastic relaxation after Fin formation in intrinsically-strained materials was presented. Reported simulations are correlated with NBD strain measurements and are consistent with length dependence of drain current in sSOI FinFETs. Stress in the Fin's channel region depends on Fin length and height. dingle layer structured Fins are much more resistant to elastic relaxation than dual layer structured Fins, due to overall structure height. On dual layer structure Fins, significant relaxation is expected for $L_{Fin}$ shorter than 750nm, while single layer structured Fins do not relax significantly until 300nm. For transistor layout, elastic relaxation effects should be modeled or circumvented. Fin Length effects need to be added to other layout effects in Spice models.

## ACKNOWLEDGEMENTS

The authors would like to thank D.K. Sadana, B.Y. Nguyen and C. Girard for useful discussions and managerial support. This work was performed at various Soitec and IBM sites.

## REFERENCES

[1] T. Ghani et al., IEDM 2003; [2] G. Eneman et al. VLSI 2005; [3] G. Eneman et al. ESSDERC 2005; [4] A. Khakifirooz et al., IEEE S3S Conf., 2013; [5] A. Khakifirooz et al., VLSI 2012; [6] K. Cheng et al., IEDM 2012; [7] C. Auth et al., VLSI, p.131, 2012; [8] S.-Y. Wu et al., IEDM 2013; [9] T. Yamashita et al., VLSI 2011; [10] J.J. Wortman et al., JAP Vol. 36, n°1, 1965; [11] K. Usuda et al., Mat. Sc. Eng. B, Vol. 124-125, 2005; [12] K. Maitra, et al., IEEE EDL, 2011; [13] A. Khakifirooz et al., 218th ECS meeting, 2010.

# Toward Robust Subthreshold Circuit Design
## – Variability and Soft Error Perspective –

Masanori Hashimoto

Dept. Information Systems Engineering, Osaka University, Japan & JST, CREST

*Abstract*—**Subthreshold circuits are drawing attention for ultra-low power application. However, subthreshold circuits have inherent problems that their performance is extremely sensitive to manufacturing and environmental variability and they are susceptible to soft errors. This paper discusses robust subthreshold circuit design from variability and soft error perspective.**

## I. INTRODUCTION

Subthreshold circuits are promising to severely energy-constrained devices with low demands for their operation speeds, such as devices for habitat monitoring, health monitoring, structural health monitoring, and biomedical equipment. Recently not only such severe low energy devices but also middle performance chips are developed [1], [2]. In addition as a new application, a cubic-millimeter wireless intraocular pressure sensor is proposed [3]. In this paper, near-threshold circuits are included in subthreshold circuits.

Subthreshold circuits have a problem that their performances are extremely sensitive to manufacturing variability and environmental variability such as temperature and supply voltage fluctuations. In addition, subthreshold circuits have been thought to be vulnerable to soft errors.

We have worked for putting subthreshold circuits robust to practical use at device, circuit and CAD levels. At device level, we constructed a transistor variability model that reproduces subthreshold circuit performance [4], and evaluated soft error immunity of subthreshold SRAM [5]–[9]. At circuit level, adaptive performance control is studied parting from conventional worst-case design [10]. For implementing and verifying adaptive control, stochastic evaluation frameworks of timing error and power consumption are developed [10], [11]. Furthermore, we propose a self-timed processor to cope with large variation in memory access time [12].

In this paper, adaptive speed control for overcoming variability is first introduced, and the soft error susceptibility of subthreshold SRAM is presented.

## II. RUN-TIME PERFORMANCE ADAPTATION

Vth variation due to manufacturing variability and temperature fluctuation significantly varies speed and power consumption of subthreshold circuits. If adding up worst-cases for each variation factor, power dissipation may increase more than 10x. We therefore devised an adaptive speed control scheme [10] (Fig. 1). The timing error predictive flip-flop (TEP-FF) causes a setup violation earlier than the main flip-flop due to the inserted delay element. This error signal is used as a

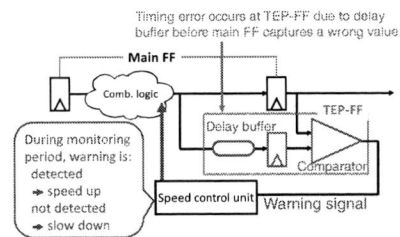

Fig. 1. Run-time adaptive speed control with TEP-FF [11].

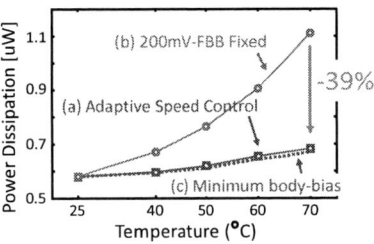

Fig. 2. Measurement result of speed adaptation (3MHz, 0.35V) [10].

warning signal indicating a shortage of timing slack, and the circuit is speeded up or down according to this signal.

This adaptive speed control was applied to a 32-bit Kogge-Stone adder. A test chip was fabricated in 65nm process. Figure 2 shows a measurement results under temperature variation. In this test chip, the circuit speed is adjusted by body-biasing. (a) corresponds to the proposed speed control, (b) is the power dissipation when 200 mV forward body-bias is given to satisfy the speed requirement at 25°C, and (c) is the power dissipation when the minimum body-bias is given at each temperature. This result shows that the power dissipation of the proposed speed control is close to (c) and the speed control is well working. Compared to conventional adaption of (b), the power dissipation is reduced by 40%. We also confirmed that 46% power reduction was possible compared to worst-case design for process and temperature.

This adaptive speed control involves a fundamental problem that timing errors cannot be completely eliminated. This is because the circuit could be slowed down excessively, if critical paths are not activated for a long time, the circuit could be slowed down excessively. However, this timing error is very difficult to evaluate in design time, since simulation is too slow for rare errors, such as an error per month. To enable design-time verification, we developed a stochastic error rate estimation method [11]. The necessary evaluation time was reduced by twelve orders of magnitude, which can guide design optimization of run-time adaptive system.

978-1-4799-7440-5/14 $31.00 © 2014 IEEE

## III. NEUTRON-INDUCED SOFT ERROR

In terrestrial environment, alpha particle and neutron are major sources for soft error, especially neutron could be dominant. This section presents measurement results of neutron-induced soft errors in 10T SRAM over a wide range of supply voltages between 1.0 and 0.3 V reported in [6], [7]. This section also mentions future trends on neutron-induced soft error.

A test chip including a 256 kb 10T SRAM was fabricated in a 65-nm bulk CMOS process with triple well structure and irradiated with accelerated spallation neutron beam. This SRAM can operate even at 0.3 V, because the cross-coupled inverters are large enough to mitigate threshold voltage variability. The size of a memory unit is 4.4 $\mu$m × 0.8 $\mu$m.

Figure 3 illustrates the dependence of the SBU and MCU rates on the supply voltage. The MCU rate was derived by dividing the number of failing bits (for example, a "2b MCU" was considered to be two errors) by the measurement period. The SBU rate dramatically increases as the supply voltage is reduced. On the other hand, the dependence of the MCU rate on the supply voltage is smaller than that of the SBU rate. Previous work [13] has shown that the MCU rate is less sensitive to the supply voltage between 1.2 and 0.7 V and concluded that this is because most neutron-induced MCUs are caused by the parasitic bipolar action. Interestingly, however, the MCU rate shown in Fig. 3 slightly increases when the supply voltage is below 0.5 V. Remind that charge sharing and parasitic bipolar action have opposite directions in terms of supply voltage. While the parasitic bipolar action is the dominant mechanism of MCUs in the super-threshold region in our design, the effect of charge-sharing becomes larger in the subthreshold region, which results in the increase in the MCU rate between 0.3 and 0.5 V, as depicted in Fig. 3.

Next, the MCU distributions in the memory cells are shown in Fig. 4. A decrease in the supply voltage also increases the probability of large-bit MCUs due to the decrease in the critical charge. 6-bit MCU was observed at 0.3V.

Figure 5 shows the simulated SEU probability including both SBU and MCU per neutron flux as a function of critical charge at the incident angles of 60° and 0°. Individual contributions from secondary H (proton), He (alpha), and heavier ions to the SEU are separated for the result of 0° in Figure 5. There is little difference between the SEU probabilities at the

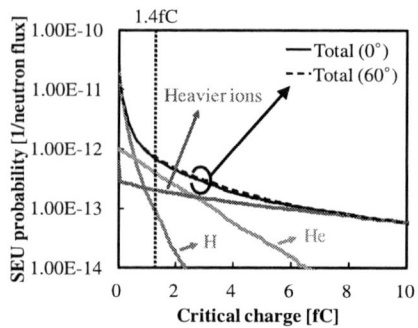

Fig. 5. Simulated SEU probability of each ion as a function of critical charge [7].

angles of 60° and 0°. On the other hand, the critical charge of our 10T SRAM in 0.4-V operation is estimated by circuit simulation to be 1.4 fC. Therefore, He and heavier ions are the dominant secondary ions causing SEUs in 0.4-V operation because these ions occupy 89 % of the SEU probability at 1.4 fC of critical charge.

At 0.4V operation, protons are not dominant, but another newer result with other 90nm bulk SRAM at 0.19V presented a dramatic SEU increase, which is explained by proton contribution [8]. Another technology direction is SOI device, and [9] reports that ultra-thin-BOX SOI is helpful to mitigate SEU, especially MCU. Thanks to this, by introducing ECC for SBU, highly reliable SRAM can be obtained.

### ACKNOWLEDGEMENTS

This work was partly supported by NEDO and LEAP.

### REFERENCES

[1] S. Jain, et al., "A 280mV-to-1.2V wide-operating-range IA-32 processor in 32nm CMOS," *ISSCC*, 2012..

[2] G. Gammie, et al, "A 28nm 0.6V low-power DSP for mobile applications," *ISSCC*, 2011.

[3] G. Chen, et al, "A cubic-millimeter energy-autonomous wireless intraocular pressure monitor," *ISSCC*, 2011.

[4] H. Fuketa, et al., "Transistor Variability Modeling and Its Validation with Ring-Oscillation Frequencies for Body-Biased Subthreshold Circuits," *IEEE Trans. VLSI Systems*, vol. 18, no. 7, 2010.

[5] H. Fuketa, et al., "Measurement and Analysis of Alpha-Particle-Induced Soft Errors and Multiple Cell Upsets in 10T Subthreshold SRAM," *IEEE Trans. Device and Materials Reliability*, vol. 14, no. 1, 2014.

[6] H. Fuketa, et al., "Neutron-Induced Soft Errors and Multiple Cell Upsets in 65-nm 10T Subthreshold SRAM," *IEEE Trans. Nuclear Science*, vol. 58, no. 4, 2011.

[7] R. Harada, et al., "Angular Dependency of Neutron Induced Multiple Cell Upsets in 65-nm 10T Subthreshold SRAM," *IEEE Trans. Nuclear Science*, vol. 59, no. 6, 2012.

[8] T. Uemura, et al., "Soft-Error in SRAM at Ultra-Low Voltage and Impact of Secondary Proton in Terrestrial Environment," *IEEE Trans. Nuclear Science*, vol. 60, no. 6, 2013.

[9] R. Harada, et al., "Measurement of Alpha- and Neutron-Induced SEU and MCU on SOTB and Bulk 0.4 V SRAMs," *NSREC*, to appear.

[10] H. Fuketa, et al., "Adaptive Performance Compensation with In-Situ Timing Error Predictive Sensors for Subthreshold Circuits," *IEEE Trans. VLSI Systems*, vol. 20, no. 2, 2012.

[11] S. Iizuka, et al., "Stochastic Error Rate Estimation for Adaptive Speed Control with Field Delay Testing," *ICCAD*, 2013.

[12] H. Fuketa, et al., "An Average-Performance-Oriented Subthreshold Processor Self-Timed by Memory Read Completion," *IEEE Trans. CAS II*, vol. 58, no. 5, 2011.

[13] T. Nakauchi, et al., "A Novel Technique for Mitigating Neutron-Induced Multi-Cell Upset by means of Back Bias," *IRPS*, 2008.

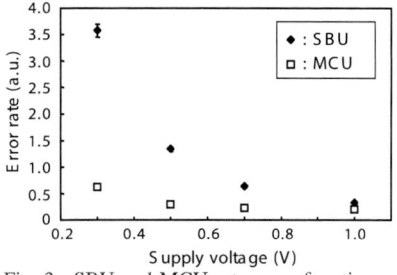

Fig. 3. SBU and MCU rates as a function of supply voltage of memory cell array [6]. SBU and MCU rates are plotted with error bars, where each error bar indicates ±3σ.

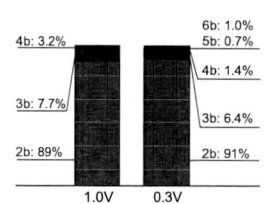

Fig. 4. Comparison of MCU distributions.

# Energy Efficiency Benefits of Subthreshold-Optimized Transistors for Digital Logic

P.J. Grossmann, S.A. Vitale and P.W. Wyatt

MIT Lincoln Laboratory, Lexington, MA

Email: grossmann@ll.mit.edu, Tel: (781) 981-1592

The minimum energy point of an integrated circuit (IC) is defined as the value of the supply voltage at which the energy per operation of the circuit is minimized. Several factors influence what the value of this voltage can be, including the topology of the circuit itself, the input activity factor, and the process technology in which the circuit is implemented. For application-specific ICs (ASICs), the minimum energy point usually occurs at a subthreshold supply voltage [1]. Advances in subthreshold circuit design now permit correct circuit operation at, or even below, the minimum energy point. Since energy consumption is proportional to the square of the supply voltage, circuit design techniques and process technology choices that reduce the minimum energy point inherently improve the energy efficiency of ICs.

Previous research has shown that optimizing process technology for subthreshold operation can improve IC energy efficiency [2]. This, coupled with the energy efficiency advantages offered by fully-depleted silicon-on-insulator (FDSOI) processes, have led to the development of a subthreshold-optimized FDSOI process at MIT Lincoln Laboratory (MITLL) called xLP (Extreme Low Power) [3]. However, to date there has not been a quantitative estimate of the energy efficiency benefit of xLP or other analagous technology for complex digital circuits.

This paper will show via simulation that the xLP process technology enables energy efficiency improvements that exceed that of process scaling by one generation. Specifically, the process is shown to improve power delay product by 57% vs. the IBM 90nm low power bulk process, and by 9% vs. the IBM 65 nm low power bulk technology at 0.3V.

## EXPERIMENTAL SETUP

The energy efficiency of three process technologies have been compared using a gate-level simulation-based technique. The process technologies considered were IBM 90nm low power bulk, IBM 65 nm low power bulk, and MITLL xLP, a 90nm subthreshold-optimized fully depleted silicon-on-insulator

This work was sponsored by the Assistant Secretary of Defense for Research & Engineering under Air Force under Air Force contract no. FA8721-05-C-0002. Opinions, interpretations, conclusions and recommendations are those of the authors and are not necessarily endorsed by the United States Government.

(FDSOI) process. The IBM process technologies offer several different transistor options with different threshold voltages; the standard threshold voltage was used in this study. Small standard cell libraries for each technology were laid out. Table 1 shows the transistor sizing of the unit inverter in each cell, used to set the relative drive strengths of other cells in the library. The lengths of the transistors in each technology were upsized by 67% of minimum to mitigate short-channel effects and improve robustness at subthreshold supply voltages. IBM PMOS widths were chosen to fit two contacts per diffusion and NMOS widths chosen to fit one contact per diffusion, rather than making a delay-based sizing choice. This represents a standard, nominal sizing for these technologies. The xLP process technology specifically targets a 1:1 P:N ratio at 0.3V, so the xLP cell library was sized to fit two diffusion contacts per diffusion for both NMOS and PMOS. In making nominal sizing choices within each process technology, rather than an iso-area comparison, the pros and cons of each process's area vs. energy efficiency tradeoffs are preserved in the results.

Table 1. Transistor Sizes for Unit Inverters in Three Process Technologies

|    | IBM 65 nm | IBM 90 nm | MITLL xLP |
|----|-----------|-----------|-----------|
| Wp | 350       | 480       | 500       |
| Wn | 150       | 200       | 500       |
| L  | 100       | 150       | 150       |

Parasitic netlists for all cells were obtained using Calibre xRC. The standard cell libraries were characterized using the parasitic netlists as input to Cadence Liberate at 0.3, 0.4, 0.5, and 0.6 V. To avoid bias toward synthesis results for any of the three processes or a particular supply voltage, the DES encryption circuit was synthesized only once, targeting a fourth process technology (IBM 180nm partially depleted silicon-on-insulator (PDSOI)) that had been characterized at 1.5V. By keeping the names of the standard cells consistent between all technologies, an iso-netlist comparison between the technologies was possible. The gate count for the DES enscryption circuit was 38024.

A set of 1000 random input vectors was applied to the DES engine as input stimulus. The probability of any input being 1 during any given clock cycle was 0.5, resulting in a net input activity factor of 0.25. For each process technology at each supply voltage, the circuit was simulated with gate-level Verilog and standard delay format (SDF) back-annotated timing using Modelsim. SDF information was regenerated for each voltage and process of interest using Cadence RTL Compiler. This step was also used to determine the maximum clock frequency of operation at each voltage for each process technology, and the maximum frequency was used in each simualtion. Value Change Dump (VCD) files obtained from each simulation were used with netlist and cell library information by Cadence RTL Compiler to obtain average power consumption over the 1000 clock cycles of simulation for each process/voltage combination. Power delay product was computed by multiplying the average power by the simulation clock period. It thus represents the average energy consumption per clock cycle for the 1000 input vectors applied. Power delay product is therefore the energy efficiency metric used to evaluate the three process technologies.

## ENERGY EFFICIENCY SIMULATION RESULTS

Figure 1 plots power delay product of the DES encryption circuit vs. supply voltage for IBM 90 nm low power bulk, IBM 65 nm low power bulk, and xLP 90nm FDSOI process technologies from 0.3-0.6V. In each process technology, this represents a voltage range spanning from subthreshold to near-threshold.

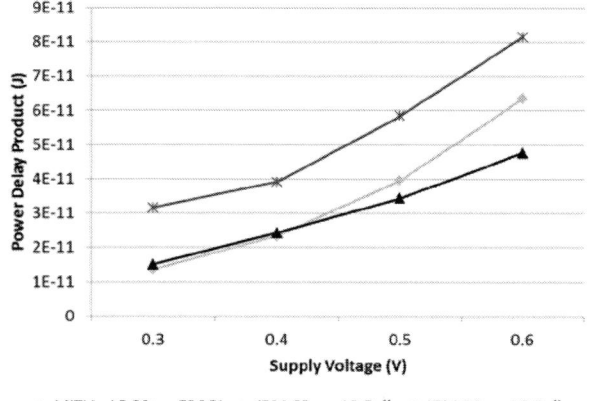

Figure 1. Power delay product vs. supply voltage for DES encryption circuit in three process technologies.

Figure 1 clearly shows that the IBM 90nm process lags behind the other two in energy efficiency. The energy efficiency of the IBM 65 nm and xLP technologies are similar. At near-threshold voltages, the IBM 65nm DES circuit is more energy efficient, while at subthreshold supply voltages, the xLP DES circuit is more energy efficient. Because the DES encryption circuit is efficient and being stimulated with high input activity in this experiment, it is most energy efficient at 0.3V. Table 2 compares the three technologies more precisely at 0.3V. Note

that the xLP implmentation compensates for increased static power with greatly improved speed vs. the bulk technologies. Overall, the xLP DES is 57% more energy efficient than the IBM 90 nm DES for which the area and capacitive loads are similar, and 9% more energy efficient than the 65nm DES despite a competitive disadvantage in area that results in net capacitances that are larger than their IBM 65nm counterparts, per the wireload models used in RTL Compiler.

Table 2. DES Encryption Circuit Energy Efficiency at 0.3V for Three Process Technologies

|  | IBM 65 nm | IBM 90 nm | MITLL xLP |
|---|---|---|---|
| Max Clock Frequency | 4.2 kHz | 15.3 kHz | 307 kHz |
| Static Power | 6 nW | 130 nW | 240 nW |
| Total Power | 64 nW | 480 nW | 4200 nW |
| PDP | 15.1 pJ | 31.5 pJ | 13.7 pJ |

## CONCLUSION

Using a simulation-based approach, the energy efficiency of the MIT Lincoln Laboratory xLP process was benchmarked against two commercial low power process technologies for a ~40,000 gate DES enryption circuit to demonstrate the benefits of subthreshold-optimized transistors for minimum energy circuit design. Because the transistors are tuned for low voltage operation, the xLP process performs best relative to other technologies at subthreshold supply voltages, where most circuits achieve their minimum energy point. With 90nm feature sizes, the xLP process at showed a 57% percent energy efficiency improvement vs. IBM 90nm technology and a 9% energy efficiency improvement vs. IBM 65 nm. Scaling the xLP process to 65 nm should provide further energy efficiency benefit.

## REFERENCES

[1] D. Blaauw and B. Zhai, "Energy efficient design for subthreshold supply voltage operation," in *Circuits and Systems, 2006. ISCAS 2006. Proceedings. 2006 IEEE International Symposium on*, 2006, p. 4 pp.-32.

[2] B. C. Paul, A. Raychowdhury, and K. Roy, "Device optimization for digital subthreshold logic operation," *Electron Devices, IEEE Transactions on*, vol. 52, no. 2, pp. 237-247, 2005.

[3] S. Vitale, P. W. Wyatt, N. Checka, J. Kedzierski, and C. L. Keast, "FDSOI Process Technology for Subthreshold-Operation Ultralow-Power Electronics," in *Proceedings of the IEEE*, 2010, vol. 98, pp. 333-342.

# Ultra-Wide Voltage Range 32-bit RISC CPU with Timing-Error Prevention in 28nm CMOS

Markus Hiienkari, Jukka Teittinen, Lauri Koskinen
University of Turku,
Technology Research Center,
Joukahaisenkatu 1C, 20520 Turku, Finland
E-mail: markus.hiienkari@utu.fi

Matthew Turnquist,
Mikko Kaltiokallio
Aalto University,
Department of Micro
and Nanosciences

Jani Mäkipää, Arto Rantala,
Matti Sopanen
VTT Technical Research Centre of Finland

*Abstract*—To minimize energy consumption of a digital circuit, logic can be operated at sub- or near-threshold voltage. Operation at this region is challenging due to device and environment variations, and resulting performance may not be adequate to all applications. This paper presents an ASIC implementation of a 32-bit RISC CPU in 28nm CMOS with wide range of adjustable voltage/frequency from 250mV/85kHz to 750mV/135MHz. The CPU employs timing-error prevention with clock stretching to enable operation with minimal safety margins while maximizing energy efficiency at a given operating point. Measurements show 3.15pJ/cyc energy consumption at 400mV, which corresponds to 39% energy savings and 83% EDP improvement compared to operation based on static signoff timing.

## I. INTRODUCTION

Near-threshold computing has caught increasing amount of attention lately. With the latest technologies, operating digital logic at sub- or near-threshold typically minimizes its energy consumption [1]. However, if operation at the minimum energy point (MEP) cannot satisfy temporal peak performance targets, dynamic voltage and frequency scaling (DVFS) is required. With energy-constrained systems, supply voltage typically varies with time (battery, solar cell etc.), creating additional challenges for ensuring efficient and reliable operation.

In this paper, we present a 32-bit microprocessor with ultra-wide supply voltage range, implemented in 28nm CMOS. It allows Ultra-Dynamic Voltage Scaling (UDVS) [2], ranging from subthreshold to nominal voltage to enable optimal energy-performance ratio for various applications. In order to minimize die characterization effort, the microprocessor voltage/frequency can be tuned using timing feedback from critical paths. To ensure correct operation with minimal safety margins, the system employs *timing-error prevention* (TEP) with adaptive clocking. TEP is a version of timing-error detection (TED) [3] [4] which has been shown to be effective in removing variation-incurred timing margins.

## II. DESIGN OF A 32-BIT RISC CPU WITH TEP

The CPU core of our system is a freely available, open-core LatticeMico32 CPU [5]. LM32 is a configurable medium-scale RISC microprocessor with 6 pipeline stages, full GCC toolchain support and sufficient performance (1.14 DMIPS/MHz, 1.83 Coremark/MHz) even for demanding sensor network applications.

We have enhanced the CPU with critical path monitoring combined with timing-error prevention. Our TEP system enables *time borrowing* (TB) on all paths, and it works as follows: When a late signal arrives, time borrowing occurs

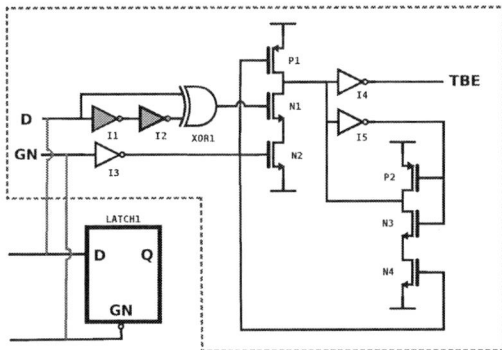

(a) The time borrow detector (TBD) circuit indicates time-borrow events (TBEs), i.e. transition at data pin (D) when the latch is transparent.

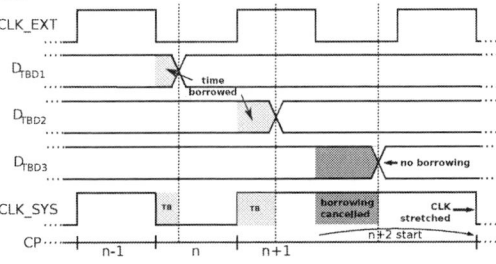

(b) A high-level illustration of adaptive clocking, where clock stretching prevents time borrow accumulation.

Fig. 1: An overview of the timing-error prevention system.

normally. *Time borrow events* (TBE) are detected with *time borrow detector* (TBD) circuits, which are integrated into critical path latches as illustrated in Fig. 1a. The circuit indicates when TB happens, and TBEs are combined and propagated to a clock control circuit, which is a state machine consisting of three flip-flops and three additional gates. It is responsible for error prevention, which is achieved by shifting the clock phase after a TBE as illustrated in Fig. 1b. Therefore, the TEP system prevents the stacking of TB, which could otherwise lead to a timing error.

Since the unmodified LM32 is a standard edge-sensitive pipeline, additional design steps are required to make it suitable for TEP. Latches and special flip-flops allow time borrowing. Pulse-latches and TB flip-flops (e.g. [6]) are the most straightforward to integrate into a given design, but they add minimum-delay constraints into fast paths and raise robust-

978-1-4799-7440-5/14 $31.00 © 2014 IEEE

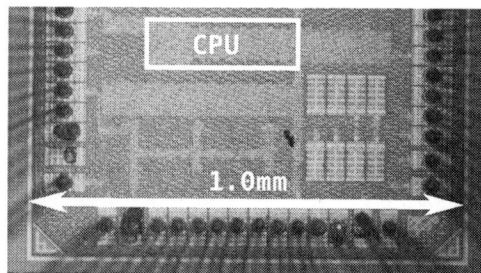

Fig. 2: Test chip microphotograph.

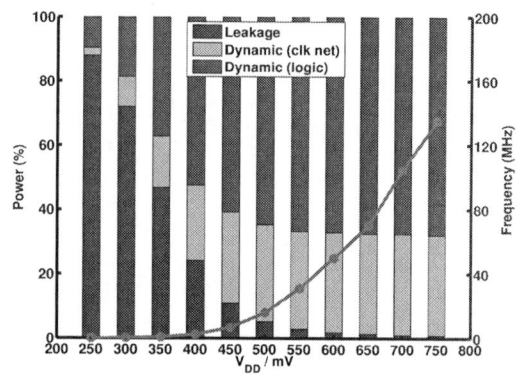

Fig. 3: Measured power distribution and performance of the CPU.

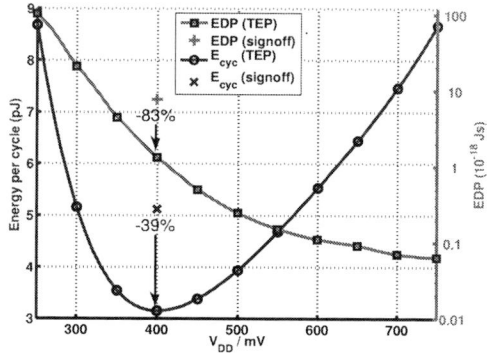

Fig. 4: Measured $E_{cyc}$ and EDP at supported voltage range, with signoff-based results as reference. A wide operation range enables an optimal tradeoff between energy and performance based on application requirements.

ness concerns at near-threshold voltages. Standard latches are readily available in foundry design kits and have no robustness issues, but the design needs to be fully transformed to operate as a dual-phase latch pipeline to avoid half-cycle min-delay requirement for critical paths.

We selected the latter alternative, and transformed RTL manually into a level-sensitive version. The transformation reduces maximum attainable clock frequency compared to original edge-sensitive version, but decreases clock network load while adding the important ability of time-borrow. On-chip memory was built from a standard cell latch array to allow wide operating range and level-sensitive operation.

## III. TEST CHIP MEASUREMENTS

The CPU was signed off with a 0.67MHz frequency target at 400mV $V_{dd}$, which resulted to lowest energy per cycle ($E_{cyc}$) in system simulations. Silicon measurements of the fabricated test chip (Fig. 2) verified correct operation at this point without TEP, but due to safety margins and inaccuracies in timing libraries, optimal energy or performance could not be achieved with static signoff-timing based operation. This was evidenced by studying timing feedback from critical paths. With the feedback and TEP active, the CPU could be run safely at 2.4MHz with the same 400mV $V_{dd}$. Voltage was scaled next to study energy/performance tradeoff at various operation points. Fig. 3 illustrates power distribution and performance of the CPU running a stress test at 250mV-750mV, while Fig. 4 shows resulting $E_{cyc}$ and energy-delay product (EDP).

The resulting frequency ranges from 85kHz to 135MHz, which caters a wide number of usage scenarios. As predicted by simulations, energy minimum (3.15pJ/cyc) is located at 400mV, while EDP minimum is found at highest voltage measured (750mV). If CPU clock was generated internally instead of an external clock, voltage and frequency could be raised even further to the nominal voltage of 1.0V. We estimate that the clock frequency at that point would be 293MHz. Compared to static signoff-based timing, the TEP system reduces $E_{cyc}$ by 39% and EDP by 83% at 400mV. Moreover, $E_{cyc}$ is relatively flat around MEP region (350mV-450mV), allowing free selection of performance target between 0.8MHz and 6.8MHz with minimal difference in energy consumption.

## IV. CONCLUSION

Ultra-Dynamic Voltage Scaling allows making an optimal tradeoff between energy and performance, but it has not been adopted widely in commercial systems due to complex design and reliability concerns. Shown here, is a simple adaptive technique, timing-error prevention. This technique helps the presented design to operate reliably with an ultra-wide voltage

range of 250mV-750mV. Compared to static signoff-based timing, the TEP system reduces energy consumption by 39% and EDP by 83%. Additionally, the technique increases robustness to ambient variability such as supply voltage ripple. This in turn allows for simpler power management design, a fact which will be proven in future implementations which encompass the system shown here.

## ACKNOWLEDGMENT

Academy of Finland Projects #124029 and #13139458.

## REFERENCES

[1] D. Bol *et al.*, "Interests and limitations of technology scaling for subthreshold logic," *Very Large Scale Integration (VLSI) Systems, IEEE Transactions on*, vol. 17, no. 10, pp. 1508 –1519, oct. 2009.

[2] B. Calhoun and A. Chandrakasan, "Ultra-dynamic voltage scaling (udvs) using sub-threshold operation and local voltage dithering," *Solid-State Circuits, IEEE Journal of*, vol. 41, no. 1, pp. 238–245, Jan 2006.

[3] K. Bowman *et al.*, "Energy-efficient and metastability-immune timing-error detection and recovery circuits for dynamic variation tolerance," in *ICICDT 2008.*, June 2008, pp. 155–158.

[4] J. Mäkipää *et al.*, "Timing-error detection design considerations in subthreshold: An 8-bit microprocessor in 65 nm cmos," *Journal of Low Power Electronics and Applications*, vol. 2, no. 2, pp. 180–196, 2012.

[5] "LatticeMico32 open, free 32-bit soft processor." [Online]. Available: http://www.latticesemi.com/en/Products/DesignSoftwareAndIP/ IntellectualProperty/IPCore/IPCores02/LatticeMico32.aspx

[6] M. Choudhury *et al.*, "TIMBER: Time borrowing and error relaying for online timing error resilience," *DATE, 2010*, pp. 1554 –1559, march 2010.

978-1-4799-7440-5/14 $31.00 © 2014 IEEE

# A Reduced-Memory FIR Filter Using Approximate Coefficients for Ultra-Low Power SoCs

Alicia Klinefelter and Benton H. Calhoun

University of Virginia, Charlottesville, VA 22903 USA, Email: {amk5vx, bcalhoun}@virginia.edu

*Abstract*-**This paper presents an ultra-low power (ULP) finite-impulse response (FIR) filter using a method that approximates filter coefficients on-chip without reliance on dedicated memory such as SRAM. In a system-on-chip (SoC) context, this method allows for full power gating of the coefficient unit without coefficient state loss, and runtime modifications of filtering specifications, such as filter order and cutoff frequency. Using trigonometric approximation methods for the sinc and resource sharing of computational units, a single coefficient is generated in five clock cycles. The approximation unit is compared against standard-cell-based memories, such as register and latch files, for energy and area, and the design is synthesized in 130nm CMOS consuming 6.9nW at 300mV and 6.5kHz.**

## I. INTRODUCTION

Supply voltage scaling into the subthreshold region of operation is becoming an increasingly attractive solution to save energy and power in cases where performance is not the driving factor. Applications with low-speed requirements, such as environmental and physiological monitoring, can operate at processing frequencies in the 10-100kHz range, while the acquired signals, such as electrocardiographs (ECGs), have sampling rates in the hundreds of Hz. This leaves hundreds to thousands of clock cycles between samples for processing, while enabling operation in the subthreshold region. In this region, where leakage energy begins to dominate, serializing logic at the cost of more clock cycles has little impact on energy consumption [1]. Flexible and programmable biomedical SoCs often require large sets of filtering coefficients during deployment for a variety of scenarios (e.g. removing 60Hz power line noise or band selection). This can lead to large amounts of coefficient storage, and for batteryless systems such as in [2], an intermittent supply can lead to complete coefficient state loss and the need for reprogramming.

Generally, an FIR filter's coefficients are stored in an on-chip SRAM or register file that must also operate in subthreshold. The SoC's data memory is rarely power-gated as it also stores chip data that may be unrelated to the filtering process. This leads to excess leakage in memory dedicated to stored coefficients when the filter is not being used. For ULP SoCs, on-chip SRAMs can consume ~60% of the total digital power, and if not power gated can dominate the digital power budget [3]. They are often the primary barrier to low-voltage operation and pose the first failure point in low-voltage designs [1]. Additionally, in the deep subthreshold region, decreased $I_{on}/I_{off}$ ratios reduce the reliability of SRAM. Standard-cell-based coefficient memories that are local to the accelerator to hold data and coefficients are an alternative, but this puts an upper bound on the order of the filter and becomes energy and area inefficient for sizes >1kb [4]. By generating the coefficients in real-time using digital, synthesized logic, design robustness is improved and the minimum supply voltage and leakage is decreased without the need for SRAM-based storage.

## II. METHODS OF APPROXIMATION

Assuming a symmetric FIR filter, the filter can be implemeted using a folded delay line shown in (1).

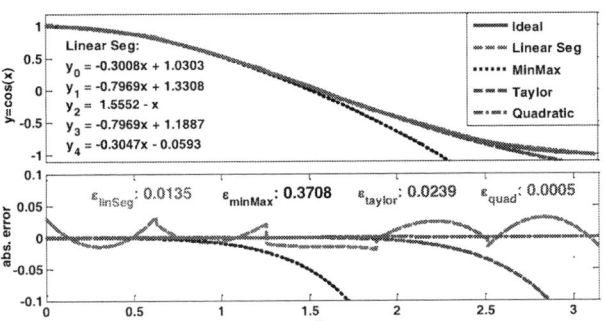

Fig. 1. Approximations (top) and absolute error (bottom) for cosine $[0, \pi]$. Average error, $\varepsilon$, is show for each method.

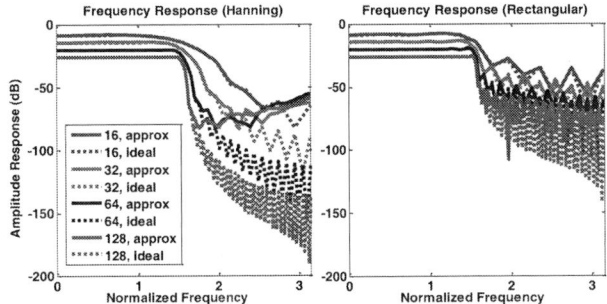

Fig. 2. Ideal and approximated frequency response functions for $f_c=0.5\pi$ using a rectangular and Hanning window.

$$y(i) = \sum_{k=0}^{M/2} h(k)[x(i-k) + x(k)] \qquad (1)$$

Here, $x$ is the sampled and delayed input data, $h$ is the array of coefficients, $M$ is the filter order, and $y$ is the filtered output. This form reduces the number of multiplications by a factor of two relative to the classical form. FIR filtering coefficients are sampled points along the sinc function, $sin(x)/x$, where windowing functions are applied to smooth the effects of truncating this infinite function. Although there are other methods of creating filter coefficients such as frequency sampling and least-squared, many of these are computationally complex requiring Fast-Fourier Transforms or optimization methods.

To generate the sinc, the sine function must be approximated in a way that isn't "too complex" to implement in hardware, but is also "accurate enough". Four methods were evaluated and the results are shown in Fig. 1. A common approximation used for trigonometric functions is the Taylor series. This series is expensive to compute for accurate results (i.e. more terms) and the resulting error is only acceptably low for small values of the input argument [5]. A 7th order approximation was used here, but it required seven multiplications, three divisions, and three additions to compute the result. An alternative method uses minmax polynomials. This method is often preferred over the Taylor polynomial method due to its distributed error over the range, which also decreases with additional terms, but suffers from the same computational complexity issue as Taylor's approximation [5]. The third method, a quadratic approximation, uses a parabola to approximate the sine/cosine function across the

978-1-4799-7440-5/14 $31.00 © 2014 IEEE

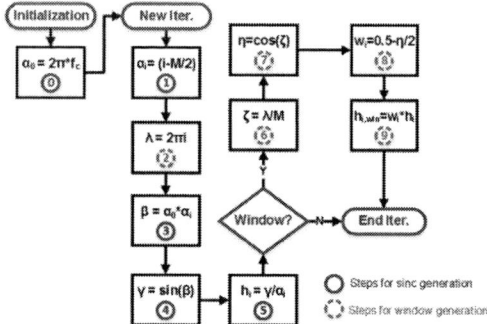

Fig. 3. Data-flow graph showing the coefficient generation algorithm using a Hanning window. Circled numbers indicate the current clock cycle (state).

[0, π] range, but requires three total multiplications per approximation. The final method uses linear segments to approximate the cosine and thus requires only one multiplication per computation. An unconstrained nonlinear optimization method was used to choose slope and intercept values that reduced the residual sum of squares of the fitted lines, and the segments are shown in Fig. 1. Five linear segments were used to achieve comparable error with the other, higher order methods, while requiring fewer hardware resources, and this was used for the final design.

## I. COEFFICIENT GENERATION

Using a low-pass filter as an example, the methodology for coefficient generation is described here. First, the filter specification is defined by providing the desired cutoff frequency, $f_c$, and the filter order. From this, an ideal impulse response function (sinc) can be generated (2).

$$h(i) = \frac{\sin(2\pi f_c(i-M/2))}{i-M/2} \qquad (2)$$

This impulse response is delayed to keep the system causal, and the sinc function is multiplied by a specified window function and sampled to obtain coefficients. Many of the most common windows such as Hanning (3), Hamming, or Blackman contain cosine functions that require an additional trigonometric computation. For this purpose, the approximation unit is dual-purposed for both sine and cosine approximations, using a shift of $\pi/2$.

$$w(i) = 0.5 - 0.5\cos(2\pi i/M) \qquad (3)$$

The approximated frequency response functions were compared with the ideal response in Fig. 2 for both rectangular ($w(i) = 1, \forall i$) and Hanning windows. Although the passband regions appear unaffected by the approximation, the stopband attenuation is slightly degraded for the approximated cases.

Since this algorithm can be mapped to hardware in a variety of ways, the specific, serialized method is shown in Fig. 3. Due to the resource-shared hardware, state order had to be carefully chosen to prevent combinational loops.

## II. RESULTS

To determine the suitability of this method in a real system, an approximation and multiply-accumulate (MAC) unit were taped out in a 130nm commercial process with chip micrograph and logical area distribution shown in Fig. 4. Fig. 5 shows the measured energy results of the approximation unit against the simulated results of 16-bit register/latch files of different

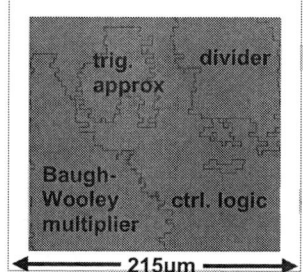

Fig. 4. Chip micrograph (left) and approximation unit division of area (right).

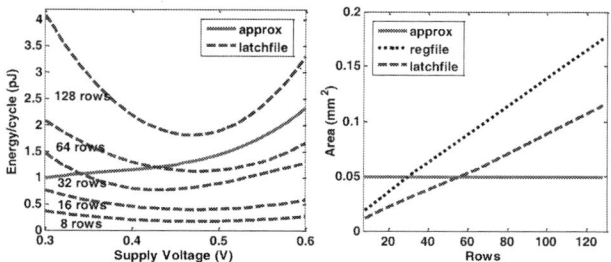

Fig. 5. Energy and area comparison of the approximation unit with 16-bit register and latch files of various depths.

depths/rows, corresponding to number of taps. From this, the point at which this method becomes more advantageous than standard-cell-based coefficient storage is seen with respect to supply voltage. Since the area of the approximation unit remains constant for any number of tap values, it becomes more area efficient than a register file at ~30 rows/taps and a latch file at ~55 rows/taps. The design was synthesized in 130nm CMOS and was tested to operate down to 300mV where it consumed an average energy of 1pJ/state.

## CONCLUSION

This work presents a methodology for generating coefficients on-chip in lieu of storing them in high-leakage SRAMs operating below the threshold voltage. This design can enable reliable, ULP operation for applications requiring large banks of coefficient data and diverse filtering operations. By evaluating a variety of trignometric approximation methods for the complexity-accuracy tradeoff, a linear segment-based approximation was selected to generate coefficients. Although this coefficient generation adds latency to the filtering operation, there is often available slack for serial processing in low-throughput, subthreshold systems.

## ACKNOWLEDGEMENTS

This work is funded by the NSF NERC ASSIST Center (EEC-1160483).

## REFERENCES

[1] Verma, N., et al., "Nanometer MOSFET Variation in Minimum Energy Subthreshold Circuits," T-ED, 2008.

[2] Zhang Y., et al., "A batteryless 19 μW MICS/ISM-band energy harvesting body area sesor node SoC for ExG applications," JSSC, 2013.

[3] Khayatzadeh M., et al., "A 0.7-V 17.4-μW 3-Lead Wireless ECG SoC," BIOCAS, 2013.

[4] Meinerzhagen, P., et al., "Towards generic low-power area-efficient standard cell based memory architectures," MWSCAS, 2010.

[5] Muller, J-M., "Elementary Functions: Algorithms and Implementation," Birkhäuser Publishing, 2005.

978-1-4799-7440-5/14 $31.00 © 2014 IEEE

# Monolithic 3D Integration: a powerful alternative to classical 2D Scaling

M. Vinet, P. Batude, C. Fenouillet-Beranger, F. Clermidy, L. Brunet, O. Rozeau, JM Hartmannn,
O. Billoint, G. Cibrario , B. Previtali, C. Tabone, B. Sklenard, O. Turkyilmaz, F Ponthenier, N. Rambal, MP. Samson[2], F. Deprat, V. Lu[2],
L. Pasini[2], S. Thuries, H. Sarhan, J-E. Michallet, O. Faynot

CEA-LETI, Minatec, [2] STMicroelectronics, 17 rue des Martyrs, Grenoble FRANCE
Email: maud.vinet@cea.fr, Tel: +33-438789900

*Abstract – Monolithic or sequential 3D Integration is a powerful technological enabler for actual 3D IC design as the stacked layers can be connected at the transistor scale. This paper reviews the opportunities brought by M3DI and highlights the applications benefiting from this small 3D contact pitch. It also presents the technological challenges of this concept and offers a general overview of the potential solutions to obtain a high performance low temperature top transistor while keeping bottom MOSFET integrity.*

## I. INTRODUCTION

For many reasons, classical CMOS scaling driven by Moore's law is being more and more questioned. In addition to the usual Power/Performance/Area (PPA) trade-off, costs considerations are opening up post 10nm technological options [1 ,2]. Extremely complex and costly engineering work required achieving 10nm and post 10nm Moore's law compatible devices [3] now favors more global circuit consideration rather than just transistor centric options [4, 5]. In this context three dimensional circuits (3D IC) have emerged as a promising option to extend further global PPA-cost benefit [6]. Among the 3D integration schemes, Monolithic 3D integration (M3DI) aims at processing transistors on top of each other sequentially. Its interest resides in providing ultra-high 3D contact density. However its implementation faces the challenge of being able to obtain a high performance top transistor processed at Low Temperature (LT) – typically at temperatures below 600°C – in order to preserve the bottom FET from any degradation during the top stacked FETs fabrication. This paper reviews the opportunities brought by M3DI and the technological optimization required to enable the integration.

## II. MONOLITHIC 3D UNIQUE TRAITS

Major asset of M3DI resides in its ability to reach a 3D contacts density, *ie* between the devices layers, close to planar contacts density. Two main reasons explain of the high density of contacts. First by construction M3DI leads to an excellent alignment precision between top and bottom transistors, process flow principle is detailed in fig.1. Alignment precision only depends on lithographic alignment capability of the stepper [7]. Then 3D contact process is very close to a standard 2D tungsten contact plug: the 3D contact is a simple contact in an oxide and its aspect ratio is kept very similar to the one in a 2D integration. Fig2 summarizes the vias density achievable either by parallel (3DIC) 3D integration or by M3DI showing that *M3DI allows a via density over of a million/mm² and up to 100 million/mm² is projected with 14nm ground rules [8].*

## II. OPPORTUNITIES

### a) Design considerations

*For CMOS logic applications, basic interest of M3DI resides in the* fact that it allows gaining IC performance without the additional cost associated to transistor scaling, fig.3. Performance gain is achieved thanks to wire length reduction.

As an example, as Field Programmable Gate Arrays (FPGAs) are one of the circuits benefiting the most of advanced technologies, we have used them to evaluate the potential gains brought by M3DI. A logic-on-memory approach is selected: memory cells in the bottom and logic in the top layer. Two intermediate metal layers are used to route the SRAM bitlines wires (in the bottom layer) and to connect the vertical vias through the two metal layers in the bottom. Hence, it is possible to adjust memory cell alignment in order to minimize the area difference to reach equilibrium between the top and bottom layers.

Results are extracted through electrical simulations and transferred in a high-level FPGA simulator.

It can be observed in fig. 4 that the area is reduced by 55%. The area benefits of M3DI can be explained as follows: the memory, corresponding to more than 45% of the initial 2D design is completely removed from the logic layer [9].

Second, due to the high granular vertical connections, replacing the memory on the bottom layer does not impose any routing congestion. Especially, the use of intermediary metal layers on the bottom layer enables very flexible memory placement while keeping high proximity to the logic layer. With careful design practice, the area overhead of the vertical connections on the logic can be completely avoided.

As for the final metric, fig.5 shows that the Energy-Delay Product (EDP) can be reduced by 47%. The improvement in the EDP is twofold: the intrinsic delays of the blocks are reduced with 3D integration due to simplified internal routing; as a consequence of shorter wire length between blocks, the capacitance of the routing wires is decreased. Therefore, the operations are carried out faster while consuming less energy which results in an improved EDP of the 3D FPGA.

It is also expected that the thermal difficulties can be overcome with the proposed partitioning. Since the memory layer holds the configuration information, there is no dynamic evolution in the values of the SRAM once it is written. Therefore, the heat generation in the bottom layer is minimized.

### b) Integration considerations

*With a CMOS over CMOS integration* (for instance like in the previous FPGA partionning example), *it is possible to optimize the memory and logic layers separately:* low leakage/high-Vt and high performance/low-Vt processes can be applied to memory and logic layers for optimal performance.

On a longer time scale, *Monolithic 3D Integration can also be a way to co-integrate different channel materials* [10] such as high mobility materials (Ge for PFET and III-V for NFET [11]) without the need to come with planar co-integration tricks. In this scheme

978-1-4799-7440-5/14 $31.00 © 2014 IEEE

CMOS co-integration is eased by independent optimization of the different layers.

### c) More Moore opportunities

Once again thanks to its high 3D via density, M3DI allows thinking of daring applications as shown in fig.6: for instance highly miniaturized imagers benefits from a 3D stacking [12]. The logic can be designed on top of the sensing photo-diode improving the sensitivity. Another field of interest is the CMOS/ MEMS co-integration [13] where bringing the computing part close the MEMS sensing part allows reducing signal attenuation and simplifying signal detection (with no signal down-mixing).

## III. M3D ENABLERS

### d) Integration analysis

Main technological challenges are listed in fig.7. In order to ensure a high quality of the top layer film, we resort to optimized low temperature molecular bonding [14]. Pristine wafer quality is preserved at the wafer scale as shown in fig.8. The most challenging development consists in the fact that the top transistor thermal budget (TB) is capped. As shown in fig.9, the acceptable top TB is a trade-off between interconnection and bottom MOSFET thermal stability and top MOSFET process steps. Analysis of the top MOSFET process flow evidences the dopant activation as the most critical TB: routinely RTA activation annealings are performed above 1000°C with a time scale around 1s.

Fig.10 presents the stability of a state-of-the art FDSOI transistors [15] providing two directions to set up top TB limit: performance degradation is observed above 500°C and it is coming from a salicide degradation. Improving silicide thermal stability [16] thus allows pushing upwards the transistor thermal stability. As shown in fig.11, the TB as experienced by the bottom MOSFETs results from a (duration, temperature) couple optimization which corresponds to different activation techniques. Basically we can either perform very short and high temperature annealing to contain the heat in the top MOSFET or resort to longer lower temperature methods. Use of laser annealing which provides high temperature activation with low in-depth thermal heat diffusion is promising [17]. This technology is very dependent of the underlying stack and requires a fine tuning of heat absorption/diffusion layers. Currently Solid Phase Epitaxy Regrowth (SPER) is the most mature way of activating dopant below 600°C. In this technique, dopant activation occurs during the recrystallization of the purposely amorphized source and drain region. Because of high activation level, provided a dose specific optimization [18], lower sheet resistance than with conventional RTA annealing can be reached as shown in fig.12. Fig. 13 provides an example of Monolithic 3D Integration where top and bottom MOSFET have been stacked with a lithographic alignment precision. 3D inverters demonstrate that both CMOS layers are functional and interconnected [19].

### e) Design tools discussion

From an IC conception perspective, taking full advantage of Monolithic 3D Integration is still challenging because of the lack of fully functional 3D place and route tools. It leads to a complexity factor for a proper benchmarking of 3D monolithic compared to planar integration. We still resort to full hand-made layout to design ICs. For instance the main parts of the FPGA namely the multiplexers, switch boxes and look-up-tables [9] were entirely customized using a 14nm FDSOI M3DI design kit.

## IV. CONCLUSION

Monolithic 3D integration allows stacking several layers of devices on top of each other with a unique connecting via density above a million/mm². For logic applications, it allows gaining performance thanks to a wire length reduction: area reduction of 55% has been shown in a FPGA case study. To enable M3DI most critical technological enablers resides in the top CMOS dopant activation. Currently Solid Phase Epitaxy Regrowth has proven its efficiency to contain thermal budget around 600°C and is expected to be effective downwards.

**ACKNOWLEDGMENT** - This work is partly funded by the ST/IBM/LETI Joint Development Program and by Qualcomm. It was also partly funded by the French Public Authorities through NANO 2017 and Equipex FDSOI11.

### References

[1] Handel Jones, IBS analyst, *High cost per wafer, long design cycles may delay 20nm and beyond*, http://electroiq.com/petes-posts/2014/01/22/high-cost-per-wafer-long-design-cycles-may-delay-20nm-and-beyond/

[2] Zvi Or Bach, Monolithic 3D Inc, Moore's Law seen hitting big bump at 14 nm, http://www.eetimes.com/author.asp?section_id=36&doc_id=1286922.

[3] Geoffrey Yeap, *Smart Mobile SoCs Driving the Semiconductor Industry, Technology Trend, Challenges and Opportunities*, IEEE Int Electron Device Meeting (IEDM), (2013).

[4] M Ritter, *Materials, Analytics & Science of Information Processing*, IBM research colloquia, Madrid (2012).

[5] M. Vinet, *Advanced CMOS roadmap*, Leti workshop @ IEDM (2013).

[6] Subramanian Iyer, *Orthogonal Scaling - the Role of Packaging and 3D Integration*, ISSCC 3D Stacking Technologies for Image Sensors and Memories Forum (2014).

[7] P. Batude, M. Vinet, A. Pouydebasque, L. Clavelier, C. LeRoyer, C.Tabone et al., *Enabling 3D monolithic integration*; Proceedings of the Electro-Chemical Society (ECS) spring meeting VOL 16 pp47 (2008)

[8] P. Batude et al., *3D sequential integration opportunities and technology optimization*, IITC-AMC proceedings, p. 373 (2014).

[9] O.Turkyilmaz et al., *3D FPGA using high-density interconnect Monolithic Integration*, IEEE Design Automation and Test Conference, DATE'14

[10] P. Batude, M. Vinet, A. Pouydebasque, C. Le Royer, B. Previtali, C. Tabone *et al*, *GeOI and SOI 3D Monolithic Cell integrations for High Density Applications*, VLSI 2009.

[11] T. Irisawa, K. Ikeda, Y. Moriyama, M. Oda, E. Mieda, T. Maeda, and T. Tezuka, *Demonstration of Ultimate CMOS based on 3D Stacked In-GaAs-OI/SGOI Wire Channel MOSFETs with Independent Back Gate*, p118, 978-1-4799-3332-7/14/$31.00 ©2014 IEEE, VLSI Symposium (2014).

[12] P Coudrain, P. Batude, X. Gagnard, C. Leyris, S. Ricq, M. Vinet et al., *Setting up 3D Sequential Integration for Back-Illuminated CMOS Image Sensors with Highly Miniaturized Pixels with low temperature Fully Depleted SOI transistors*, IEEE Int Electron Device Meeting (IEDM), pp 271 (2008).

[13] P. Batude,T. Ernst,J. Arcamone,G. Arndt,P. Coudrain, P.-E. Gaillardon, *3D sequential integration: a key enabling technology for heterogeneous co-integration of new function with CMOS*, JETCAS 2012 714-722.

[14] L .Brunet et al, Direct Bonding: a Key Enabler for 3D Monolithic Integration, ECS 2014.

[15] C. Fenouillet-Beranger, B. Previtali, P.Batude, F. Nemouchi, M. Cassé et al, *FDSOI bottom MOSFETs stability versus top transistor thermal budget featuring 3D monolithic integration*, ESSDERC 2014.

[16] F. Nemouchi, E.Bourjot, V. Carron, Y. Morand, J.P. Barnes, P. Batude et al., *W implantation for Ni0,9Pt0,1-silicide and -germano-silicide morphology enhancement featuring CMOS 3D monolithic integration*, MAM 2014

[17] J. Venturini et al., *Laser Thermal Annealing: Enabling ultra-low thermal budget processes for 3D junctions formation and devices*, IWJT 2012 p57-62.

[18] L Pasini et al, *Insights In Accesses Optimization for nFETLow Temperature Fully Depleted Silicon On Insulator Devices*, IWJT 2014.

[19] P. Batude, M. Vinet, B. Previtali, C. Tabone, C. Xu. Et al, IEDM proceedings, p. 151 (2011)

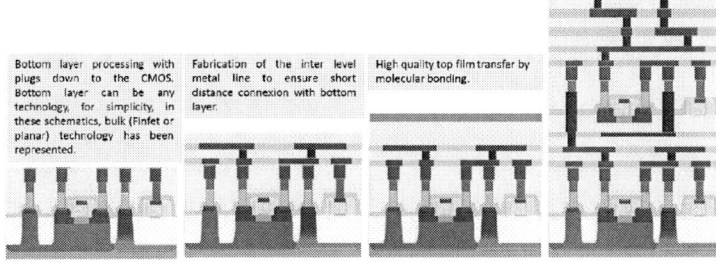
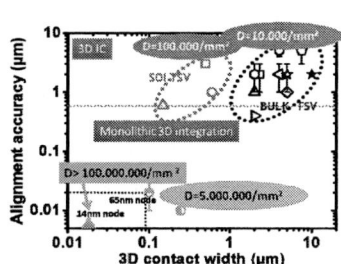

Fig. 1: Process flow principle of Monolithic 3D Integration. By resorting to a unique alignment flow throughout the whole process, layers are stacked on top pf each other within a lithographic alignment precision.

Fig. 2: 3D via density between CMOS layers depending on the 3D integration scheme. M3DI allows reaching over a million/mm² as opposed to parallel 3D IC where typically via density is below 100 thousand /mm² [8].

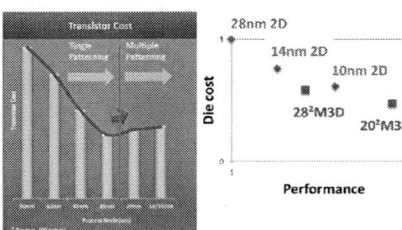

Fig. 3: Cost vs performance opportunities for Monolithic 3D Integration. By stacking older generations M3DI allows containing the cost increase associated to double patterning introduction.

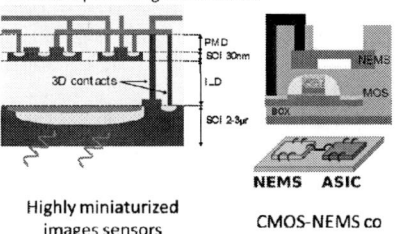

Fig. 4: Area gain for memory-logic partitioning of FPGA [9].

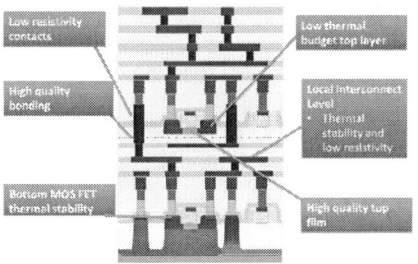

Fig. 5: Energy x Delay Gains for memory-logic partitioning of FPGA [9].

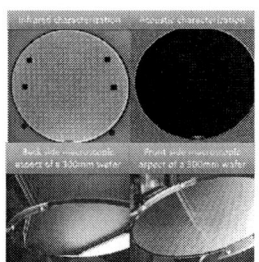

Fig. 6: More than Moore opportunities. Image sensors benefit from M3DI by partionning photo-diode and CMOS layer for signal treatment. NEMS and CMOS co-integration benefits from M3DI from a strong increase of signal to noise ratio [12, 13].

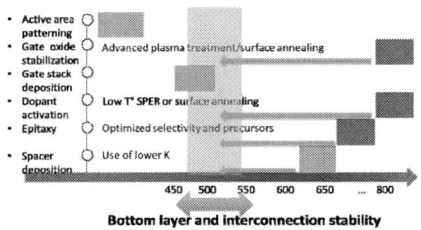

Fig. 7: Main technological challenges for M3DI. Thermal stability is required for CMOS bottom layer and local interconnects level to be compatible with the low thermal budget of the top CMOS layer.

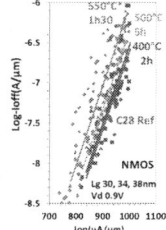
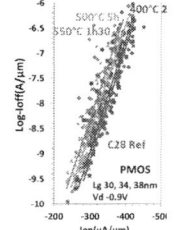

Fig. 8: Top – bonding characterization of 200mm wafers after bonding showing a high bonding quality at the wafer scale. Bottom – 300mm macroscopic characterization of wafers after bonding [14].

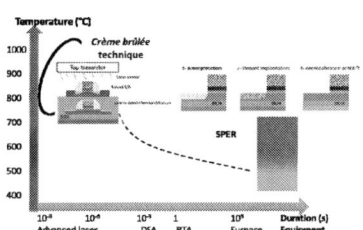

Fig. 9: Top thermal budget analysis and solutions to bring the process temperature of the critical thermal steps in the 500°C range.

Fig 10: Study of a state-of-the-art FDSOI MOSFET technology. Performance degradation is observed above 500°C and is attributed to silicide degradation [15].

Fig 11. Top thermal budget dopant activation strategies

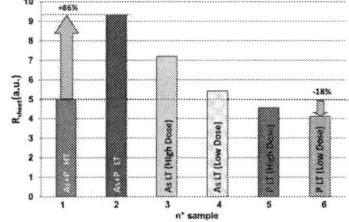
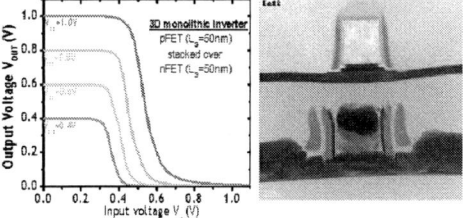

Fig. 12: R$_{sheet}$ comparing SPER at 600°C (Low temperature – LT) and RTA (1050°C) activated implanted Si layers for several doses [18].

Fig. 13: Electrical results of a 3D monolithic inverter featuring P over TFET. M3DI cross-section highlighting the alignment accuracy between top and bottom layers [19].

# Design Challenges and Solutions for Ultra-High-Density Monolithic 3D ICs

Shreepad Panth[1], Sandeep Samal[1], Yun Seop Yu[2], and Sung Kyu Lim[1]

[1]Dept. of Electrical and Computer Engineering, Georgia Institute of Technology, Atlanta, USA

[2]Hankyong National University, Korea

*Abstract*—**Monolithic 3D ICs (M3D) are an emerging technology that offers an ultra-high-density 3D integration due to the extremely small size of monolithic inter-tier vias. We explore various design styles available in M3D and present design techniques to obtain GDSII-level signoff quality results for each of these styles. We also discuss various challenges facing each style and provide solutions to them.**

## I. INTRODUCTION

Monolithic 3D ICs (M3D) is an emerging technology that offers orders of magnitude higher integration density than conventional through-silicon-via (TSV) based 3D ICs. This is because it utilizes a sequential stacking process, which eliminates the need for tier alignment. This enables the monolithic inter-tier vias (MIVs) to be the same size as regular local vias ($< 100nm$) [1].

This ultra-high-density enables several design styles, as shown in Figure 1. First, with respect to SRAM, the PMOS and NMOS of the bit-cell can be split onto multiple tiers. This gives us the opportunity to tune the PMOS and NMOS process separately. Next, a similar separation can be done for standard cells themselves, and this is known as transistor-level M3D. This design style has both intra-cell and inter-cell MIVs. Another design style is gate-level M3D, where the standard cells themselves are 2D, but they are placed in a 3D space, and interconnected using MIVs. This design style has only inter-cell MIVs. Finally, the coarsest level of integration in block-level M3D, where each functional block is 2D, and the 2D blocks are floorplanned onto a 3D space. In this design style, the MIVs are limited to the whitepsace between blocks. We now discuss each of these design styles in detail.

## II. MONOLITHIC 3D SRAM

Monolithic 3D offers a unique optimization opportunity for SRAM designs [2]. We can split the PMOS and NMOS onto different tiers, which allows us to optimize the process of each type of transistor independently. We pick a state-of-the art 6T SRAM cell as our 2D baseline. This is designed in a 22nm node, and it has an area of $0.1 \mu m^2$. The default 6T SRAM cell has a 2 PMOS and 4 NMOS (2P4N) configuration. The obvious choice is to blindly split up this bit-cell into two tiers, but we observe that it only gives us a 33% footprint reduction due to the imbalance in the PMOS and NMOS count. We therefore explore various alternative design options to give us a larger footprint reduction.

The first option we explore is the same 2P4N configuration, but with different sizing. We are able to obtain a footprint reduction of 44% with the same static noise margin (SNM) as 2D, but slightly worse write stability. Next, we explore a 3P3N configuration, replacing one pass transistor with an NMOS. The footprint reduction in this case is 45%. Using a single-ended read technique, we are able to achieve a high SNM margin. Lastly, we explore an 8T bit-cell, changing the conventional 2P6N configuration to 4P4N for area

This research is supported by Intel Research, Qualcomm Research, and the Center for Integrated Smart Sensors (CISS-2012366054194).

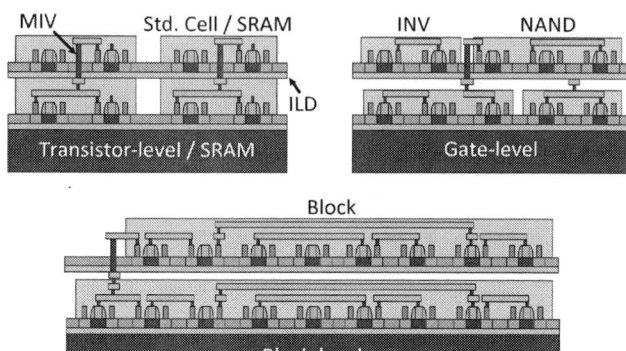

Fig. 1. Various design styles available for monolithic 3D ICs.

balance. This gives us a 40% footprint reduction under the same read margin, write margin and access time as 2D.

## III. TRANSISTOR-LEVEL MONOLITHIC 3D ICS

Transistor-level M3D is similar to the SRAM case in the sense that the PMOS and NMOS are split onto multiple tiers. In this design style, each standard cell is redesigned such that its PMOS and NMOS are on different tiers [3], [4], [5]. As in the case of SRAM, the advantage of doing this is that the PMOS and NMOS can be optimized separately.

We begin by constructing a library of 66 monolithic 3D standard cells using a cell folding technique. When compared with 2D, we observe a footprint reduction of about 40% because of an imbalance between PMOS and NMOS sizes. We re-characterize the cells taking into account the new cell internal parasitics. The advantage of this design style is that we can utilize existing 2D P&R tools to perform all the design steps for us. The standard cells have pins on different metal layers, and the router is capable of connecting all these pins together, inserting inter-cell MIVs in the process.

Since the total number of pins remain the same and the footprint is reduced, there is a 1.7-2× increase in the pin density of the chip, which causes several routability issues. We explore several interconnect options to mitigate this impact. Our first comparison set is the default case with one metal layer on the bottom tier (1BM), three additional metal layers on the top tier (3TM), and three additional metal layers on the bottom tier (4BM). We observe that the 4BM case leads to a significant increase in the cell internal parasitics, which increases the cell delay and power by up to 9.86% and 15.65% respectively. Overall, we observe that the 3TM case gives best results with up to a 22% reduction in the total power of the chip. We also explore other options like utilizing two intermediate and two global metal layers instead of three intermediate metal layers, and this gives us a further 2.8% power benefit.

We also study the benefit of this design style at more advanced and future technology nodes such as $22nm$ and $7nm$. We observe that at the $22nm$ and $7nm$ nodes, we get an additional 4% and 23% power benefit respectively.

## IV. Gate-level Monolithic 3D ICs

In this design style, existing 2D standard cells are placed onto a 3D space [6], [3], [7]. The advantage of this design style is that it offers reuse of existing standard cells, a 50% or more footprint area reduction, and since each tier has equal number of metal layers as 2D designs, no increase in pin density.

We propose a design flow based on "shrunk 2D" gate placement that leverages existing commercial 2D placers. This approach halves the footprint area and doubles the placement capacity in each place-ment bin to give an initial placement. The shrunk 2D placement is then partitioned with local area balance in each placement bin to give us a gate-level M3D design. We demonstrate that it can give us up to 30% HPWL savings when compared with 2D ICs. We also propose a commercial-router driven MIV insertion algorithm that improves the routed wirelength (WL) by up to 16.6% and the power delay product (PDP) by up to 6.1%. Next, we propose a routability-driven partitioner that utilizes the fine-grained nature of MIVs to reduce routing congestion. Our approach helps give us an additional 4% WL and 4.33% PDP benefit. We also demonstrate that using multiple MIVs per 3D net can help give us a 8.43% WL benefit and 2.25% power benefit. Next, we propose techniques for utilizing a commercial tool for timing optimization and clock-tree-synthesis (CTS). We demonstrate that keeping the clock backbone on a single tier gives us 29.82% clock power reduction compared to the case where we have one separate clock tree per tier. Overall, we demonstrate on the OpenSparc T2 design that M3D can give a 15.57% power benefit compared to commercial-quality 2D designs. We also demonstrate that this benefit rises to 16.08% when utilizing dual-$V_t$ libraries.

We also explore power delivery network (PDN) issues in M3D [8]. Since the top metal layers need to be used for both PDN as well as MIV landing pads, we demonstrate that PDN increases the M3D WL by 20.5% compared to only a 7.1% increase in 2D. This in turn increases the net power and temperature, reducing the benefit of M3D. We propose PDN optimization techniques that reduce the routed WL by up to 8% and the maximum temperature by up to 5% while still meeting the original IR-drop budget.

## V. Block-level Monolithic 3D ICs

Block-level monolithic 3D ICs utilize existing functional 2D IP blocks and floorplan them onto a 3D space [9], [10]. This design style can be used for SoC-level integration, and it also has the benefit of IP reuse.

We present a simulated annealing framework for M3D floorplan-ning, which uses a weighted sum of wirelength and area as the cost function. We also present a router-driven MIV insertion algorithm that inserts MIVs into the whitespace between blocks. We first demonstrate that in the case of a perfect manufacturing process, we can close the gap to the ideal block-level implementation by up to 50% w.r.t. both power and performance. The ideal block-level implementation is obtained by designing the chip assuming perfect inter-block interconnects. This is the best possible block-level design for a given benchmark.

However, during the manufacturing process of the top tier, we need take care not to damage the underlying interconnects and transistors, which can be achieved by using tungsten on the bottom tier. We model the impact of the tungsten interconnects and present a variation-aware floorplanning scheme that improves the performance and power by up to 12.6% and 10.6% respectively. Finally, we demonstrate that

even under such performance variations, we can still close the gap to the ideal block-level implementation by up to 50% w.r.t. performance and 36% w.r.t power.

The increase in power density associated with 3D ICs mean that thermal-aware design methodologies have become necessary [11]. We first study the thermal properties of monolithic 3D ICs and observe that the extremely thin tiers leads to negligible lateral thermal coupling. In addition, the absence of a bonding layer means that heat is not trapped in a given tier, and that the vertical thermal coupling is very high. In addition, the small size of MIVs mean that they do not serve as a conduction path, and their location need not be optimized for thermal reasons.

These properties enable us to develop a non-linear multi adaptive regression spline (MARS) model to quickly estimate the temperature of a monolithic 3D IC. We demonstrate a modeling error of $< 5\%$ when compared to GDSII-level FEA simulations. In addition, our model is extremely fast, and is $10^5$ times faster than prior quick-thermal approaches. This extremely quick computation means that our model can be used within a simulated annealing floorplanning framework. We modify our simulated annealing framework to include the temperature in the cost function. The non-modified floorplanner is first run until a certain area and wirelength target are met. Next, the temperature term in the cost function is introduced, and the area and wirelength serve as constraints instead of objectives. Using this approach, we demonstrate up to a 22% reduction in the maximum temperature of the chip, without affecting other design metrics such as wirelength and area.

## VI. Conclusion

We have explored several design styles that are available for monolithic 3D ICs – SRAM, transistor-level, gate-level, and block-level. For each design style, we have presented design flows to obtain GDSII-level signoff-quality power and performance results. We have enumerated various challenges facing M3D, and techniques to overcome them. Overall, ultra-high-density monolithic 3D ICs offers significant benefit over 2D ICs.

## References

[1] P. Batude et al., "Advances in 3D CMOS Sequential Integration," in *Proc. IEEE Int. Electron Devices Meeting*, 2009, pp. 1–4.

[2] C. Liu and S. K. Lim, "Ultra-High Density 3D SRAM Cell Designs for Monolithic 3D Integration," in *Proc. IEEE Int. Interconnect Technology Conference*, 2012.

[3] C. Liu and S. K. Lim, "A Design Tradeoff Study with Monolithic 3D Integration," in *Proc. Int. Symp. on Quality Electronic Design*, 2012.

[4] Y.-J. Lee, P. Morrow, and S. K. Lim, "Ultra High Density Logic Designs Using Transistor-Level Monolithic 3D Integrationn," in *Proc. IEEE Int. Conf. on Computer-Aided Design*, 2012.

[5] Y.-J. Lee, D. Limbrick, and S. K. Lim, "Power Benefit Study for Ultra-High Density Transistor-Level Monolithic 3D ICs," in *Proc. ACM Design Automation Conf.*, 2013.

[6] S. Panth, K. Samadi, Y. Du, and S. K. Lim, "Placement-Driven Par-titioning for Congestion Mitigation in Monolithic 3D IC Designs," in *Proc. Int. Symp. on Physical Design*, 2014.

[7] S. Panth, K. Samadi, Y. Du, and S. K. Lim, "Design and CAD Methodologies for Low Power Gate-level Monolithic 3D ICs," in *Proc. Int. Symp. on Low Power Electronics and Design*, 2014.

[8] S. K. Samal et al., "Full Chip Impact Study of Power Delivery Network Designs in Monolithic 3D ICs," in *Proc. IEEE Int. Conf. on Computer-Aided Design*, 2014.

[9] S. Panth, K. Samadi, Y. Du, and S. K. Lim, "High-Density Integration of Functional Modules Using Monolithic 3D-IC Technology," in *Proc. Asia and South Pacific Design Automation Conf.*, 2013.

[10] S. Panth, K. Samadi, Y. Du, and S. K. Lim, "Power-Performance Study of Block-Level Monolithic 3D-ICs Considering Inter-Tier Performance Variations," in *Proc. ACM Design Automation Conf.*, 2014.

[11] S. Samal et al., "Fast and Accurate Thermal Modeling and Optimization for Monolithic 3D ICs," in *Proc. ACM Design Automation Conf.*, 2014.

978-1-4799-7440-5/14 $31.00 © 2014 IEEE

# 3D Memory with Shared Lithography Steps:
# The Memory Industry's Plan to "Cram More Components onto Integrated Circuits"

Deepak C. Sekar

Rambus Labs

*Invited Paper*

**In his 1965 paper titled "cramming more components onto integrated circuits" [1], Gordon Moore predicted the number of components on a chip would increase exponentially with time due to continual reduction of feature sizes. That paradigm has continued successfully for the past 50 years, but cracks are starting to appear. Lithography is becoming prohibitively expensive and component quality is expected to degrade significantly beyond the 7nm node. To lower cost per bit further without relying on feature size scaling, monolithic 3D flash memories are being introduced where lithography steps are shared among multiple memory layers. In this paper, I review the flash memory industry's direction and describe 3D concepts that can scale other parts of the memory hierarchy.**

## I. INTRODUCTION

Systems for most enterprise and consumer markets use a memory hierarchy consisting of memory and storage components (Fig. 1). Memory components such as DRAM are relatively fast - the CPU waits for the requested data. Storage components such as NAND flash are slow and cheap - the CPU does not wait for data and changes the process or thread.

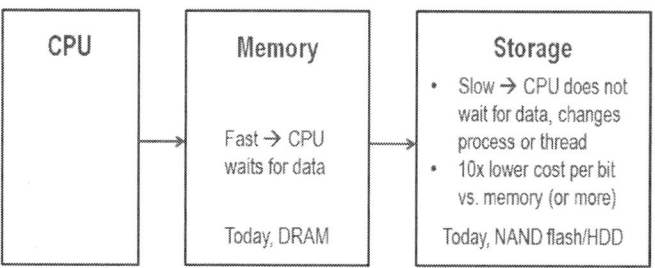

Figure 1: Memory hierarchy.

Historically, we have seen exponential improvements of cost per bit for memory and storage ICs with time. This was driven by shrinking feature sizes every 1-3 years. However, this paradigm seems to be hitting some severe challenges now.

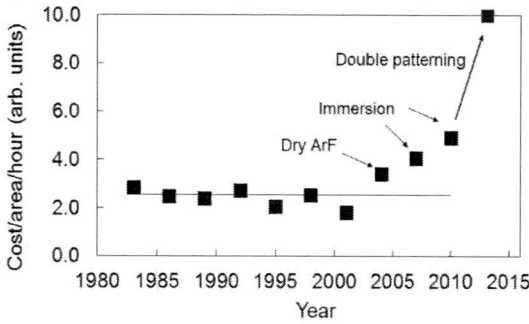

Figure 2: Lithography cost no longer follows historical trends.

As Fig. 2 indicates, lithography cost for scaling ICs to smaller feature sizes no longer follows historical trends [2]. For memory and storage ICs which are driven by cost per bit, this can be a show-stopper. Component quality, which I define as

the ability of components to meet an application's performance, power and density requirements, is considered an important challenge beyond the 10nm and 7nm nodes too [3]. DRAMs face difficult challenges with stacked capacitor scaling [3], while NAND flash faces difficult challenges with few electron effects and cell-to-cell interference [3]. It is clear that alternative paradigms to scale cost per bit for memory and storage ICs are required.

## II. 3D STORAGE WITH SHARED LITHOGRAPHY STEPS

A new approach that allows continuous cost per bit reduction has recently gained traction for NAND flash memories. Innovative architectures have been developed that allow multiple 3D stacked memory layers to be patterned with a common set of lithography steps [4][5][6]. Fig. 3 shows an example, which uses charge trap NAND flash cells.

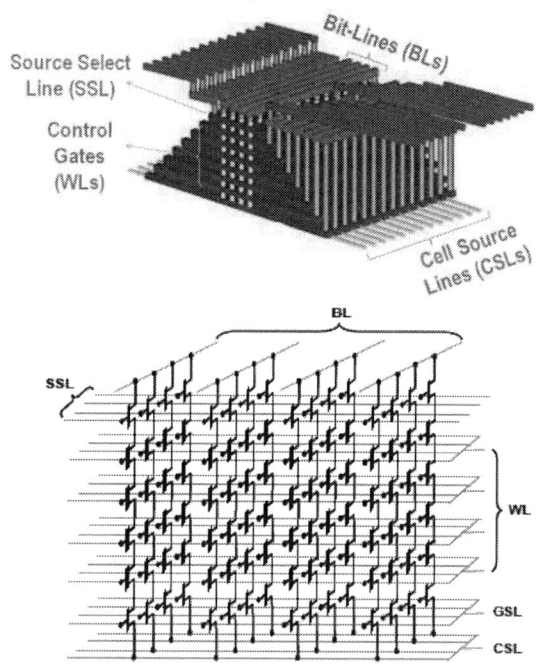

Figure 3: A 3D NAND flash architecture which uses shared lithography steps [5].

The process flow in Fig. 4 depicts how shared lithography steps can be used for defining such an architecture. Multiple oxide/nitride layers are deposited, following which shared lithography steps are used to form NAND flash memory strings with metal gates and charge trap dielectrics. Polysilicon is used as the channel material, but I-V characteristics are acceptable since the grain size is large compared to cell dimensions, and because the device structure is "gate-all-around", with excellent electrostatic control.

Fig. 5 shows the architecture in Fig. 3 allows continuous cost per bit reduction by periodically increasing the number of

978-1-4799-7440-5/14 $31.00 © 2014 IEEE

memory layers. This puts the burden of cost per bit reduction on processes such as high aspect ratio etch and deposition, which have more room for improvement vis-à-vis lithography. Compared to 2D NAND, one can use larger dimension cells but still get similar densities (Fig. 6). The use of larger cell sizes improves performance and endurance characteristics [6], as indicated by the threshold voltage distributions in Fig. 6. With commercial 3D NAND introduced recently [8], and several generations of improvement possible, it certainly looks like this "post-feature-size scaling" paradigm is here to stay.

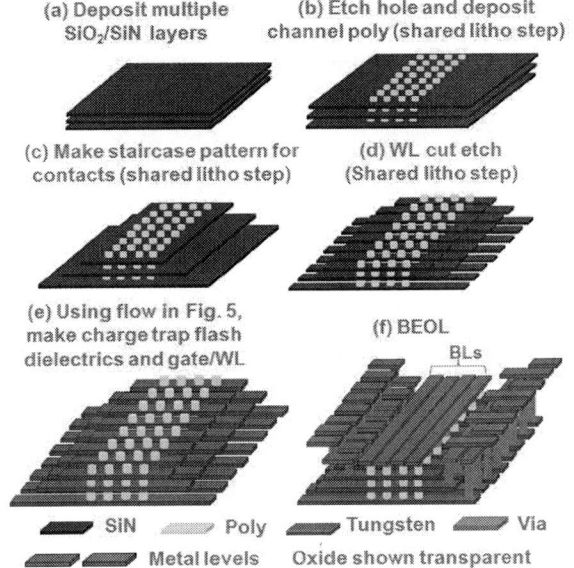

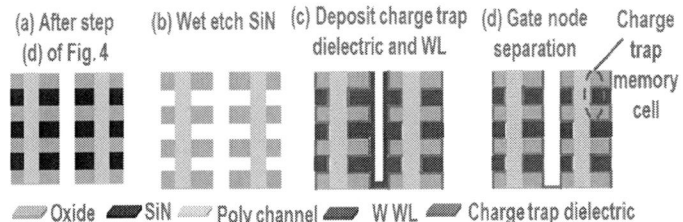

Figure 4: Process flow for 3D NAND flash [5].

Figure 5: Charge trap dielectric and electrode definition [5].

| Memory Layers | Die capacity | |
|---|---|---|
| 8 | 16Gb | |
| 24 | 128Gb [6][7] 3D: 40nm, 133mm$^2$ 2D: 16nm, 173mm$^2$ | |
| 192 | 1Tb | |

Figure 6: Scaling, distributions for 3D NAND [6].

## III. 3D MEMORY WITH SHARED LITHOGRAPHY STEPS

Just like for *storage*, if one can have monolithic 3D architectures with shared lithography steps for *memory*, it would be quite beneficial. However, the high aspect ratio capacitor needed for conventional DRAM makes this difficult. Fig. 7 shows an approach that is being researched by several organizations [9][10]. Emerging memories such as Resistive RAM (RRAM), Conductive Bridge RAM (CB-RAM) and others lend themselves to 3D stacking with shared lithography

steps easier than capacitor-based DRAM. The "memory" layer of future systems could potentially be composed of two types of memory: a small amount of conventional DRAM and a large amount of 3D Resistive Memory made with shared lithography steps, as shown in Fig. 7.

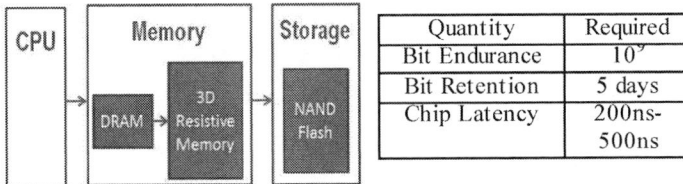

| Quantity | Required |
|---|---|
| Bit Endurance | 10$^9$ |
| Bit Retention | 5 days |
| Chip Latency | 200ns-500ns |

Figure 7: A paradigm for continuing memory scaling.

The International Technology Roadmap for Semiconductors [3] gives requirements for this application (Fig. 7). A low-power memory device that can meet all the requirements in Fig. 7 is yet to be developed. However, there is promise. RRAMs are known to switch at less than 10ns [9], so the chip latency requirements in Fig. 7 are feasible. Endurance required at the bit-level is 10$^9$ cycles. As Fig. 8 indicates [11], endurance for RRAM has been improving as the technology matures, indicating 10$^9$ cycles may be feasible. Several avenues exist to improve endurance further. (1) RRAMs are conventionally optimized for 10 year retention. For the "memory application" in Fig. 7, memory devices and programming algorithms can be optimized for 5 day retention, which can improve endurance significantly. (2) Endurance fail modes for RRAM are still largely unknown. As endurance fail modes are understood, countermeasures can be developed. (3) Fast wear-leveling algorithms [12] can help too.

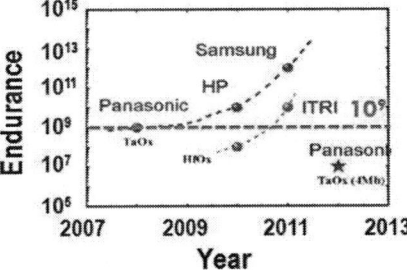

Figure 8: Endurance for RRAM has steadily improved [11].

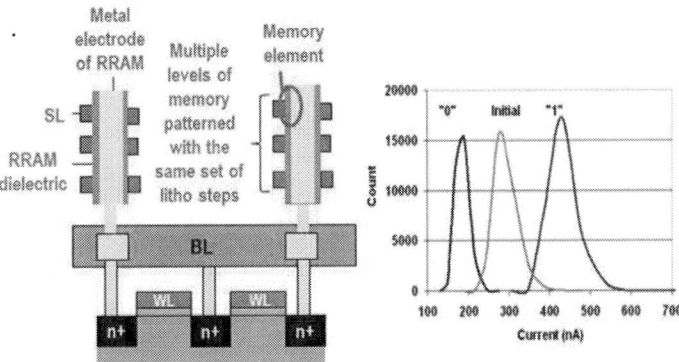

Figure 9: Vertical crosspoint memory with shared litho steps.

Several architectures have been pursued for building monolithic 3D resistive memories with shared lithography steps. Fig. 9 shows a commonly pursued architecture, a vertical crosspoint memory, also called a 1T-many R memory

[13]. Select transistors on the substrate are shared among multiple memory layers patterned with shared lithography steps. The key challenge with this architecture is that not every memory device has a selector of its own, leading to "sneak leakage paths" when memory arrays are built. Ways to reduce "sneak paths" include designing and smartly biasing memory devices with non-linear I-V curves [14], as well as circuit schemes to detect and compensate sneak paths. Fig. 9 shows ON and OFF state distributions of a 130nm crosspoint memory chip [14] from Rambus that successfully tackled sneak path issues, albeit with access times slower than Fig. 7.

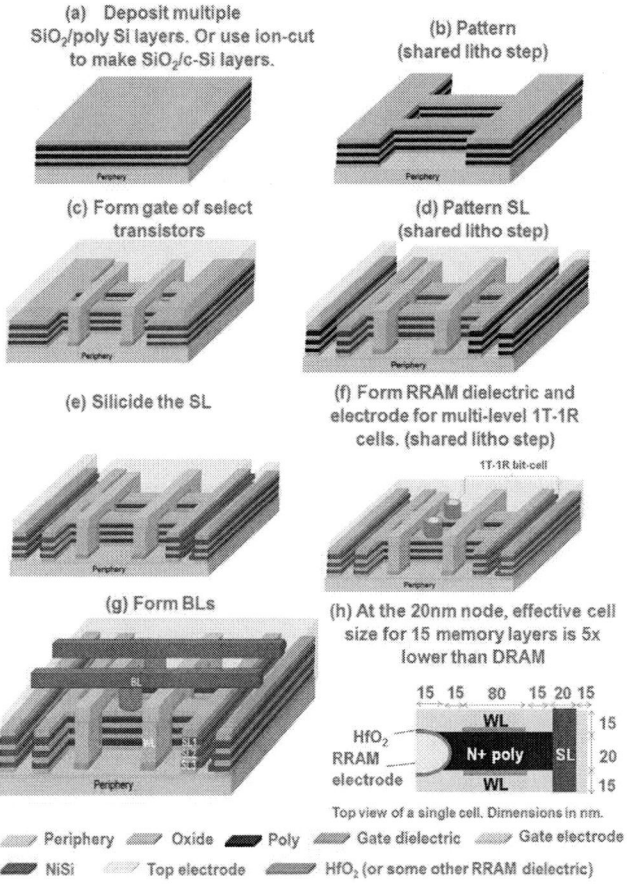

Figure 10: An alternative 3D architecture [15] with shared lithography steps and multilayer 1T-1R cells.

The biggest challenge with vertical crosspoint memories is that it is difficult to develop a low-power memory cell *with non-linear I-V characteristics* which can meet all the requirements in Fig. 7. The multi-layer 1T-1R architecture in Fig. 10, invented by the author and Zvi Or-Bach [15], could be used to tackle this problem. It uses shared lithography steps and creates double gated junction-free transistors as selectors. Sneak paths are eliminated since each RRAM device has a selector of its own. Compared to DRAM, significant improvements in density, scalability and refresh power are possible (Fig. 10(h)).

Unlike the concepts in Section II, the monolithic 3D memory concepts in Section III are still in the research stage, with several unsolved challenges. It should be an exciting area for innovation for the next 5-10 years.

## IV. SUMMARY

For the past 50 years, integrated circuits have seen tremendous improvements in cost, performance and power by scaling feature sizes of components every few years. We are now reaching feature sizes where lithography cost and component quality are posing serious limitations. In this paper, I described the strategy the memory industry is developing to tackle these limitations and move forward. It involves using monolithic 3D technology with lithography steps shared among multiple memory layers. 3D NAND flash chips built with these concepts are being introduced to the marketplace now [8], revealing the feasibility of this strategy. I described ways one could extend these monolithic 3D concepts beyond just storage chips, and showed ways to apply them to memory.

## ACKNOWLEDGEMENTS

The author would like to thank Gary Bronner (Rambus) and Zvi Or-Bach (MonolithIC 3D Inc.) for helpful discussions.

## REFERENCES

[1] Moore, Gordon E., "Cramming more components onto integrated circuits, Electronics, volume 38, number 8, April 19, 1965.

[2] H. Levinson, "The insertion of EUV lithography into high-volume manufacturing and other options", Workshop at the IEEE Intl. Interconnect Technology Conference, 2014.

[3] International Technology Roadmap for Semiconductors (ITRS). Available at http://public.itrs.net

[4] Tanaka, H.; et al., "Bit Cost Scalable Technology with Punch and Plug Process for Ultra High Density Flash Memory," VLSI Technology, 2007 IEEE Symposium on , vol., no., pp.14,15, 12-14 June 2007.

[5] Jaehoon Jang; et al, "Vertical cell array using TCAT (Terabit Cell Array Transistor) technology for ultra high density NAND flash memory," VLSI Technology, 2009 Symposium on, vol., no., pp.192,193, 16-18 June 2009 *(references)*

[6] Ki-Tae Park; et al., "Three-dimensional 128Gb MLC vertical NAND Flash-memory with 24-WL stacked layers and 50MB/s high-speed programming," Proc. of the ISSCC, Feb. 2014

[7] Helm, M.; et al. "A 128Gb MLC NAND-Flash device using 16nm planar cell," Solid-State Circuits Conference Digest of Technical Papers (ISSCC), 2014 IEEE International , vol., no., pp.326,327, 9-13 Feb. 2014

[8] K. Parrish, "Samsung launching branded SSD with 3D V-NAND", Tom's Hardware, July 2014.

[9] Hyung Dong Lee; et al., "Integration of 4F$^2$ selector-less crossbar array 2Mb ReRAM based on transition metal oxides for high density memory applications," VLSI Technology (VLSIT), 2012 Symposium on , vol., no., pp.151,152, 12-14 June 2012

[10] Burr, G.W.; Kurdi, B.N.; Scott, J.C.; Lam, C.H.; Gopalakrishnan, K.; Shenoy, R.S., "Overview of candidate device technologies for storage-class memory," IBM Journal of Research and Development , vol.52, no.4.5, pp.449,464, July 2008.

[11] Summarized by Panasonic, International Memory Workshop, 2012.

[12] Qureshi, M.K.; Karidis, J.; Franceschini, M.; Srinivasan, V.; Lastras, L.; Abali, B., "Enhancing lifetime and security of PCM-based Main Memory with Start-Gap Wear Leveling,", Proc. MICRO, 2009

[13] Hong Sik Yoon; et al., "Vertical cross-point resistance change memory for ultra-high density non-volatile memory applications," VLSI Technology, 2009 Symposium on , vol., no., pp.26,27, 16-18 June 2009

[14] Chevallier, C.J.; et al., "A 0.13μm 64Mb multi-layered conductive metal-oxide memory," Proc. ISSCC, 2010

[15] D. C. Sekar, Z. Or-Bach, US Patent 8581349, Filed 2011, Issued 2013.

# Monolithic 3D Integration in a CMOS Process Flow

E.A. Fitzgerald[1,2], S.F. Yoon[1,3], C.S. Tan[1,3], T. Palacios[1,2], X. Zhou[1,3], L.S. Peh[1,2], C.C. Boon[1,3], D.A. Kohen[3], K.H. Lee[3], Z.H. Liu[3], P. Choi[1,3]

[1]Low Energy Electronic Systems, SMART
Singapore
[2]Massachusetts Institute of Technology
Cambridge, MA
[3]Nanyang Technological University
Singapore

*Abstract*—**We describe a 3D integration process flow in which the vertical distance from the CMOS layer to the novel device layer is 100-1000 nm. This short distance effectively defines the process flow as a silicon CMOS process flow and allows for the use of silicon infrastructure in process and design. Progress has been made in demonstrating various pieces of III-V device integration into a foundry 0.18 µm process on 200 mm wafers.**

*Keywords—CMOS; III-V; integration; 3D; GaN; InGaAs; HEMT; LED*

## I. INTRODUCTION

Forms of "3D Integration" such as through-silicon-vias (TSV) have been evolving in research, sometimes at substantial scale, since the 1970s. As Moore's Law slows down, and ends for many market participants, new methods of producing circuit-level innovation are desired, and 3D integration is a potential route for innovation. Many methods and processes for 3D integration have been developed, yet a universal platform-level adoption like integrated CMOS has not appeared.

Potential roadblocks for an evolution to a true platform for innovative use in multiple applications lie in technology and market adoption. Technologically, many 3D integration methods result in interconnect densities and lengths that are incremental improvements over packaging technologies; therefore CMOS-level adoption and impact is elusive and 3D technologies appear more akin to packaging solutions. Adoption is also difficult because a truly universal design platform needs to exist at a level of abstraction above the current design-level, thus limiting broad use and evolving 3D solutions beyond specialized packaging cases.

In SMART LEES, we are pursuing a form of 3D integration that brings the vertical distance between CMOS and the other technology plane to approximately a micron. In so doing, the definition between 3D integration and monolithic CMOS fades, as the process is effectively a CMOS process flow that enables the integration of new devices. Such a method alleviates both the technological and adoption limits mentioned above. Interconnect densities can be approximately the same as that for the interconnect between silicon devices, and by placing the new device models into the CMOS design

kit, circuit designers can use the same design platforms and tools already present in the industry. Initially, we are integrating a variety of III-V devices (HEMTs or LEDs) based on different III-V materials, thus creating an array of new design platforms for investigating novel integrated circuits. However, we have downstream research on future integrated platforms involving thin film batteries and photovoltaics/detectors.

Our current approach is an evolution of our previous work which became the monolithic process flow in the DARPA COSMOS program [1]-[2]. In that case, we designed a CMOS process flow that is similar to a BiCMOS flow. The CMOS front end transistors are built first in series, followed by the InP heterojunction bipolar transistors, and finally the back-end which creates the interconnects for the CMOS but also for the InP devices, thus creating a monolithic integrated circuit containing CMOS and InP. However, if InP epitaxy is initiated in this process directly on the silicon substrate, the InP will be of poor quality and thus bring no advantage to the CMOS platform. We therefore integrated a closer-matched lattice constant to InP into the silicon wafer itself before the start of the CMOS process. We termed this wafer process SOLES (silicon on lattice-engineered substrate) and used Ge as the inserted template since it is acceptable to silicon process facilities currently. In effect, we were performing "3D integration" at the substrate level immediately, before device fabrication. SOLES wafers are fabricated using epitaxy on silicon wafers, and bonding thin silicon on top of that engineered substrate (i.e. SOI with a lattice-mismatched layer underneath the oxide). The CMOS process temperatures in the front-end were slightly modified to prevent melting of the Ge template.

Although proving that monolithic integration was not as impossible as was currently thought at the time, the process revealed some shortcomings from an adoption perspective. The wafers are made in research-volume levels, and therefore perfection (e.g. no small flakes at edges, reproducible performance) is not achievable. Although not important for demonstration when using experimental silicon CMOS facilities, such wafers would not be allowed into a real silicon foundry. On the other hand, without a market producing revenue, there is no method of maturing wafer fabrication. In addition, this method inherently assumes that the front-end of the CMOS is modified for each potential platform. Exploring many combinations in early commercialization as different technologies need to be explored suggests designing many CMOS processes, thus increasing market and prototype exploration costs.

## II. SMART LEES MONOLITHIC PROCESS

To alleviate the short-comings mentioned in the last section, we have developed the following innovative flow:

*1) Front-end foundry silicon CMOS processing using SOI wafer*

978-1-4799-7440-5/14 $31.00 © 2014 IEEE

*2) Double-bond and transfer CMOS on top of a lattice-engineered wafer (e.g. GaN/Si)*

*3) Open vias to engineered substrate*

*4) Fabricate III-V devices*

*5) Re-insertion into foundry for back-end interconnects*

This method transfers completed front-end CMOS on top of the engineered wafer, thus being "3D-like". However, the distance between the CMOS-level and III-V device level is the thickness of the oxide equivalent in an SOI wafer, approximately 100-1000 nm. Note also that devices have not yet been fabricated in the III-V layer until after the CMOS transfer. This aspect retains the automatic alignment of devices through photolithographic processes as opposed to needing to align pre-fabricated devices on different planes (although such a process is a derivative of this flow if necessary).

The SOI wafers inserted at the start of the flow are of high quality and acceptable to foundries, yet have an oxide etch stop present for aiding the CMOS transfer. Also, the foundry CMOS is never altered, and therefore the only variants to the process are located within the III-V processing, thus concentrating resources on the novel devices inserted into the process. III-V devices are modeled, and those models are inserted into the foundry PDK. Again, most of the design kit is unaltered, and only the innovative piece needs to be added.

An advantage of this innovative flow is also that various platforms can be created at low cost, design kits developed, and novel circuits explored. The reason is that the choice of engineered substrate to attach in step 2) defines the platform. Everything else remains the same.

This flow is also attractive from a commercial perspective. An entire CMOS process does not need to be redeveloped for each design platform. Thus, the process is a new path to innovate in silicon: the foundry retains its ability to focus on high volume manufacturing without altering process, yet new design platforms can be explored by designers to find novelty and market whitespace. High-value but lower volume circuits can be manufactured in this fashion, and presumably if very high volume is needed, the process would likely evolve into a specific CMOS flow with an engineered substrate upfront (i.e. the COSMOS process described in the Introduction), to minimize costs.

### III. SAMPLE RESULTS

As the SMART LEES program is now 2.5 years old, we have constructed the infrastructure required to perform steps 2)-4) listed above, and have already demonstrated key aspects of the flow at the materials, process, device, and circuit level. Even though we have not yet run a full integrated flow, results at the circuit-design level are possible since early III-V device models can be inserted into the design kit even before the full integrated process is ready. Such parallel exploration at the circuit level is important since it can provide feedback into the platform development while research is occurring.

Our infrastructure was built for 200 mm wafer flows. 200 mm has CMOS that is advanced enough for many analog-

Research supported by the National Research Foundation of Singapore through CREATE SMART

mixed signal devices (0.18 μm technology continues to be heavily used in these markets) and therefore combinations of such CMOS with III-V devices offers novel disruptive platforms on inexpensive trailing-edge silicon manufacturing infrastructure. In addition, research and development costs are lower for 200 mm wafer processing as various novel platforms and circuits are explored.

### A. Materials

A variety of engineered substrates have been demonstrated in order to begin developing various design platforms. Fig. 1 is a cross-section transmission microscope (XTEM) image of the InGaAs HEMT engineered substrate. This platform will host high-speed, low noise InGaAs HEMTs in the CMOS design kit. Fig. 2 is an example of a GaN/Si engineered substrate with a silicon layer transferred onto it (the donor wafer at the top of the image has not been removed yet). And Fig. 3 is a demonstration of our double-bond and transfer process for a

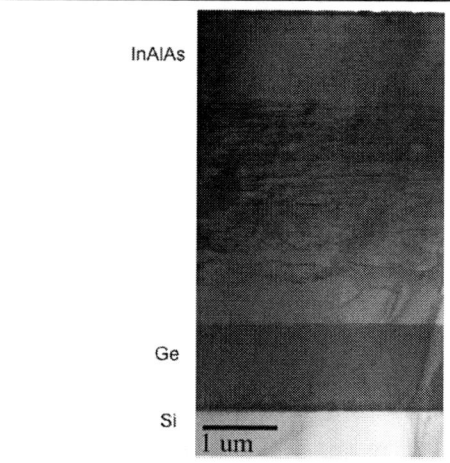

Figure 1: XTEM of the InAlAs and Ge metamorphic buffer on silicon. The structure shifts the lattice constant from that of silicon to that of InP while retaining low enough threading dislocation density for InGaAs HEMT performance

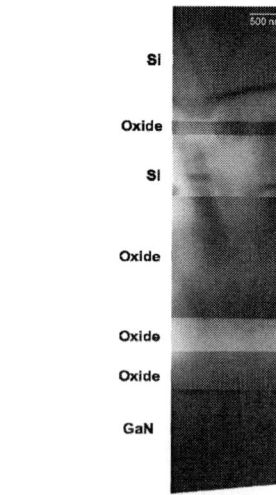

Figure 2: XTEM of a silicon layer from a host SOI wafer bonded on top of a GaN/Si engineered substrate. After removing the top handle wafer, this is the process that will be used to transfer front-end CMOS onto the engineered GaN/Si.

foundry-processed CMOS wafer.

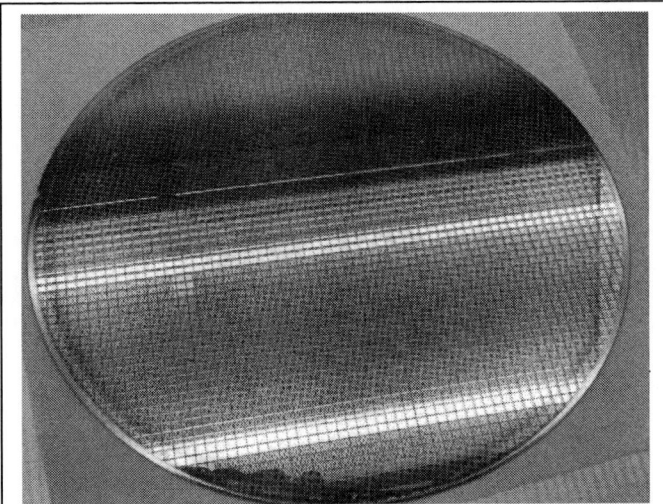

Figure 3: Double-bond transfer of a foundry-processed front-end onto a host 200 mm substrate. The structure is void-free.

### B. Devices

III-V devices have been fabricated on silicon. Fig. 4 shows device characteristics from a GaN/Si HEMT. Our goal is to initially integrate standard HEMT devices that add value to the silicon platform as our device sub-group continues to push

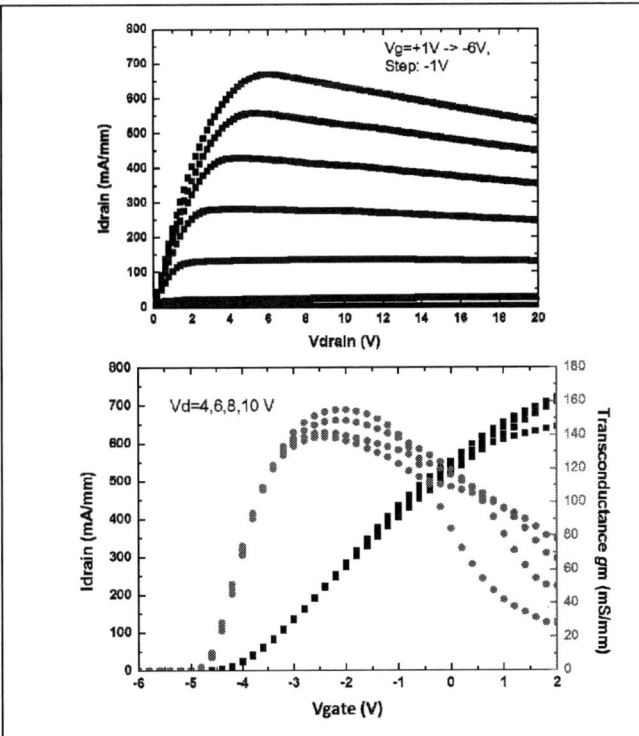

Figure 4: DC and transfer characteristics of GaN/Si HEMTs ($L_g = 2$ μm). Second and third generation devices will be tailored to meet circuit designer needs for particular applications.

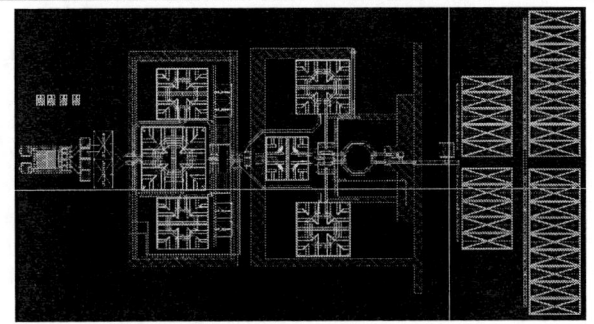

Figure 5: A GaN+CMOS power amplifier design using the integrated foundry PDK.

such devices to higher performance for future integration.

For both the InGaAs and GaN HEMT, innovation has appeared in the processes that are compatible with the silicon CMOS process flow. For example, low temperature ohmic contacts to the source and drain were developed for the GaN HEMT, as the normal non-gold-containing contacts are processed at too high a temperature for the front-end of the silicon CMOS.

### C. Modeling and Circuits

We have constructed a GaN HEMT+CMOS design kit by adding GaN HEMT models to a foundry 0.18 μm PDK. Fig. 5 is an example of a design for a GaN+CMOS power amplifier that will be included in early circuit demonstrations. Such platforms allow our circuit designers to give feedback to the device and materials researchers as they develop the novel design platforms.

### ACKNOWLEDGMENT

The authors on this abstract are SMART LEES researchers that contributed to aspects of the data included in this abstract. We gratefully acknowledge the team work of the various other SMART LEES researchers who contributed to this interdisciplinary effort.

### REFERENCES

[1] C.L. Dohrman, K. Chilukuri, D.M. Isaacson, M.L. Lee, and E.A. Fitzgerald, "Fabrication of silicon on lattice-engineered substrate (SOLES) as a platform for monolithic integration of CMOS and optoelectronic devices", Materials Science and Engineering B, v. 135, 235, (2006).

[2] T.E. Kazior, J.R. LaRoche, D. Lubyshev, J.M. Fastenau, W.K. Liu, M. Urtega, W. Ha, J. Bergman, M.J. Choe, M.T. Bulsara, E.A. Fitzgerald, D. Smith, D. Clark, R. Thompson, C. Drazek, N. Daval, L. Benaissa, E. Augendre, "A high performance differential amplifier through the direct monolithic integration of InP HBTs and Si CMOS on silicon substrates", 2009 IEEE MTT-S International Microwave Symposium Digest (MTT), p. 1113, (2009).

978-1-4799-7440-5/14 $31.00 © 2014 IEEE

# Monolithic 3D Integration Advances and Challenges: From Technology to System Levels

Mohammad Sadegh Ebrahimi, Gage Hills, Mohamed M. Sabry, Max M. Shulaker, Hai Wei, Tony F. Wu, Subhasish Mitra, and H.-S. Philip Wong

Department of Electrical Engineering and Department of Computer Science, Stanford University

The exploration of three-dimensional (3D) integrated circuits (ICs) and systems has been triggered by the ever-increasing demand for high-performance and highly energy-efficient computing systems to meet current and emerging application needs. 3D-ICs overcome the limitations of traditional processing technologies; such ICs can potentially improve the energy-delay- product (EDP) metrics and form factors of resulting computing systems by significant amounts compared to planar 2D-ICs. Thus, 3D-ICs can enable higher performance, better energy efficiency, and enhanced system functionality [1].

Most 3D integration concepts are based on stacking heterogeneous technologies (logic, memory, analog-mixed circuits, RF circuits...) layered on top of each other, with vertical through-silicon vias (TSVs) [2] connecting those layers. TSVs are generally large (e.g., 5 µm diameter and 20 µm inter-TSV pitch [2]). Such large TSV dimensions severely limit the density of vertical connections between various layers. **Monolithic** 3D-ICs overcome the vertical connectivity challenge of TSV-based 3D-ICs through the use of nano-scale interlayer vias (ILVs) [3]. ILVs have the same pitch and dimensions as tight-pitched metal layer vias, which are orders-of-magnitude smaller than TSVs. With high-density vertical connectivity, monolithic 3D-ICs can enable significantly higher on-chip data transfer bandwidth compared to TSV-based 3D-ICs. This enhancement makes monolithic 3D-ICs suitable for big-data applications [30].

This paper highlights recent technological advances in monolithic 3D integration in Section I. Section II discusses system architecture questions and challenges.

## I. TECHNOLOGICAL ADVANCES

Monolithic 3D integration requires low-temperature processing of both the front-end of line (FEOL) and back-end of line (BEOL) to preserve the performance of active and passive devices on the lower layers [4]. However, today's silicon technology generally requires high-temperature annealing for transistor fabrication, making it difficult to fabricate monolithic 3D circuits. Replacement of high-temperature dopant activation with solid phase epitaxy on the upper layers of circuits has allowed transistors to be fabricated on top of other transistors [5]. Methods to transfer single crystal layers to the substrate have also been proposed [6]. While advances in silicon-based electronics continue to be made, alternative technologies for logic and memory are being explored. Carbon nanotube transistors (CNFETs), which are amenable to low-temperature 3D monolithic integration, are highly promising candidates for building energy-efficient digital logic due to their superior electrostatic and transport properties [7], [8].

CNFETs are projected to improve the energy-delay product (EDP, a measure of energy efficiency) of 2D digital ICs by an order of magnitude vs. silicon-CMOS [9]. Sub-10 nm CNFETs have been demonstrated with significantly less short channel effects than previously predicted [8].

CNFETs can be fabricated using low-temperature CNT transfer techniques below the traditional thermal budget limit of 400°C [11], [12]. CNTs can be transferred onto multiple circuit layers to build CNT-based monolithic 3D ICs [13]. Multiple transfers onto each layer of circuits have also shown to linearly increase on-currents of CNFETs [12]. Thus, it is natural for CNFETs to provide a unique opportunity to achieve monolithic 3D integration.

Monolithic CNFET 3D-ICs have been demonstrated by building fully-complementary logic circuits spanning multiple logic layers in a

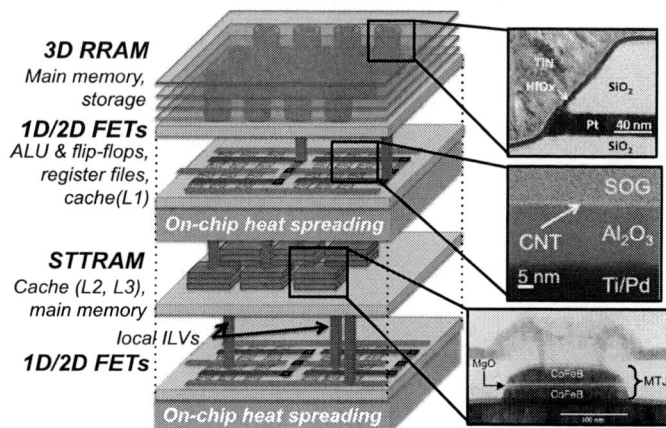

Figure 1. Monolithically integrated computing platform with CNFET-based logic [13], STTRAM-based cache [25], RRAM main storage [26], and on-chip thermal solutions. The left half of the figure includes the TEM figures of the different technologies.

VLSI-compatible manner [13]. The imperfection-immune paradigm [14], which includes VLSI-compatible metallic-CNT removal [15], mis-positioned CNT-immune circuit design [16], and aligned-active layouts [17], is essential for fabricating CNFET-based logic circuits on each layer, enabling massive monolithic 3D integration. Recent experimental demonstrations of CNFET digital systems include all-digital capacitive sensor-interface subsystems [18], [19], fully-programmable processors [20], CNFET circuit integration up to sub-20 nm channel lengths [10], monolithic 3D integrated circuits using CNFETs and CNT interconnects [21], air-stable complementary logic using P-type CNFETs and N-type CNFETs [22], and monolithic 3D integration of CNFETs on top of silicon CMOS [23].

Monolithic 3D enables tight integration of logic and memory. SRAM, DRAM, and Flash memory technologies are becoming increasingly difficult to scale from performance and storage capacity standpoints [24]. Emerging non-volatile memories such as STTRAM, RRAM, CBRAM, and PCRAM are promising candidates for replacing these existing technologies. STTRAM is a promising candidate for system main memory, with read latencies comparable to SRAM but with 90% less leakage [25]. RRAM, CBRAM, and PCRAM can enable massive on-chip storage [26], [27], [28]. The processing temperature of such emerging RAM technologies can be as low as 80°C [29], making them attractive for monolithic 3D integration.

Figure 1 shows a possible computing platform with traditional silicon-based or CNFET-based circuits on the first layer. Subsequent logic layers are fabricated using CNFETs. This platform combines emerging memory technologies (e.g., STTRAM, RRAM), and integrated thermal solutions. The CNFET logic and memory layers can be interleaved in a flexible fashion, enabled by low-temperature fabrication. ILVs allow for massive connectivity between layers of logic circuits and/or memory. Such a platform can additionally enable the integration of sensors and computing into a single-chip integrated system, allowing local processing of sensor data (followed by subsequent transfer of "processed information" rather than raw data).

## II. SYSTEM ARCHITECTURE QUESTIONS

With significant technology-level advances in monolithic 3D integration (Sec. 1), it is essential to understand system architecture-level implications of 3D-ICs. The ultimate question is: *"which applications can benefit the most from monolithic 3D-ICs and by how much?"*

One such application domain can be big-data (massive structured or unstructured data volume and rate) analytics, which represents an important class of emerging applications [30]. Big-data analytics can impose stringent system-level requirements with respect to low-latency processing, high-bandwidth transfer, and energy-efficient storage of massive amounts of data. Monolithic 3D-ICs can potentially integrate massive amounts of on-chip non-volatile memory (possibly at the terabyte scale) which itself can translate into significant energy consumption and access latency benefits (compared to off-chip storage) [6]. Moreover, high-density ILVs with nano-scale dimensions can improve inter-layer communication bandwidth by 10,000× compared to today's TSV-based 3D-ICs [3]. This significant gain in bandwidth can provide further benefits by alleviating some of the major memory bottlenecks faced by conventional computing systems, where complex memory hierarchies are used to "hide" memory latency limitations. However, significant research is necessary to quantify such benefits. New memory technologies have read/write and retention/endurance characteristics different from conventional SRAM, DRAM, and Flash. Hence, we expect strong interactions between device-level optimization of new memory technologies and application-level requirements for memory hierarchy design.

The above observations lead us to the next question: *"With monolithic 3D, should various architectural blocks be re-designed and how?"*

Conventional computer architectures often assume that memory access latencies are orders of magnitudes longer than the time spent on actual computations. Accordingly, various architectural techniques are used to hide memory latencies [32]. With low-latency and high-bandwidth memory accesses in monolithic 3D-ICs, it may be possible to simplify the design of various micro-architectural blocks (and, hence, further improve energy efficiency). Memory interfaces are also expected to undergo redesign because they can now take advantage of improved memory bandwidth and latency, as well as the non-volatile nature of new memory technologies.

The above questions only scratch the surface of various design questions that must be addressed: *Should the design of various building blocks, such as memory arrays, fundamentally change? What are the native computation elements? What block sizes are required for optimized data transfer? How to design scalable monolithic 3D architectures such that the application-level performance and energy efficiency benefits can scale with the number of (parallel) computational units?*

Monolithic 3D integration requires deep understanding of variability, yield, and reliability. To that end, we have created a preliminary framework for CNFET circuit design [38], based on hardware calibrated device model [39]. This software runs 100 times faster than existing approaches, and enables us to quickly explore the large space of interplay among carbon nanotube variations, energy, delay, noise margin, and circuit-level functional and performance yield. Monolithic 3D integration of logic and memory requires detailed study of ways to manage yield, variability, and reliability at the device, circuit, and architecture levels. We have also developed SPICE device models for emerging memories to enable such detailed studies [41].

In addition, one must pay attention to heat generation, dissipation, and possible cooling mechanisms for monolithic 3D-ICs. The use of energy-efficient technologies, such as transistors based on carbon nanotubes and layered transition metal dichalcogenides [34], can result in lower heat generation. However, the large number of layers projected for monolithic 3D-ICs can result in a substantial increase in on-chip heat flux. In addition, today's technologies use thin-film inter-layer dielectrics (ILD). The use of ultra-thin ILD significantly reduces the thermal conductivity of the dielectric material [35]. As a result, monolithic 3D-ICs may require advanced cooling technologies, within the layers and around the peripheries. A possible solution would be to incorporate materials with high thermal conductivity, such as graphene, CNT bundles, or novel 2D materials [36], to aid in heat removal from the layers furthest from the heat sink. Another promising approach is to include advanced convective structures such as metallic nanomesh [37] to handle the heat flux of monolithic 3D-ICs. These nanomesh structures can use phase-change materials [40] to overcome spatial and temporal thermal hotspots. In addition, advanced chip-external technologies are required to remove the heat from the chip.

## CONCLUSION

Recent technological advances in monolithic 3D integration lay the foundation for highly efficient next-generation computing systems. These advances, however, can only be utilized to their full potential if system architectures are properly optimized by utilizing monolithic 3D-IC technology for target applications. Thus, a multidisciplinary research framework from system architecture to technology layers, and several experimental demonstrations are required.

## ACKNOWLEDGMENTS

This work is supported in part through the NCN-NEEDS program, which is funded by the National Science Foundation, contract 1227020-EEC, and by the Semiconductor Research Corporation, and through Systems on Nanoscale Information fabriCs (SONIC), one of the six SRC STARnet Centers, sponsored by MARCO and DARPA, and IARPA TIC (Program Director: Dennis Polla), as well as the member companies of the Non-Volatile Memory Technology Research Initiative (NMTRI) at Stanford University. We gratefully acknowledge discussions with Professors J. Bokor, J. Rabaey, C. Kozyrakis, L. Pileggi, K. Goodson, M. Asheghi, and E. Pop at UC Berkeley, Stanford University, and Carnegie Mellon University.

## REFERENCES

[1]  P. G. Emma et al., *IBM J. Res. Dev.,* 2008.
[2]  Z. Xu et al., *IEEE TSM,* 2013.
[3]  S. Panth et al. *ASPDAC,* 2013.
[4]  S. Wong et al., *VLSI-TSA,* 2007.
[5]  P. Batude et al., *IEDM,* 2011.
[6]  "Monolithic3D", www.monolithic3d.com
[7]  J. Appenzeller, *Proc. IEEE,* 2008.
[8]  A. D. Franklin et al., *Nano letters,* 2012.
[9]  Chang, L. et al., IEDM Short Course, 2012
[10] M. M. Shulaker et al., *ACS Nano,* 2014.
[11] N. Patil et al., *Symp. VLSI Tech,* 2008.
[12] M. M. Shulaker et al., *Nano letters,* 2011.
[13] H. Wei et al., *IEDM,* 2013.
[14] J. Zhang, TCAD, 2012
[15] N. Patil et al., *IEDM,* 2009.
[16] N. Patil et al., *TCAD,* 2008.
[17] J. Zhang et al., *DAC,* 2010.
[18] M. M. Shulaker et al., *ISSCC,* 2013.
[19] M. M. Shulaker et al., *JSSC,* 2014.
[20] M. M. Shulaker et al., *Nature,* 2013.
[21] H. Wei et al., *IEDM,* 2009.
[22] H. Wei, et al., *IEDM,* 2011.
[23] M. M. Shulaker et al., *Symp. VLSI Tech,* 2014.
[24] S. Hong, *IEDM,* 2012
[25] C. W. Smullen et al., *HPCA,* 2011.
[26] H. –S P. Wong, et al., *Proc. IEEE,* 2012.
[27] H. -S P. Wong, et al., *Proc. IEEE,* 2010.
[28] R. Fackenthal et al., *ISSCC,* 2014.
[29] H. Jeong, et al., *Nanotechnology,* 2010.
[30] P. Russom, *TDWI Best Practices Report,* 2011.
[31] S. L. Kim, *Elsevier,* 2013.
[32] K. Olukotun et al., *ASPLOS,* 1996.
[33] David Brooks et al., *Comput. Archit. News,* 2000.
[34] S. McDonnell et al., *ACS Nano,* 2014.
[35] E. Pop, *Nano Research,* 2010.
[36] E. Pop et al. *MRS Bulletin,* 2012.
[37] M. T. Barako et al., *ITHERM,* 2014.
[38] G. Hills et al., *DAC,* 2013.
[39] J. Luo et al., *TED,* 2013.
[40] S. Lingamneni et al., *ITHEM,* 2014.
[41] Z. Jiang et al., *SISPAD,* 2014.

# Performance Assessment of ULP Analog/RF MOSFET Architectures

## Dipankar Ghosh and Abhinav Kranti

Low Power Nanoelectronics Research Group, Electrical Engineering Discipline,
Indian Institute of Technology Indore, India. Email: akranti@iiti.ac.in

Ultra low power (ULP) operation is increasingly in demand for future System-on-Chip (SOC) circuits using nanoscale analog/RF transistors. In this work, we analyze the analog/RF performance metrics of different ULP MOSFET architectures and quantify the advantages and challenges associated with underlap inversion-mode (INV), Junctionless (JL), lateral tunnel FET (LTFET) and vertical tunnel FET (VTFET) in terms of gain and bandwidth.

## INTRODUCTION

Nanoscale MOSFETs for ULP subthreshold operation require enhanced gate controllability and lower parasitics [1-2]. As shown in fig. 1a-d, several architectures have gained considerable interest due to their ability to suppress short channel effects (such as underlap INV [3] and JL MOSFETs [4]) and improved switching from off-to-on state (such as LTFET and VTFET [5-9]). While much attention has been devoted to digital performance of these devices, ULP analog/RF metrics have not been investigated in detail.

## RESULTS AND DISCUSSION

MOSFETs are analyzed (fig. 2a) with ATLAS simulator [10] using Lombardi mobility model [11] and non-local Hurkx band-to-band tunneling model for tunnel FETs (TFETs) [6, 8] with device parameters as given in Table 1. The transfer characteristics of TFETs are compared with available experimental data published in the literature [12-13] for nMOS and pMOS TFETs (fig. 2b-c). As the focus is on ULP operation, the devices have been compared for the current range 0.1 $\mu$A/$\mu$m to 10 $\mu$A/$\mu$m (fig. 2a).

At a drain current ($I_{ds}$) level of 10 $\mu$A/$\mu$m (fig. 3a-b), JL and INV underlap devices exhibit a cut-off frequency ($f_T$) of 200 GHz which is 15× higher than that achieved by TFETs (7-9 GHz). For the same current level, intrinsic voltage gain ($A_{VO}$) for JL, INV underlap and VTFET device is ~40 dB which is nearly 3 times higher than exhibited by LTFET (15 dB). The highest gain exhibited by VTFETs over the entire operating region suggests the usefulness for high gain ULP operation. A lower transconductance ($g_m$) is expected in TFETs due to high tunneling junction resistance [7]. JL and INV devices achieve the same gm values for the current range shown in fig. 3c. The lower $g_m$ is also reflected in the lower mobility of 20 cm²/Vsec and 5 cm²/Vsec for LTFET and VTFET, respectively, extracted at mid-gate position of the conduction channel at $I_{ds}$ of 10 $\mu$A/$\mu$m. Due the underlap design in INV mode and JL devices, relatively higher mobility of 230 cm²/Vsec were extracted at mid-gate positions at the same current level. The high values of

carrier mobility in JL and INV devices correlates well with the higher $g_m$ values for INV and JL MOSFETs (fig. 3c).

Total gate capacitance ($C_{gg}$) of JL and INV underlap device (≈ 0.35 fF/$\mu$m) is 4 times lower than LTFET (1.17 fF/$\mu$m) and 2 times lower than VTFET (0.7 fF/$\mu$m) (fig. 3d). Significant proportion of charges are provided by drain in LTFET [7] and drain field influences conduction band (fig. 4a) which results in high gate-to-drain capacitance ($C_{gd}$) and $C_{gg}$ values, with lower gate-to-source capacitance $C_{gs}$. Valence band (not shown here) also shows a similar trend. On the contrary, VTFET exhibits higher $C_{gs}$ due to enhanced tunneling but lower $C_{gd}$ as channel potential is not affected by the drain voltage (fig. 4a) [8-9]. Lower values of $C_{gg}$ in JL and INV underlap devices is due to the reduction of inner fringing capacitance [14-15]. Lower $C_{gg}$ (~ 0.35 fF/$\mu$m) exhibited by LTFET, JL and INV underlap devices as compared to VTFET (0.7 fF/$\mu$m) at $I_{ds} = 0.1$ $\mu$A/$\mu$m also correlates with lower leakage current as shown in fig. 2a.

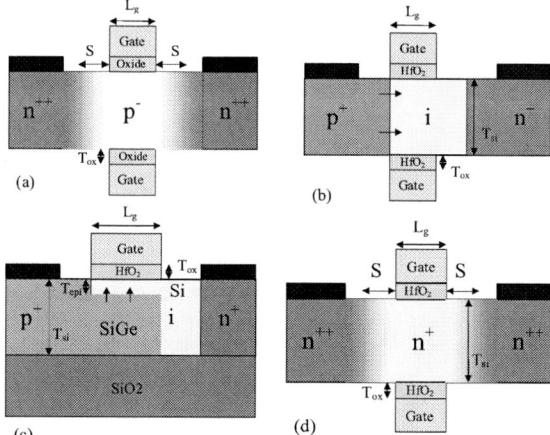

**Fig. 1:** Schematic diagram of (a) Inversion mode (INV) underlap MOSFET (spacer width ($s$) = 24 nm and doping gradient = 3 nm/decade), (b) Lateral TFET, (c) Vertical TFET and (d) Junctionless (JL) MOSFET. Solid arrows indicate tunneling direction for electrons in (b) and (c).

Transconductance generation efficiency ($g_m/I_{ds}$) (fig. 4b) for INV underlap and JL devices (25 to 30 V⁻¹) is found to be 10× higher than VTFET (2.5 V⁻¹) and twice higher than LTFET (10 V⁻¹) due to lower $g_m$ values. Simultaneous improvement in $g_m/I_{ds}$ and $f_T$ for underlap INV and JL devices signifies the balance achieved between power efficiency and bandwidth [16] useful in ULP operational transconductance amplifiers (OTAs). High Early voltage ($V_{EA}$), which indicates enhanced gain and suppressed SCEs, is 15 times higher for VTFET (60 V) than that for

978-1-4799-7440-5/14 $31.00 © 2014 IEEE

JL and INV devices (~3 V) and "2 order" higher than LTFET at $I_{ds}$ = 10 μA/μm (fig. 5a). $V_{EA}$ for VTFET is high due to reduced influence of drain field on band bending and tunneling (fig. 4a and 5b). LTFET exhibits low $V_{EA}$ and gain beyond $I_{ds}$ of 1μA/μm. JL and INV underlap exhibit moderately high $V_{EA}$ and $A_{VO}$, due to reduced drain electric field provided by underlap region (fig. 5b). Significantly higher ($V_{EA}$) results in high gain ($A_{VO}$ = $g_m/I_{ds}$ x $V_{EA}$) for VTFET over the entire current range, despite a degraded $g_m/I_{ds}$.

## CONCLUSION

A comparative analysis of ULP analog/RF performance metrics have been presented. TFETs will be useful in high-gain ULP circuits with limited bandwidth requirement. Underlap INV and JL devices, achieving a balance between gain and bandwidth, are most suitable for ULP analog/RF applications. JL MOSFETs do not suffer from any significant degradation in metrics despite higher channel doping.

## ACKNOWLEDGEMENT

This work is supported by Science and Engineering Research Board, Department of Science and Technology, Government of India, under Grant SR/S3/EECS/0130/2011.

## References

[1] www.itrs.net
[2] J.-P. Colinge *et al.*, *Solid–State Electronics*, 48, 897–905, 2004.
[3] A. Kranti *et al.*, *IEEE TED*, 54, 3308-3316, 2007.
[4] J.-P. Colinge *et al.*, *Nature Nanotechnology*, 5, 225-229, 2010.
[5] K. Boucart *et al.*, *IEEE TED*, 54, 1725-1733, 2007.
[6] C. Anghel *et al.*, *APL*, 96, 122104, 2010.
[7] A. Tura *et al.*, *IEEE TED*, 57, 1362-1368, 2010.
[8] A. Rajoriya *et al.*, *IEEE TED*, 60, 2626-2633, 2013.
[9] R. Asra *et al.*, *IEEE TED*, 58, 1855-1863, 2011.
[10] ATLAS Users Manual, Silvaco, 2010.
[11] C. Lombardi *et al.*, *IEEE TCAD*, 7, 1164-1171, 1988.
[12] A. Walke *et al.*, *IEEE TED*, 61, 707-715, 2014.
[13] F. Mayer *et al.*, *In Proc. IEDM*, 2008.
[14] M.S. Parihar *et al.*, *IEEE TED*, 60, 1540–1546, 2013.
[15] D. Ghosh *et al.*, *IEEE EDL*, 33, 1477-1479, 2012.
[16] B. Murmann *et al.*, *IEEE TED*, 53, 2160-2167, 2006.

| Parameters | INV | LTFET | VTFET | JL |
|---|---|---|---|---|
| $L_g$ (nm), $V_{ds}$ (V) | 20, 0.5 | | | |
| $T_{ox}$ (nm) | 3 nm HfO$_2$ ($\kappa$ = 21) | | | |
| $T_{si}$, $T_{epi}$(nm), Ge mole (%) | 10, NA, NA | 10, 2, 30 | | 10, NA, NA |
| Source doping ($\times 10^{20}$ cm$^{-3}$) | ($N_d$) 5 | ($N_a$) 2 | ($N_a$) 1 | ($N_d$) 5 |
| Channel doping ($\times 10^{17}$ cm$^{-3}$) | ($N_d$) 0.01 | ($N_d$) 1 | | ($N_d$) 10 |
| Drain doping ($\times 10^{20}$ cm$^{-3}$) | ($N_d$) 5 | ($N_d$) 0.2 | ($N_d$) 1 | ($N_d$) 5 |
| Spacer (nm) | 24 | NA | | 24 |

**Table 1:** Parameters used in the analysis. Underlap INV and JL devices were designed with spacer width ($s$) = 24 nm and S/D doping gradient (at the gate edge) of 3 nm/decade.

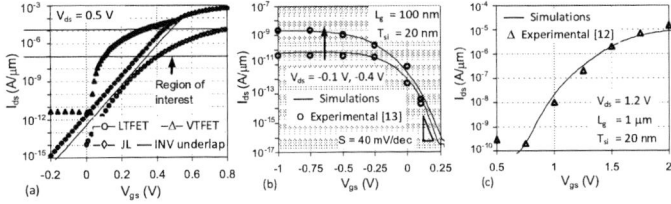

Fig. 2. (a) Simulated $I_{ds}$-$V_{gs}$ characteristics for INV underlap, JL, LTFET and VTFET devices. Comparison between simulation and experimental results for (b) LTFET [13], and (c) VTFET [12].

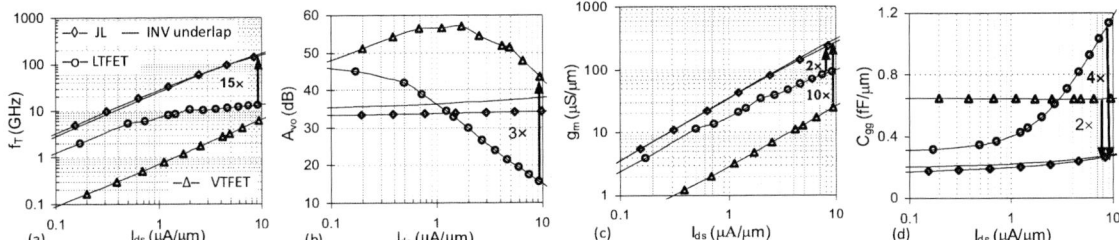

Fig. 3. Dependence of (a) cut-off frequency ($f_T$), (b) intrinsic voltage gain ($A_{VO}$) (c) transonducatnce ($g_m$) and (d) total gate capacitance ($C_{gg}$) for different devices on $I_{ds}$. Notations are same as in fig. 3a.

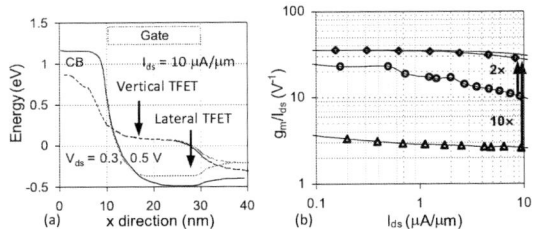

Fig. 4. (a) Conduction band (CB) variation with drain bias for vertical and lateral TFETs. (b) Dependence of $g_m/I_{ds}$ on $I_{ds}$ for dfiferent devices. Notations are same as in fig. 3a.

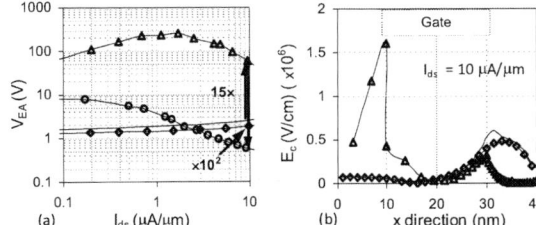

Fig. 5. (a) Dependence of $V_{EA}$ on $I_{ds}$ for different devices. (b) Variation of electric field along the channel direction ($x$). Notations are same as in fig. 3a.

978-1-4799-7440-5/14 $31.00 © 2014 IEEE

# On the Cryogenic Performance of Ultra-Low-Loss, Wideband SPDT RF Switches Designed in a 180 nm SOI-CMOS Technology

Adilson S. Cardoso, Partha S. Chakraborty, Anup P. Omprakash, Nedeljko Karaulac,
Prabir Saha, and John D. Cressler

School of Electrical and Computer Engineering, Georgia Institute of Technology, Atlanta, GA U.S.A
Email: cardosoa@gatech.edu / Tel: (404) 874-6403

**Abstract:** The RF cryogenic performance of ultra-low-loss, wideband (DC to 40 GHz) single-pole double-throw (SPDT) RF switches implemented in a 180 nm SOI CMOS technology is reported for the first time. Results show that the switch insertion loss (IL), isolation (ISO), small- and large-signal linearity all improve as the temperature decreases. DC characterization of individual transistors was performed and analyzed to provide insight into the mechanisms underlying the observed changes in the RF switches.

**1. Introduction:** RF switches are ubiquitous building blocks utilized in a wide variety of space-based electronic systems (e.g., satellites, deep-space remote sensing electronics, and spacecraft navigation systems) [1]. Therefore, it is important to investigate and understand the operational integrity of the RF switches while operating in extreme environments where radiation and cryogenic temperatures are commonplace. This information is required to provide mission architects and circuit designers with a better understanding of how system operation can be influenced by cryogenic temperatures, and more importantly, how to leverage the circuit response for cryogenic operation. This paper presents the first full RF characterization of the FET-based SPDT switch at cryogenic temperatures. The switch has the lowest reported insertion loss (< 1.25 dB at 40 GHz) on a silicon-based technology operating at 78 K. Although [2] has reported a comparable IL at 296 K (room temperature), that SPDT was designed in a 45 nm SOI CMOS technology. The radiation robustness of these switches were reported in [3].

**2. SPDT Switch Design:** SOI CMOS technologies are naturally suited for low-loss FET-based switches since the isolated bulk reduces the loss through the substrate [4]. The circuits were designed and fabricated in a commercially-available 180 nm twin-well SOI CMOS (IBM 7RF-SOI) process on high-resistivity ($\approx 1000$ $\Omega$-cm) substrate with 3 metal layers [5]. The detailed design and RF performance of the SPDT at ambient temperature ($\approx 300$ K) was previously presented in [5] (which did not include the cryogenic results). A $\pi$-matching network consisting of series inductor $L_X$, shunt capacitor $C_X$, and the parasitic capacitance from $M_1$ and $M_2$ (denoted as $C_P$ in Fig. 1 (a)) was realized to improve the SPDT input matching and IL at higher frequencies. A photomicrograph of the fabricated circuit is shown in Fig. 1 (b).

**3. Measured Results and Analysis:** As illustrated in Fig. 2, the switch IL (on-state) improves with decreasing temperature; < 1.25 dB at 78 K, 1.73 dB at 150 K, and 2.25 dB at 300 K over the bandwidth of 40 GHz. The IL is < 0.13 dB from DC to 20 GHz at 78 K. These are significant results, since the IL is typically the primary specification for applications, as it can impact the overall system performance. The reduction in IL can be explained by the extracted ON-resistance ($R_{ON}$)

and carrier mobility ($\mu_e$) results shown in Figs. 3(a) and 3(b), respectively. As temperature is reduced, the $R_{ON}$ decreases due to an increase in $\mu_e$. The $R_{ON}$ and the $\mu_e$ parameters were extracted over temperature from DC measurements using similar sized devices as in the SPDT. Fig. 4 depicts the measured ISO (off-state) at varying temperatures. The ISO results follow a similar trend as the IL demonstrating an improvement in isolation with decreasing temperature; a larger improvement is observed at lower frequencies, from DC to 20 GHz ($\approx 4.5$ dB difference between 300 K and 78 K). Above 20 GHz, the ISO enhancement is only $\approx 1.8$ dB. At high frequencies the parasitic capacitances of the OFF devices (e.g., when $RF_{OUT1}$ is enabled, $M_1$ and $M_3$ are ON while $M_2$ and $M_4$ are OFF) begin to oppose and cancel out the added benefit of cooling (reduction in IL) from the devices that are ON. The ISO improvement with decreased temperature is most likely due to a decrease in the parasitic source-drain junction capacitances as temperature is lowered. The ISO degrades as frequency increases, due to the increased parasitic coupling to the substrate [6]. The measured input and output reflection coefficients, $S_{11}$ and $S_{22}$ are shown in Fig. 5(a) and (b), respectively. Below 10 GHz, the worst-case variation is about 5 dB from 300 K to 78 K. Since the $S_{11}$ and $S_{22}$ values are high (> 20 dB) the temperature-induced variations do not impact the overall switch performance. At high frequencies (> 10 GHz) the influence of temperature on $S_{11}$ and $S_{22}$ is small, and no discernable trend is observed. As anticipated, the $S_{11}$ results demonstrate that the designed $\pi$-matching network is not too sensitive to temperature changes. The power handling capability of the SPDT was also characterized over-temperature. Fig. 6(a) shows the output power ($P_{OUT}$) and IL ($S_{21}$) as the input power is swept at a fixed frequency (8 GHz). The large-signal linearity, $P_{1dB}$ (input-referred) was extracted from Fig. 6(a) results. As temperature decreases, the $P_{1dB}$ improves from $\approx 11$ dBm at 300 K to $\approx 12.6$ dBm at 78 K (as shown in Fig. 6(b)). Small-signal linearity measurements were performed using two-tone RF signals at 8 GHz with a 10 MHz frequency spacing. The third-order intercept points (TOI) IIP3 and OIP3 were extracted, and as expected, the results follow the same trend as $P_{1dB}$; both IIP3 and OIP3 increase with decreasing temperature. This observed improvement in linearity is due to the increase in $\mu_e$ (as shown in Fig. 3(b)). The $P_{1dB}$ and TOI expressions for FET (previously derived in [7]) confirm that the observed improvement in large- and small-signal linearity is due to the increase in $\mu_e$ as temperature decreases. In (1) and (2) the parameter $v_{sat}$ represents the saturation velocity, "L" the transistor gate length, "$R_s$" the input impedance, and $\mu_1$ is the effective carrier mobility,

978-1-4799-7440-5/14 $31.00 © 2014 IEEE

$$P_{1dB} = \frac{\left(1+\frac{\mu_1 V_{od}}{2v_{sat}L}\right)^4}{2R_S\left(\frac{\mu_1}{2v_{sat}L}\right)^2\left[V_{od}\left(1+\frac{\mu_1 V_{od}}{4v_{sat}L}\right)+\frac{6.88v_{sat}L}{\mu_1\left(1+\frac{\mu_1 V_{od}}{2v_{sat}L}\right)^2}\right]} \quad (1)$$

$$IIP3 = \frac{8}{3}\frac{v_{sat}L}{\mu_1 R_S}V_{od}\left(1+\frac{\mu_1 V_{od}}{4v_{sat}L}\right)\left(1+\frac{\mu_1 V_{od}}{2v_{sat}L}\right)^2 \quad (2)$$

Where,

$$\mu_1 = \mu_0 + 2\theta v_{sat}L \quad (3)$$

$$V_{od} = V_{GS} - V_{TH} \quad (4)$$

In (3) and (4), $\mu_0$ is the low-field mobility and $\theta$ is the mobility degradation factor due to the applied perpendicular electric field [7].

**4. Summary:** The RF performance of SPDT switches designed in a 180 nm SOI-CMOS technology has been demonstrated at cryogenic temperatures. All key performance metrics like IL, ISO, large- and small-signal linearity improve with decreasing temperature. For this technology, operation of the SPDT at low temperatures can be leveraged to meet challenging system requirements, particularly for applications that require low-loss switches operating in cryogenic environments.

**Acknowledgement:** The authors are grateful to T. Albert and B. Offord for their contributions and support.

**References:**
[1] J. D. Cressler, *Proc. IEEE TNS*, p. 1992, 2013.
[2] M. Parlak *et al., Proc. IEEE CSICS*, p. 1, 2011.
[3] A. S. Cardoso *et al., IEEE TNS*, vol. 61, no. 2, p. 756, 2014.
[4] A. B. Joshi *et al., Proc. IEEE SOI*, 2012.
[5] A. S. Cardoso *et al., Proc. IEEE RWS*, p. 22, 2014.
[6] F. Gianesello *et al., Proc. IEEE SOI*, 2010.
[7] T. Soorapanth *et al., Proc. Workshop on Design of Mixed-Mode Integrated Circuits & Applications*, pp.81-84, 1997.

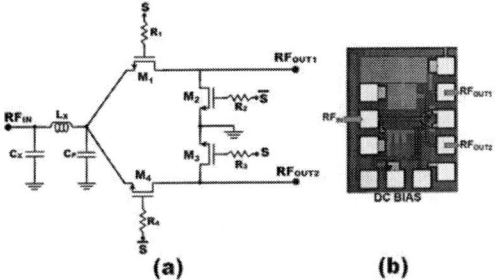

Fig. 1. (a) Schematic diagram of the SPDT RF switch, and (b) photomicrograph of the fabricated SPDT circuit.

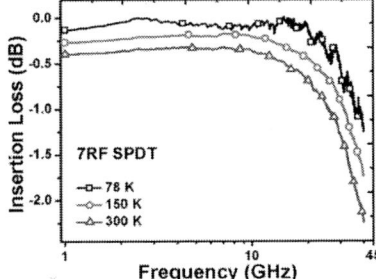

Fig. 2. Measured insertion loss for SPDT at 300 K (blue triangle), 150 K (red circle), and 78 K (black square) traces.

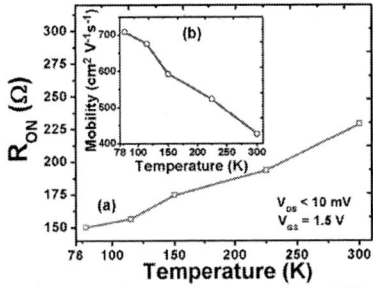

Fig. 3. (a) Extracted ON-resistance ($R_{ON}$) and (b) mobility as a function of temperature.

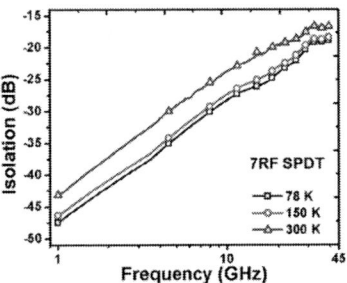

Fig. 4. Measured ISO for SPDT at 300 K (blue triangle), 150 K (red circle), and 78 K (black square) traces.

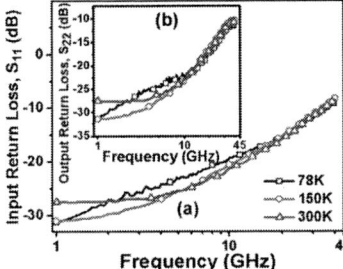

Fig. 5. (a) Measured SPDT input return loss, $S_{11}$ and (b) output return loss, $S_{22}$ at 300 K (blue triangle), 150 K (red circle), and 78 K (black square) traces.

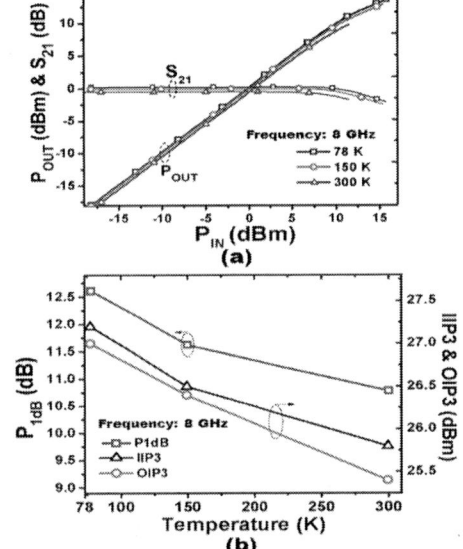

Fig. 6. (a) Measured SPDT output power ($P_{OUT}$) and insertion loss ($S_{21}$) as a function of input power ($P_{IN}$) at 300 K (blue triangle), 150 K (red circle), and 78 K (black square) traces. (b) Large-signal linearity ($P_{1dB}$ in blue square) and small-signal linearity, IIP3 (black triangle) and OIP3 (red circle) as a function of temperature.

978-1-4799-7440-5/14 $31.00 © 2014 IEEE

# Analog Performance of Short-Channel Asymmetric Self-Cascode of Junctionless Nanowire nMOS Transistors

M. de Souza, R. T. Doria, R. Trevisoli, and M. A. Pavanello

Electrical Engineering Department, Centro Universitário da FEI, São Bernardo do Campo, Brazil

michelly@fei.edu.br

## I. Introduction

Junctionless nanowire transistors (JNTs) were proposed and demonstrated to circumvent the technological challenge of realizing ultra-sharp junctions in transistors with extremely downscaled channel length ($L$) [1, 2]. JNTs are composed by a heavily doped silicon nanowire surrounded by gate stack. A schematic representation of a junctionless n-type transistor is presented in Fig. 1(top), indicating the channel length, fin width ($W_{fin}$) and height ($H_{fin}$). Apart from the advantages related to the improvement of short-channel effects (SCE) [3] in comparison to inversion mode (IM) MuGFETs, JNTs have already shown to present improved analog performance [4, 5]. However, it also presents some drawbacks, such as high threshold voltage ($V_{TH}$) sensitivity with $W_{fin}$ [6] and high series resistance ($R_S$) [7]. A common approach to improve the output conductance ($g_D$) of MOSFETs is to associate them in self-cascode (SC) configuration, as represented in Fig. 1(bottom) [8]. Recently, it has been shown that by using lower $V_{TH}$ for M$_D$, further improvements can be achieved in comparison to the symmetric $V_{TH}$ self-cascode (S-SC) [9]. In that work, different $V_{TH}$ were obtained through the reduction of doping concentration in M$_D$ channel region.

This work presents the analog performance of asymmetric $V_{TH}$ self-cascode (A-SC) structure implemented with short-channel Junctionless Nanowire SOI transistors. The $V_{TH}$ of M$_D$ has been varied by changing the device fin width. Important analog parameters such as the transconductance ($g_m$) and output conductance are presented. An evaluation of the junctionless A-SC impact over the performance of common-drain and common-source amplifiers is also presented. These two single stages amplifiers are important basic blocks in many analog circuits, and are schematically represented in Fig. 2(A) and (B), respectively [10, 11]. In both topologies, the gate terminal of the transistor serves as the input for the circuit, whereas the source is the output in the former and the drain in the later. While the common-drain (CD) amplifier is the classical approach for an unit-gain buffer (also known as source-follower), common-source (CS) amplifiers are expected to have the largest possible intrinsic voltage gain.

## II. Devices Characteristics

The devices used in this work were fabricated at CEA-Leti, following the process described in [12]. Transistors with 50 parallel fins with $L$=50 nm and $W_{fin}$ ranging between 20 nm and 440 nm were combined to configure the SC structures, maintaining the width of M$_S$ ($W_{fin,S}$) as 20 nm. Also single junctionless transistors (ST) with $W_{fin}$=20 nm and $L$=50 nm and 100 nm were measured for comparison purposes. Table I presents the $V_{TH}$ values of the different devices. From this table, it is possible to see the strong dependence of $V_{TH}$ on $W_{fin}$. However, in the case of the SC, $V_{TH}$ is kept constant ($V_{TH}$=0.12V) and close to the threshold voltage of single transistor with $W_{fin}$=20nm, independent on the width of M$_D$ ($W_{fin,D}$) since the overall $V_{TH}$ is given by $W_{fin,S}$.

## III. Self-Cascode Measurements and Discussion

The independence of self-cascode threshold voltage with $W_{fin,D}$ change can be seen in the curves of Fig. 3, that present the drain current ($I_{DS}$) vs. the gate voltage ($V_{GF}$) of single transistor and self-cascode configurations. One can also note the degradation of subthreshold slope (SS) with channel shortening (SS of 61.5, 62.9 and 75.3 mV/dec for single transistors with $L$=10μm, 100nm and 50nm, respectively). Also, self-cascode devices present the same SS as the transistor used as M$_S$.

The $I_{DS}$ and $g_D$ vs. the drain voltage ($V_{DS}$) curves of ST and SC junctionless devices are presented in Fig. 4 at gate voltage overdrive ($V_{GT}=V_{GF}-V_{TH}$) of 200 mV. As expected, for single transistors, $L$ reduction increases $I_{DS}$ level but worsens $g_D$. Although the S-SC (i.e. $W_{fin,S}=W_{fin,D}$) still presents lower $I_{DS}$ than the single transistor with $L=L_S+L_D$=100nm, their $g_D$ are practically the same. The reduction of $V_{TH,D}$ (increase of $W_{fin,D}$) in the SC promoted the increase of $I_{DS}$, without penalty for $g_D$. In the case of SC with wider M$_D$, the drain current level is larger than that of the ST with $L$=100 nm, with improved output conductance.

The larger $I_{DS}$ seen in Figs. 3 and 4 at the same $V_{GT}$ is associated to the different saturation voltages ($V_{Dsat}$) among the devices. As can be seen in Fig. 5(left), by widening M$_D$ (reducing $V_{TH}$), there is an increase of the voltage at the intermediate node, $V_X$ in Fig. 1(bottom), due to the decrease of M$_D$ resistance at a given $V_{GT}$. Therefore, larger voltage reaches the drain of M$_S$, increasing $I_{DS}$. From Fig. 5(right) it is possible to see $V_X$ variation with $V_{DS}$. At low drain bias, $V_X$ is very close to the applied $V_D$ for all SC. However, the lower $V_{TH,D}$, higher is the $V_{DS}$ where $V_X$ saturates. After this saturation, further increase of $V_D$ is absorbed by M$_D$, which is the reason for the reduction of $g_D$ with $W_{fin,D}$ increase.

978-1-4799-7440-5/14 $31.00 © 2014 IEEE

The $g_m/I_{DS}$ ratio is an important tool for the design of analog circuits. The correspondence between $g_m/I_{DS}$, $V_{GT}$ and $I_{DS}$ for the measured devices configured as common-source amplifier is shown in Fig. 6. For the same $V_{GT}$ (or $I_{DS}$), ST and SC structures have similar efficiency for converting $I_{DS}$ in $g_m$, except for the ST with $L=100$ nm, which presents larger $g_m/I_{DS}$ at low $V_{GT}$, due to its smaller $SS$.

The $g_m$ and $g_D$ curves as a function of $g_m/I_{DS}$ ratio are presented in Fig. 7, at $V_{DS}=1$V. At low $g_m/I_{DS}$, the $g_m$ of any SC is smaller than that of the single transistor with $L=50$nm, due to its larger effective length, as reported for inversion-mode transistors [9, 13]. However, as $W_{fin,D}$ is increased, $g_m$ increases and approaches that of a single transistor with $L=100$nm, due to $V_X$ increase, caused by the reduction of $M_D$ resistance. At high $g_m/I_{DS}$ values, $g_m$ of all ST and SC devices tends to a similar value. Despite of some $g_m$ degradation (maximum of 42% in the case of S-SC in comparison to ST with $L=50$nm), $g_D$ is strongly improved by the adoption of the asymmetric self-cascode configuration, especially at higher $g_m/I_{DS}$ ratio (low $V_{GT}$), showing the potential of this structure for low voltage applications. Higher $V_{GT}$ values (lower $g_m/I_{DS}$) make the A-SC to work closer to saturation voltage, degrading $g_D$, and no clear advantage over single transistors can be seen. For $g_m/I_{DS} > 4$V$^{-1}$, even the S-SC presents smaller $g_D$ than the ST with $L=100$nm. At $g_m/I_{DS} = 15$V$^{-1}$, that biases the devices at moderate conduction level, the $g_D$ reduction provided by the A-SC with $W_{fin,D}=290$ nm is about 4 and 14 times in comparison to the ST with $L=100$ nm and 50 nm, respectively. Despite of some $I_{DS}$ level reduction that can be found, this $g_D$ reduction results in larger Early voltage ($V_{EA}\cong I_{DS}/g_D$) for the self-cascode transistor for $g_m/I_{DS} > 2$V$^{-1}$ as shown in Fig. 8. It is shown that $V_{EA}$ is larger in any SC configuration than ST, and improves as $W_{fin,D}$ is increased up to 190 nm. Further $M_D$ widening has shown to promote no $V_{EA}$ increase. Similar trend is then seen in the intrinsic voltage gain ($A_V=g_m/g_D=V_{EA}\times g_m/I_{DS}$), presented in Fig. 9 as a function of $g_m/I_{DS}$ ratio. It is shown that, depending on the configuration, the A-SC can provide larger $A_V$ than the longer device. In the case of $W_{fin,D}=190$ or 290nm, the $A_V$ increase is 17 and 7 dB in comparison to ST with $L=50$nm and 100nm, respectively, at $g_m/I_{DS} = 15$V$^{-1}$.

The studied single transistors and self-cascode transistors were also configured as common-drain amplifiers (unit-gain buffer) [13]. Fig. 10 presents $V_{OUT}$ and $A_V$ vs. $V_{IN}$ at current bias of 10 nA. The gain of the unity-gain amplifier is given by $A_V=g_m/(n.g_m+g_D)$, with $n$ being the body factor, that is directly related to the subthreshold slope. As can be seen, for the single transistors, as $L$ is shortened, $A_V$ is reduced. This degradation is related both to $g_D$ degradation and short-channel effects, that increase $SS$. However, the improvement of $g_D$ in the self-cascode devices minimizes $A_V$ degradation due to the body factor ($SS$) increase, resulting in gain values

closer to the gain of longer single transistor. This improvement can be seen in Fig. 11 that presents the gain as a function of a wide range of bias current.

The reduction of $V_{TH}$ with $L$ also allows for higher $V_{OUT}$, as shown in Fig. 10. Indeed, devices with same $V_{TH}$ present similar $V_{OUT}$ vs. $V_{IN}$ characteristics. The input voltage range ($V_{IR}$) has been extracted from $V_{OUT}$ vs. $V_{IN}$ curves, as the difference between the maximum value of $V_{IN}$ before saturation of $V_{OUT}$ (or $V_{DD}$, when saturation is not reached) and the minimum $V_{IN}$ for $V_{OUT} \geq 0$. $V_{IR}$ has been extracted for several current biases and has shown little dependence with $I_{bias}$. The obtained results are presented in Fig. 12. For ST, the increase of $L$ has shown to reduce $V_{IR}$, due to $V_{TH}$ increase. However, despite of presenting similar $V_{TH}$ than the ST with $L=50$nm, SC structures have reduced $V_{IR}$ since the SC configuration causes an increase in the saturation voltage.

## IV. Conclusions

This work presented an evaluation of self-cascode association of short-channel junctionless nanowire transistors, by means of experimental results, comparing data of this configuration to single transistors. Even though the self-cascode transistors have shown to reduce dc drain current and transconductance level with respect to single device, due to their longer effective channel length, this effect becomes pronounced with the widening of the transistor close to the drain ($M_D$), with consequent threshold voltage reduction. Both symmetric and asymmetric self-cascode configurations do not degrade the efficiency of converting current into transconductance ($g_m/I_{DS}$ ratio) in relation to single transistor with same length. On the contrary, the increase of width of $M_D$ is capable of reducing the output conductance, reaching values smaller than those obtained for longer single transistors, especially at low gate voltages. The combination of these characteristics results in improved performance for the self-cascode configured as common-source amplifier if compared to symmetric self-cascode or individual longer transistors, reaching a voltage gain increase of up to 17 dB at $g_m/I_{DS} = 15$V$^{-1}$. When used as common-drain amplifiers, the asymmetric self-cascode junctionless transistors has shown to improve the electrical characteristics (voltage gain and input voltage range) in comparison to a short-channel single transistor, whereas had not shown advantages over a longer single device.

## Acknowledgments

This work was supported by FAPESP, CAPES and CNPq. The devices were fabricated by CEA-LETI in the framework of the European project SQWIRE under Grant Agreement N° 257111. The authors thank I. Ferain and J.-P. Colinge for supplying the JNTs.

## References

[1] J.P. Colinge et al., Proc. of IEEE Intern. SOI Conf., 1(2009).
[2] B. Sorée et al., J Comput Electron, 7, 380(2008).

978-1-4799-7440-5/14 $31.00 © 2014 IEEE

[3] C.W. Lee *et al.*, *Solid State Elec.*, **54**, 97(2010).

[4] R.T. Doria *et al*, *IEEE Trans. Electron Devices*, **58**(8), 2511(2011).

[5] R.T. Doria *et al.*, *Journal of Integrated Circuits Systems*, **6**, 114(2011).

[6] R.D. Trevisoli *et al.*, *Semiconductors Science & Technology*, **26**, 105009(2011).

[7] R.T. Doria *et al.*, *Journal of Integrated Circuits Systems*, **7**, 121(2012).

[8]C. Galup-Montoro *et at.*, *IEEE Journal of Solid-State Circuits*, **29**, 1094(1994).

[9] M. de Souza *et al.*, *Proc. of IEEE Intern. SOI Conf.* 1(2011).

[10] K. R. Laker, W. M. C. Sansen. Design of Analog Integrated Circuits and Systems. 1st ed. McGraw-Hill; 1994.

[11] B. Razavi, Design of Analog CMOS Integrated Circuits, McGraw-Hill, 2002.

[12] S. Barraud *et al.*, *IEEE Electron Dev. Letters*, **33**, 1225(2012).

[13] R. T. Doria *et al.*, *ECS Transactions*, **49**(1), 215(2012).

[13] W. Sansen., *IEEE Trans. On Circ. Systems II*, **46**, 315 (1999).

**Table I.** *Threshold voltage of single transistors, extracted at* $V_{DS}=50mV.$

| $L$ [nm] | 50 | | | | | 100 |
|---|---|---|---|---|---|---|
| $W_{fin}$ [nm] | 20 | 40 | 190 | 290 | 440 | 20 |
| $V_{TH}$ [V] | 0.14 | 0.08 | -0.13 | -0.14 | -0.17 | 0.28 |

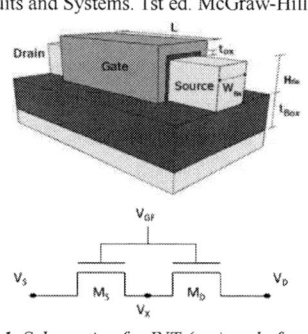

**Fig.1.** *Schematic of a JNT (top) and of a self-cascode (bottom).*

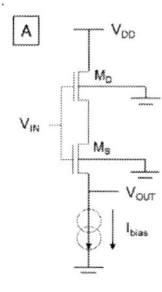

**Fig. 2.** *Schematic representation of Self-Cascode Common-Drain (A) and Common-Source (B) Amplifiers.*

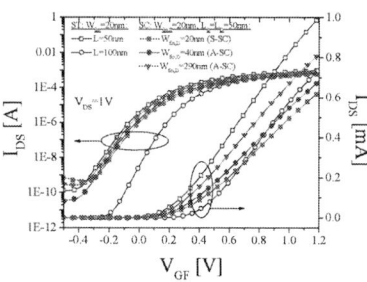

**Fig. 3.** *Drain current as a function of gate voltage for different ST and SC transistors at* $V_{DS}=1V.$

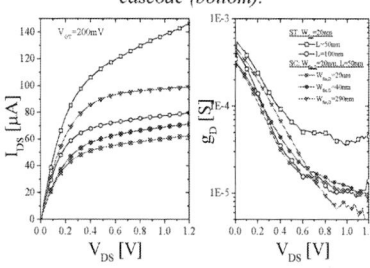

**Fig.4.** *Drain current and output conductance of ST and SC, biased at* $V_{GT}=200mV.$

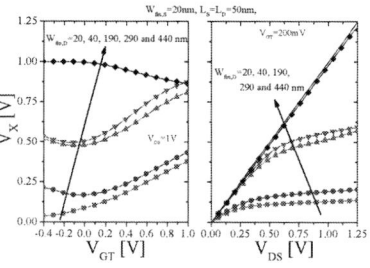

**Fig.5.** $V_X$ *vs.* $V_{GT}$ *and* $V_{DS}$ *for self-cascode JNTs.*

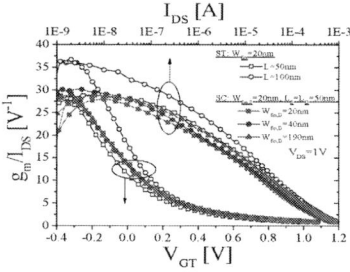

**Fig.6.** $g_m/I_{DS}$ *vs.* $V_{GT}$ *and* $I_{DS}$ *for JNT ST and SC.*

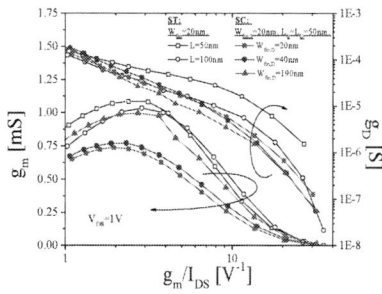

**Fig. 7.** *Transconductance and output conductance vs.* $g_m/I_{DS}$ *ratio for junctionless ST and SC transistors* ($V_D=1V$).

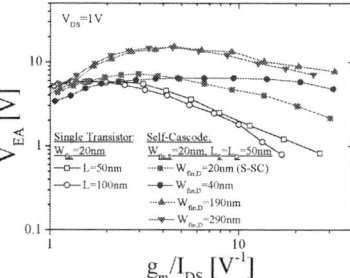

**Fig.8.** *Early voltage vs.* $g_m/I_{DS}$ *ratio of different junctionless ST and SC transistors.*

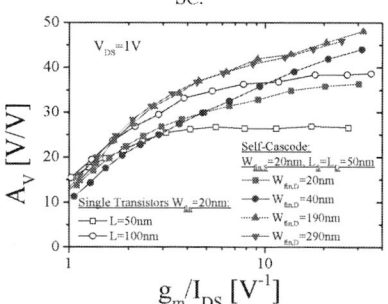

**Fig.9.** *Intrinsic voltage gain vs.* $g_m/I_{DS}$ *ratio of different junctionless ST and SC transistors.*

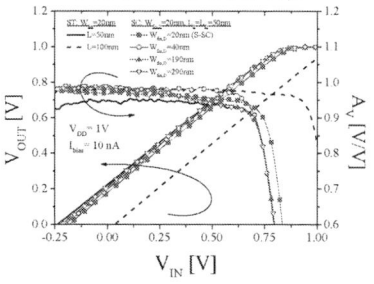

**Fig. 10.** *Output voltage and voltage gain of common-drain amplifiers with JNT single transistors and self-cascode, biased at 10nA.*

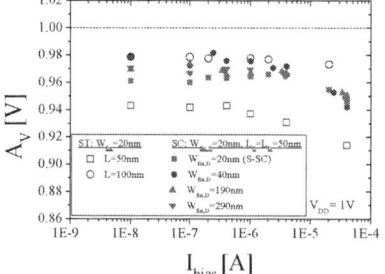

**Fig. 11.** *Voltage gain as a function of bias current of buffers junctionless SC transistors*

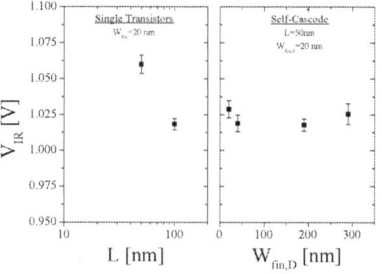

**Fig. 12.** *Input range of buffers JNT with single transistors and self-cascode*

978-1-4799-7440-5/14 $31.00 © 2014 IEEE

# Multi-Threshold Design Methodology of Stacked Si-Nanowire FETs

Yi-Bo Liao[1] and Meng-Hsueh Chiang[2]

[1]Institute of Microelectronics, Department of Electrical Engineering, National Cheng Kung University, Tainan 701, Taiwan
[2]MS Degree Program on Nano-IC Engineering, Department of Electrical Engineering, National Cheng Kung University, Tainan 701, Taiwan.
E-mail: mhchiang@mail.ncku.edu.tw, Phone: +886-6-275-7575 ext. 62418

*Abstract*- **A multi-threshold design methodology of stacked silicon nanowire MOSFETs is proposed. A flexible doping scheme is demonstrated for high-performance and low-operating power designs integrated together on a same substrate. With additional channel length adjustment, low standby power is further achieved.**

## I. INTRODUCTION

While non-planar FinFETs are emerging for mass production [1]-[4], gate-all-around (GAA) MOSFETs provide even better gate control and scalability in sub-10nm regime. However, the drive current of each GAA device is limited by reduced channel area, unlike conventional planar case with arbitrary channel width for producing desired drive current. As a result, a multi-fin/wire layout is needed, but on the other hand the total transistor area is increased. The vertically stacked-wire concept provides another approach to improve drive current at same footprint. Using Si/SiGe superlattice epitaxy and in-situ doping process for stacked wires, desired threshold voltage ($V_t$) can be met following conventional $V_t$ adjustment, but the challenge is how multi-$V_t$ can be implemented without significant additional cost. While conventional multi-$V_t$ approach via channel length adjustment has become insufficient, this work proposes a stacked approach with 3 wires to be feasible for a wide range of $V_t$. Furthermore, a design window is demonstrated for potential SoC application with high-performance (HP) and low-operating power (LOP) designs integrated together on a same substrate.

## II. STACKED-WIRE STRUCTURE

Based on epitaxial Si/SiGe superlattice growth, stacked-wire structure can be made [5][6]. A conceptual 4-wire GAA structure is shown in Fig. 1. Following ITRS roadmap at 11.9 nm node [7], the dimension of nominal device for simulation is designed with $L_g$ = 10 nm, EOT = 0.62 nm, and off-state current ($I_{off}$) ≤ 100 nA/μm at $V_{DD}$ = 0.68 V. Both channel width ($W_{Si}$) and height ($H_{Si}$) are 10 nm based on electrostatic scale length of 3.3 nm [8]. A raised-source/drain is assumed with *in-situ* doping ($2 \times 10^{20}$ cm$^{-3}$) and 1-D Gaussian doping profile along channel direction. 3-D TCAD simulator is employed with Fermi-Dirac statistics, drift-diffusion transport using the Philips unified mobility model, and density gradient quantization model [9]. For off-state current ($I_{off}$) normalization, effective channel width ($W_{eff}$) is calculated as $W_{eff} = n \times 2 \times (H_{Si} + W_{Si})$ where n is the number of wires. Constant current method ($W_{eff}/L_g \times 10^{-7}$ A) is used for $V_t$ extraction.

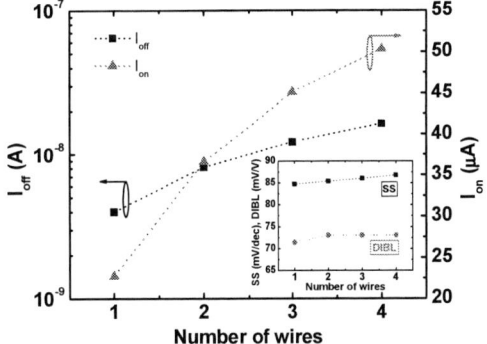

Fig. 2. $I_{off}$ and $I_{on}$ characteristics of different stacked wires. Inset shows the corresponding SS and DIBL.

Fig. 2 shows $I_{off}$ and $I_{on}$ characteristics of different stacked wires. Both $I_{off}$ and $I_{on}$ show an increasing trend with added wires. However, the increasing trend is not linear; the increase in current from 3 to 4 wires is marginal due to increased series source/drain resistance. Short-channel effect (SCE) related properties such as subthreshold swing (SS) and drain-induced barrier lowering (DIBL) are shown in the inset. Considering the complexity in technology with limited etching capability, building 4 stacked wires requires a

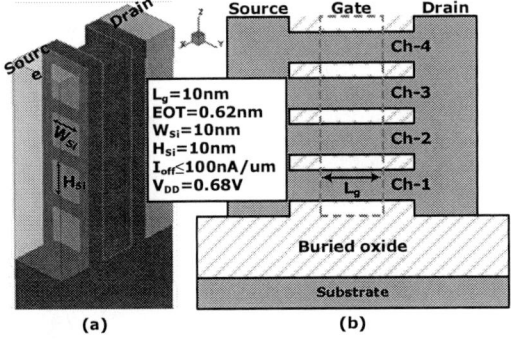

Fig. 1. The stacked-wire structure of GAA MOSFETs in (a) 3D view and (b) X-Z plane cross section view.

978-1-4799-7440-5/14 $31.00 © 2014 IEEE

challenging aspect ratio of more than 5. In addition, when aforementioned parasitic resistance is accounted for, the stacked number of 3 is suggested.

## III. THRESHOLD VOLTAGE ADJUSTMENT

Threshold voltage doping scheme for stacked wires is far different than for conventional approach, especially when multiple layers of transistors are integrated on the same substrate. Leaving the channel undoped has an advantage in mobility and is expected to relieve the issue of random dopant fluctuation, but it does not meet the need for multi-$V_t$ design being commonly used in SoC application. Instead, different gate work functions (or gate materials) will be needed for different $V_t$'s and hence such undoped approach would be even more complicate. How to implement a multi-$V_t$ design in a stacked wire configuration is proposed in this work. Threshold voltage is a function of gate oxide thickness, gate work function, and channel doping concentration. Among them, doping scheme is most commonly used to set needed Vt. Fig. 3 shows $V_t$ and $I_{off}$ in different channel doping concentrations for a single wire, as noted "Ch-1" in Fig. 1(b). Inset shows simulated $I_{DS}$-$V_{GS}$ characteristics in various channel doping levels. $V_t$ changes from 170 mV to 310 mV as $N_{a,Ch-1}$ being increased from $10^{18}$ to $10^{19}$ cm$^{-3}$. For low doping, $V_t$ is insensitive as the depletion charge is insignificant.

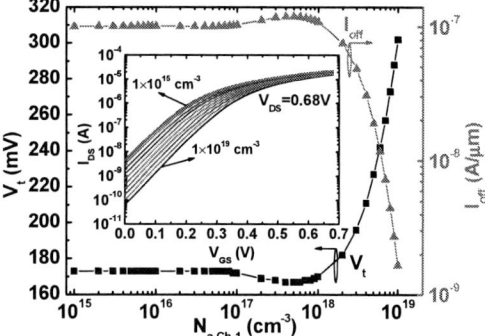

Fig. 3. $V_t$ and $I_{off}$ characteristics in different channel doping concentrations. Inset shows simulated $I_{DS}$-$V_{GS}$ characteristics.

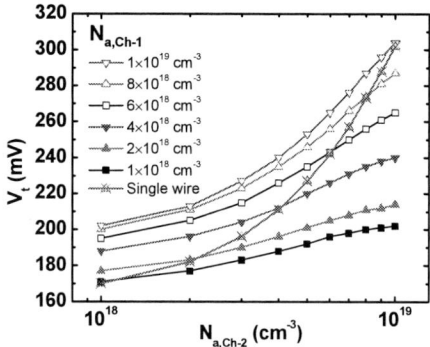

Fig. 4. Predicted $V_t$ vs. channel doping concentration for 2 stacked wires.

During epitaxy process, the *in-situ* doped channel is implemented for each of the stacked wires. Fig. 4 shows $V_t$

adjustment by changing $N_{a,Ch-2}$ for 2 stacked wires at different given $N_{a,Ch-1}$'s. The identical doping case ($N_{a,Ch-1} = N_{a,Ch-2}$) is equivalent to a single wire (also shown in Fig. 4), but it gives near twice higher current. When $N_{a,Ch-1}$ is low, Ch-1 becomes predominant and $V_t$ is nearly unchanged. For 3 stacked wires, Fig. 5 shows the predicted $V_t$ in channel doping variations of top and middle wires (Ch-2 and Ch-3) with $N_{a,Ch-1} = 10^{15}$ and $10^{19}$ cm$^{-3}$, respectively. $V_t$ adjustment becomes insufficient even with heavily doped Ch-2 and Ch-3 when Ch-1 is kept undoped, as shown in Fig. 5(a). On the other hand, a wide range of $V_t$ is possible when Ch-1 is heavily doped, as shown Fig. 5(b). For LOP application, low off-state current is required and hence all 3 wires must be heavily doped.

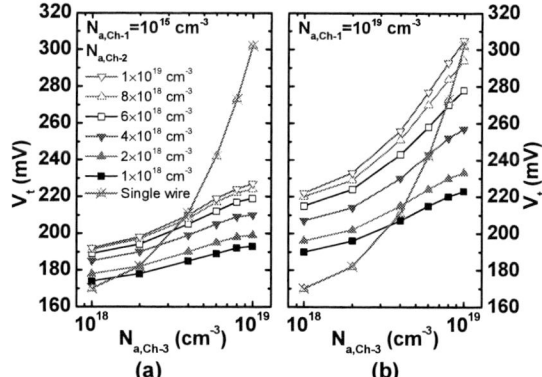

Fig. 5. Predicted $V_t$ in different channel doping variations for 3 stacked wires.

## IV. INSIGHT OF USING CHANNEL DOPING SCHEME

Doped stacked-GAA MOSFETs provide flexible options for $V_t$ adjustment. However, how to implement such a multi-$V_t$ scheme based on a same substrate is another issue. In addition, whether a design window exists for a wide range of $V_t$ is unknown. This section demonstrates a design methodology of multi-$V_t$ for HP and LOP applications according to their respective requirements as suggested in ITRS [7]. Different wire doping combinations gives different $I_{on}$ versus $I_{off}$ characteristics, as shown in Fig. 6 where $I_{off}$'s of 100 nA/µm, and 5 nA/µm are labeled for HP and LOP designs, respectively.

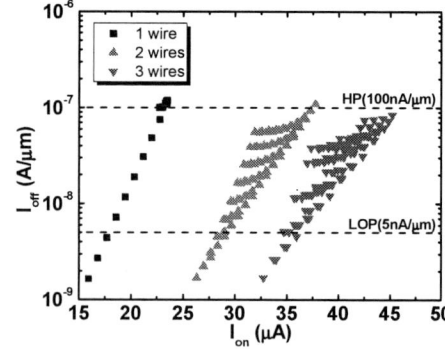

Fig. 6. $I_{on}$ vs. $I_{off}$ characteristics for different stacked cases where $I_{off}$ is normalized to effective width.

For a certain application with required $I_{off}$, we can choose the stacked-wire configuration based on needed $I_{on}$. Next, we should determine the $V_t$ value for a certain design. Fig. 7 shows $I_{off}$ versus $V_t$ characteristics. As LOP design requires low $I_{off}$, $V_t$ is higher than 277 mV. $V_t$ for HP application falls in the range between 173 mV to 277 mV. Though the stacked-wire structure provides flexible doping combinations, we need a design guideline for technology to follow.

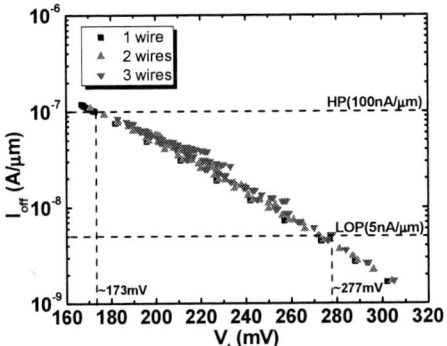

Fig. 7. $V_t$ vs. $I_{off}$ characteristics for different stacked cases.

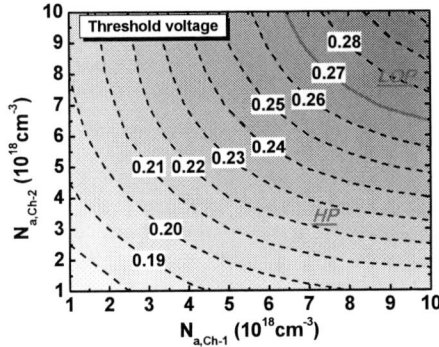

Fig. 8. The doping scheme window for 2 stacked wires.

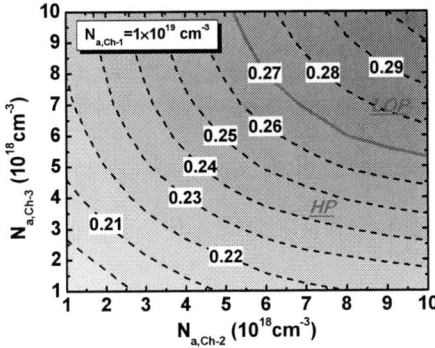

Fig. 9. The doping scheme window for 3 stacked wires with $N_{a,Ch-1} = 1 \times 10^{19}$ cm$^{-3}$.

For integration of different designs with different $I_{off}$ requirements, a design window should exist for all of them. Based on $I_{on}$ and $I_{off}$ requirements for a given application, we first choose the type of stacked wires and then decide the doping scheme. Fig. 8 shows the predicted $V_t$ from combinational doping levels of 2 stacked wires. The right-top corner can be used for LOP application with both doping levels higher than $7.5 \times 10^{18}$ cm$^{-3}$. A fairly large window is applicable to HP design. Fig. 9 shows doping scheme windows for 3 stacked wires with different given $N_{a,Ch-1}$'s. High $N_{a,Ch-1}$ gives a more flexible doping range for $N_{a,Ch-2}$ and $N_{a,Ch-3}$ for further $V_t$ adjustment. The design of 3 stacked wires provides a more flexible doping scheme and is suggested for SoC design. For low standby power (LSTP), further channel length adjustment could achieve another level of $I_{off}$ requirement, as shown in Fig. 10.

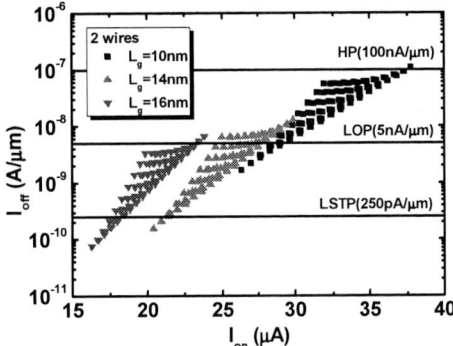

Fig. 10. Channel length adjustment for three $I_{off}$ levels of SoC application.

## V. CONCLUSION

The multi-threshold design methodology of stacked silicon nanowire MOSFETs has been proposed. The stacked approach with 3 wires provides a flexible doping scheme for a wide range of $V_t$ and hence is suggested for SoC design when considering technology requirement and parasitic resistance.

### REFERENCES

[1] C. Auth et al., "A 22nm high performance and low-power CMOS technology featuring fully-depleted tri-gate transistors, self-aligned contact and high density MIM capacitors," in Proc. VLSI Symp. Tech., pp. 131-132, 2012.

[2] C.-H. Jan et al., "A 22nm SoC platform technology featuring 3-D tri-gate and high-k/metal gate, optimized for ultra low power, high performance and high density SoC application," in IEDM Tech. Dig., pp. 3.1.1-3.1.4, 2012.

[3] K. J. Kuhn et al., "The ultimate CMOS device and beyond," in IEDM Tech. Dig., pp. 8.1.1-8.1.4, 2012.

[4] S.-Y. Wu et al., "A 16nm FinFET CMOS technology for mobile SoC and computing applications," in IEDM Tech. Dig., pp. 9.1.1-9.1.4, 2013.

[5] T. Ernst et al., "Novel 3D integration process for highly scalable Nano-Beam stacked-channels GAA (NBG) FinFETs with HfO2/TiN gate stack," in IEDM Tech Dig., pp. 1-4, 2006.

[6] C. Dupre et al., "15nm-diameter 3D stacked nanowires with independent gates operation: ΦFET," in IEDM Tech. Dig., pp. 1-4, 2008.

[7] International Technology Roadmap for Semiconductors, 2011 (http://public.itrs.net/).

[8] Y.-B. Liao et al., "6-T SRAM cell design with gate-all-around silicon nanowire MOSFETs," in Proc. VLSI-TSA, pp. 137-138, 2013.

[9] Sentaurus Device, User Guide, Synopsys Inc. ver. H-2013.03, Mar. 2013.

# Analog Building Block Design in 14nm FinFET using Inversion Coefficient*

A. Wang, V. Dhawan, and C.-J. R. Shi

Department of Electrical Engineering, University of Washington, Seattle WA 98195

{ailiw,vandana1,cjshi}@uw.edu, Tel: (206)-221-5291

*Supported by Semiconductor Research Corporation (SRC) under Task 1836.095 and the State of Washington Joint Center for Aerospace Techology Innovation (JCATI).

*Abstract*---**This paper presents the characterization of inversion coefficient and technology current for 14nm FinFET. Analog performance parameters, their variability and controlability, including transconductance effieincy, intrinsic gain, gain-bandwidth product, flicker noise and current mismatching, are then characterized in terms of inversion coefficient. These characterized relations are used for sizing and dynamically compensating a set of analog building blocks including differential pairs, current mirrors, PTAT current generator, and re-generative structures. Post-layout simulation results using IBM's 14nm SOI FinFET process are presented.**

## I.  INTRODUCTION

14nm FinFET has been reported to provide a set of advantages for analog design over bulk CMOS at 20nm [1]. However analog design at 14nm FinFET especially in the subthrehold region presents a lot of challenges, including complex coupled tradeoffs among analog performances, discrete transistor width sizing in terms of number of fins, complex device parastics,  and strong effect due to process voltage temerapture (PVT) variations.

In this work, a systematic methodolgy is presented for 14nm FinFET analog design. The methodology is focused on how to compensate analog performance varaibility due to PVT variations. The foundation is the characterization of analog performance parameters, their varaibility, and their controlability in terms of a concept called inversion coefficient [2] and technology current [3].  The methodology consists of the following steps:

(1)  Characterization of technology current
(2)  (Performance Map) Characterization of analog performance parameters including transconductance efficiency, instrisic gain, gain-bandwidh product, flick noise, current mismatcing in terms of the inversion coefficient, the number of Fins (transistor W), and the Fin length (L)
(3)  (Variability Map) Charaterization of the variability of analog performance parameterss due to process, voltage and temperature variations
(4)  (Controlabilty Map): Characterization of the controlability of analog performance paramters in terms of controllable deisgn paramters:

inversion coefficient, number of Fins (W), and the Fin length (L)

(5)  (Design) The design of analog block for performance parameters and their control  in terms of FinFET parameter using the characterized mappings of performance, variability and controlability

This methodology has been demonstrated on the design and tape out of a set of basic analog functional building blocks such as differential pairs, current mirrors, re-generative latches. The results of PTAT current generator and dynamic comparator as such examples are included in this sumary.

## II.   14NM FINFET TECHNOLOGY CURRENT

Inversion coefficient IC, is a normalized measure of MOS drain current $I_D$ and represents the degree of MOS channel inversion. The inversion coefficient $IC$ is given by [2]

$$IC = \frac{I_D}{I_0(\frac{W}{L})} \qquad (1)$$

where $I_D$ is the drain current, $I_0$ is the technology current, and $W/L$ is the aspect ratio. The key step here is to determine the technology current $I_0$. The graphical determination of $I_0$ is based on [2] and [3]. This abstract presents the calculation of $I_0$ for 14nm FINFET technology. If we model the $I - V$ curve of the MOSFET in saturation at the constant $V_{DS}$ as the $\alpha$ power-law model

$$I_{Dsat^\alpha} = B \cdot \frac{W}{L} \cdot (V_{GS} - V_T)^\alpha \qquad (2)$$

The intersect point of the asymptote in weak inversion and the strong inversion for $\alpha$-law trans-conductance efficiency $g_{m^\alpha}/I_{Dsat^\alpha}$ is at the center of moderate inversion, that is $IC = 1$. Technology current is then evaluated using a  tangent with slope $-1/\alpha$ to the $g_m/I_D$ plot in strong inversion.

The transconductance efficiency versus the drain current is shown in Fig. 1. From this plot, $I_0 = 1.346uA$ for device length of 40nm, and I0=0.917 μA for another device with length of 200nm. I0 is the drain current $I_D$ normalized to the aspect ratio of the device.

978-1-4799-7440-5/14 $31.00 © 2014 IEEE

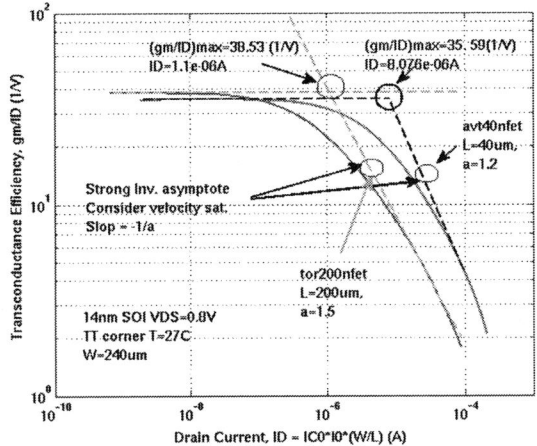

Fig. 1. Calculation of technology current $I_0$ from intersection of $g_{m\alpha}/I_{Dsat\alpha}$ with the tangent of slope$=-1/\alpha$ for two FINFET analog devices with FIN length of 40nm and 200nm.

## III. DESIGN AND SIMULATION RESULTS OF A PTAT CURRENT GENERATOR

The proportional to abstract temperature (PTAT) current generator is a building block widely used in temperature monitor systems, where a reference current varying as a linear function of temperature is required to interface sensors to the analog processing domain. A CMOS based PTAT current generator is shown in Fig. 2. Using weak inversion concepts derived in [4], if M1 and M2 are in weak inversion, the current through resistor R is given as:

$$I_{PTAT} = \frac{\eta \cdot V_T}{R} \ln\left[\frac{S_2}{S_1}\right] \quad (3)$$

where S2 and S1 are the aspect ratios of M2 and M1, respectively.

For I_PTAT=500nA, width of M1 is chosen 10X of minimum width (allowed by technology), and M1 is biased in weak inversion. In order to realize the PTAT with a reasonable sized resistor, if we choose a higher values of current (5μA), then the width of the weak inversion MOSFETs scales up by 100x and inversion cofficient is chosen to be 0.01. As the width of M2 is 8X of M1, that implies a multiplier of 800 for M2. Though the size of the resistor scales down by 10X, realizing such a large size FINFET is an overkill on area.

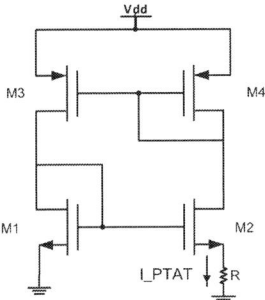

Fig. 2. Weak inversion CMOS based PTAT current source.

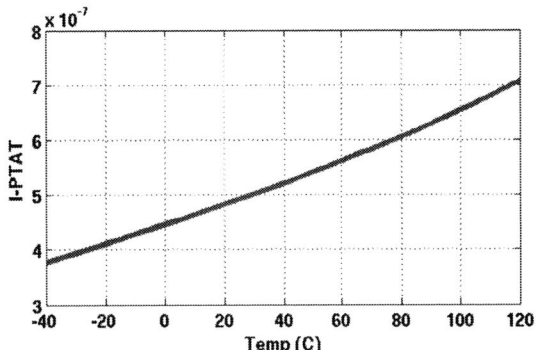

Fig. 3. I_PTAT varies linearly with temperature.

PTAT with I_PTAT = 500nA was designed in the 14nm SOI technology. The absolute value of the current versus the tempreture is shown in Fig. 3.

## IV. DESIGN AND SIMULATION RESULTS OF A DYNAMIC COMPARATOR

The comparator---a decision making circuit---is a fundamental building block for analog-to-digitial converters (ADCs), and many mixed-signal systems. Dynamic comparator is widely used in high-speed ADCs due to its low power consumption (no static power consumption) and fast speed (positive feedback in the regenerative latch). However, designing a high speed, high accuracy, low power consumption and low offset comparator is very challenging. We have designed and tapped out a single-ended "Lewis-Gray" comparator [5]. The concept of inversion coefficient has been used to size the transistors and optimize the comparator performance.

The comparator as shown in Fig. 4 operates as follows: During the reset phase (CLK=0, M9, M10 are on, and M5, M6 are off), the output nodes are charged to VDD. During the evaluation phase (CLK=1, M9, M10 are off, and M5, M6 are on), the output nodes start to discharge at a rate proportional to the difference of input voltages, once one of the nodes decreases about $|V_{thp}|$,

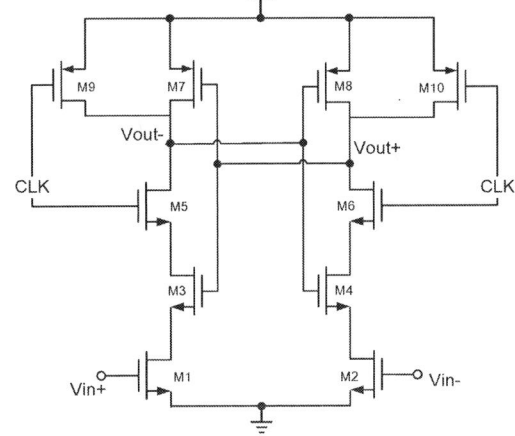

Fig. 4. Single-ended "Lewis-Gray" comparator.

978-1-4799-7440-5/14 $31.00 © 2014 IEEE          53

then the regenerative latch pulls one node to VDD, and the other node to ground.

The sizes of the input NMOS differential pair (M1, M2) and PMOS pair (M7, M8) in the cross-coupled inverters are critical for the comparator performance. Fig. 5 plots the IC of the input transistor (M1) and regenerative transistor (M7) during circuit operation. We observe that IC changes dynamically during circuit operation for this decion makign circuit, where IC is a constant for linear circuits. Noticing when the input transistor IC equals the re-generative transistor IC, the re-generative process (evaluation) starts, and the evaluation phase determines the comparator power and delay. We thus define the IC of the Lewis-Gray comparator to be the the inversion coefficient of M1 and M7 when they intersect with each other at the evaluatoin phase.

The delay (measured when the difference of comparator outputs is 50% of VDD) and average power consumption versus the common mode voltage for different Number of Fins (NF) of transistor M1 and M7 are given in Fig. 6. Fig. 7 shows the variation of power delay product with the comparator IC.

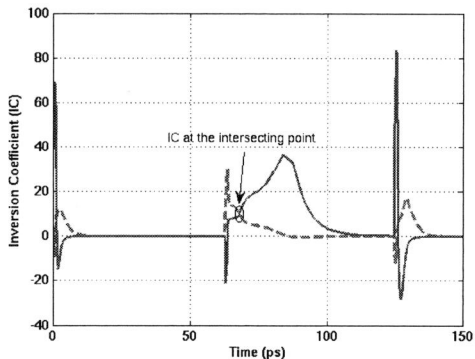

Fig. 5. The plot of inversion coefficients of M1 and M7 during the evaluation phase.

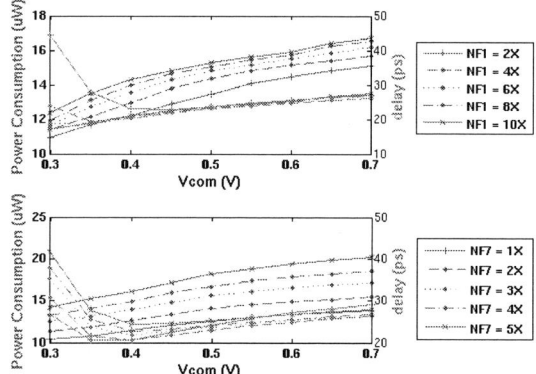

Fig. 6. Power consumption and delay versus $V_{com}$ for different NF1 and NF7.

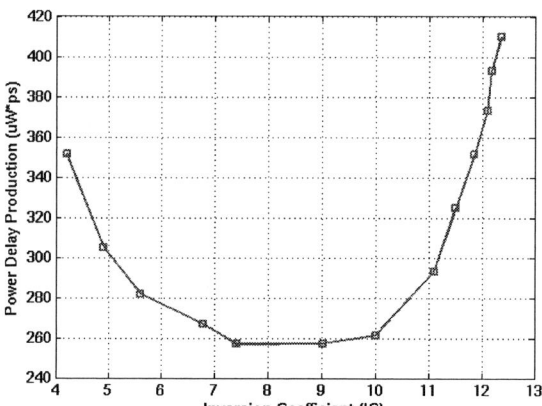

Fig. 7. Variation of power delay product with IC.

TABLE I.
SUMMARY OF COMPARATOR PERFORMANCE.

| Item | Value |
|---|---|
| Technology | 14nm SOI Finfet |
| Supply voltage | 0.8V |
| Power dissipation | 12.9uW |
| Delay@$\Delta V_{in} = 10mV$ | 20.54ps |
| Delay/log($\Delta V_{in}$) | 10.14ps |
| Energy efficiency@8GHz | 1.612fJ |

The simulation results for the designed dynamic comparator are summarized in Table I.

## V.    CONCLUSIONS

This paper has presented and demonstrated a systematic methodology for analog design in 14nm FinFET. The technology current and inversion coefficient provide a foundation for co-relating analog performance paramters to the sizes. We presented a methodology for compensating various performance parameters with respect to process, voltage and temperature variations. With this methodology, a set of basic building blocks have been designed and fabricated using IBM 14nm SOI FinFET process.

## REFERENCES

[1] V. Subramanian, B. Parvais, J. Borremans, et al., "Device and circuit-level analog performance trade-offs: a comparative study of planar bulk FETs versus FinFETs," in Electron Device Meeting, *IEEE IEDM Technical Digest*, pp. 898-901, 2005.

[2] D. M. Binkley, C. E. Hopper, et al. "A CAD methodology for optimizing transistor current and sizing in analog CMOS design," *IEEE Trans. on CAD*, vol. 22, no. 2 pp. 225-237, Feb. 2003.

[3] W. R.-Torres and Z. S. Roth, "Determination and Study of MOSFET Technology Current," *Canadian Journal on Electrical and Electronics Engineering*, vol. 4, no. 2, pp. 75-82, April 2013.

[4] E. Vittoz and J. Fellrath, "CMOS Analog Integrated Circuits Based on Weak Inversion Operation," *IEEE J. Solid-State Circuits*, vol. SC-12, no.3, June 1977.

[5] T. B. Cho and P. R. Gray, "A 10b, 20Msample/s, 35mW pipeline A/D converter," *IEEE J. Solid-State Circuits*, vol. 30, no. 3, pp. 166-172, Mar. 1995.

# 28 nm FD SOI Technology platform RF FoM

B. Kazemi Esfeh[1], V. Kilchytska[1], V. Barral[2], N. Planes[3], M. Haond[3], D. Flandre[1], J.-P. Raskin[1]

[1]ICTEAM, Université catholique de Louvain, 1348 Louvain-la-Neuve, Belgium
[2]CEA-Leti, MINATEC Campus, 17, rue des Martyrs, 38054 Grenoble Cedex 9, France
[3]ST-Microelectronics, 850 rue J. Monnet, 38926 Crolles, France
Email: babak.kazemiesfeh@uclouvain.be, Tel: (+32) 10472153

**Abstract** – This work provides a detailed study of 28 nm fully-depleted silicon-on-insulator (FD SOI) ultra-thin body and buried oxide (BOX) (UTBB) MOSFETs for high frequency applications. RF figures of merit (FoM), i.e. the current gain cut-off frequency ($f_T$) and the maximum oscillation frequency ($f_{max}$), are presented for different transistor geometries. The parasitic gate and source/drain series resistances, as well as capacitances and their effect on RF performance are analyzed.

**Keywords: FD-SOI MOSFET; UTBB; RF FIGURES OF MERIT.**

## INTRODUCTION

UTBB FD SOI technology is a well-known promising approach to satisfy ITRS requirements on device downscaling [1]. 28 nm FD SOI platform was demonstrated to provide good electrostatic features, leading to high immunity to short channel effects (SCE) and digital FoM. Many other advantages widely discussed in literature include thermal properties improvement, variability reduction thanks to undoped channel and good threshold voltage control by back gate scheme [2]. However, up to know only few studies were devoted to investigation of RF FoM of these innovative devices [3, 4]. [4] mentions achievable $f_T$ and $f_{max}$ values without deep analysis and does not include extraction of parasitic capacitances and resistances of great importance for RF FoM of advanced devices.
This work firstly provides detailed analysis of $f_T$ and $f_{max}$ versus gate length ($L_g$) and finger width ($W_f$). Secondly, the transconductance ($g_m$), parasitic gate resistance ($R_g$), source and drain resistances ($R_d$ and $R_s$), total gate capacitance ($C_{gg}$) and its intrinsic ($C_{ggi}$) and extrinsic ($C_{gge}$) parts are extracted and their effect on RF FoM is analyzed. Particular attention is paid to distinguishing intrinsic ('useful') and extrinsic ('parasitic') components.

## EXPERIMENTAL DETAILS

UTBB FD SOI devices were fabricated at ST-Microelectronics [2]. Si body, BOX and equivalent gate oxide thicknesses are 7 nm, 25 nm and 1.3 nm, respectively. The channel is strained and rotated by 45° from the <100> plane. The ground-plane implantation under the BOX is well-type. Multi-finger devices are designed and embedded in CPW (Coplanar Waveguide) pads for RF characterization. For the measured devices $L_g$ ranges from 25 nm to 0.5 µm. For 30 nm gate length, $W_f$ ranges from 5 µm to 0.3 µm. S-parameters are measured in a frequency range from 45 MHz up to 110 GHz under saturation ($V_{ds}$ =1 V) and cold ($V_{ds}$ = 0 V) conditions for different applied gate voltages ($V_{gs}$). The CPW feed line pads are de-embedded thanks to a dedicated open structure for each device. Therefore, effect of interconnections on extrinsic capacitances is eliminated from S-parameter measurements. $f_T$ and $f_{max}$ are extracted from the measured S-parameters extrapolating $H_{21}$ and MAG to 0 dB at $V_{ds}$= 1 V and $V_{gs}$ at the maximum $g_m$. The total capacitances are obtained from measured S-parameters in deep depletion regime ($V_{ds}$ = 0 V and $V_{gs}$ < $V_{threshold}$) [5]. The parasitic gate resistance is extracted in strong inversion ($V_{gs}$ > $V_{threshold}$) under cold condition ($V_{ds}$ = 0 V) [6].

## RESULTS AND DISCUSSION

According to the MOSFET small-signal equivalent circuit (Fig. 1) [3], $f_T$ and $f_{max}$ are expressed by [3], [7-8]:

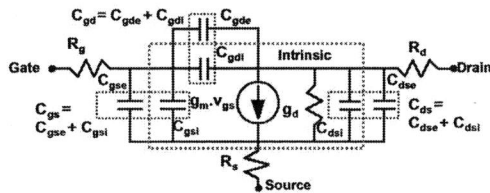

Fig.1. Small-signal equivalent circuit used for modeling the RF behavior of UTBB MOSFETs [3].

$$f_T \frac{g_m}{2\,\pi\,C_{gs}\left(1+C_{gd}/C_{gs}\right)+\left(R_s+R_d\right)\left(C_{gd}/C_{gs}\,\left(g_m+g_d\right)+g_d\right)} \quad (1)$$

$$\frac{g_m}{2\,\pi\,C_{gg}}$$

$$f_{max} \frac{g_m}{4\,\pi\,C_{gs}\,\left(1+C_{gd}/C_{gs}\right)\sqrt{g_d\,\left(R_s+R_s\right)+1/2\,C_{gd}/C_{gs}\,\left(R_s\,g_m+C_{gd}/C_{gs}\right)}} \quad (2)$$

$$\frac{f_T}{2\sqrt{\left(R_s+R_g\right)g_{ds}+2\pi f_T\,R_s\,C_{gd}}}$$

where $C_{gg}$, $g_m$ and $g_{ds}$ are the total gate capacitance, gate transconductance and channel conductance respectively.
Since $g_m \sim N_f W_f/L_g$ and $C_{gg} \sim N_f L_g W_f$, $N_f$ being the number of fingers, according to Eq.1 (in ideal case) $f_T$ is expected to increase with length reduction as $1/L_g^2$ and be independent of $W_f$. Fig. 2 shows the $f_T$ and $f_{max}$ variations with $L_g$ and $W_f$. Both $f_T$ and $f_{max}$ increase with $L_g$ scaling down (Fig. 2a), but $f_T$ follows a $1/L_g$ trend (and an even weaker one in shortest devices) i.e. attenuated comparing with ideally predicted $1/L_g^2$. This is due to velocity saturation, parasitic $R_s$, $R_d$ and extrinsic $C_{gg}$ effects (which will be discussed below). Similar trends were observed in other advanced devices [3]. It is important to point out that in case of devices

shorter than 90 nm, $f_{max}$ becomes smaller than $f_T$ (Fig. 2a). According to Eq. 2, this can be due to the increase of $R_g$ with $L_g$ reduction. Fig. 2b evidences the $f_T$ independence on $W_f$. This trend fits our expectations thus suggesting that in this $W_f$ range strong parasitic effect at the finger perimeter (reported in [3]) does not appear. In addition, one can see that $W_f$ reduction leads to increase of $f_{max}$ (Fig. 2b). This can be a result of the gate resistance reduction in narrow-finger devices.

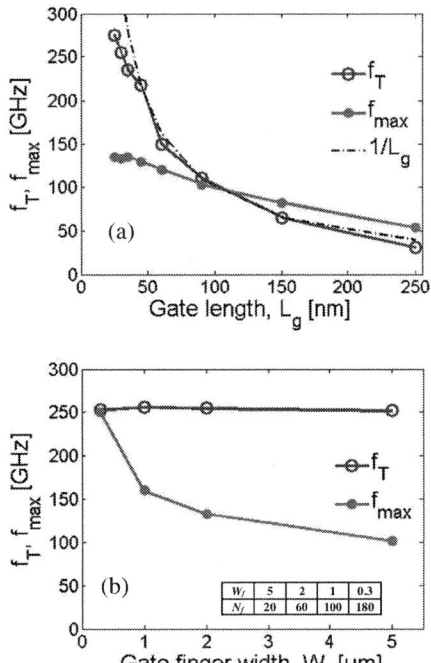

Fig. 2. $f_T$ and $f_{max}$ versus (a) gate lengths ($L_g$) for $W_f = 2$ μm, $N_f = 60$, (b) gate finger widths ($W_f$) for $L_g = 30$ nm. $N_f$ is shown as inset Table in the figure.

These results evidence that $f_T$ and $f_{max}$ dependence on $L_g$ and $W_f$ deviates from theoretical expectation and is dominated by the effect of parasitic elements. To understand the observed trends, complete equivalent circuit elements (both intrinsic and extrinsic, denoted 'i' and 'e') were extracted and analyzed.

Fig. 3 shows the dependence of extracted equivalent circuit elements on $L_g$. From Fig. 3a, one can see that $L_g$-dependence of both $g_{me}$ (as measured) and $g_{mi}$ (after $R_{sd}$ withdrawal) is much weaker than ideal $1/L_g$. This can be related to the velocity saturation effect. $R_{sd}$ effect can be seen through higher $g_{mi}$ w.r.t. $g_{me}$ values. Furthermore, difference between $g_{mi}$ and $g_{me}$ increases slightly with $L_g$ reduction, pointing out stronger $R_{sd}$ effect on short-L devices. Next to that, Fig. 3b shows that $C_{ggi}$ decreases proportionally with $L_g$ reduction, whereas $C_{gge}$ stays almost unchanged. Furthermore, $C_{ggi}$ dominates over $C_{gge}$ in long devices, whereas $C_{gge}$ is

higher than $C_{ggi}$ in shortest ones. This results in sub-linear $C_{gg}(L_g)$ dependence and can explain the fact that $f_T(L_g)$ becomes even weaker than $1/L_g$ in shortest devices. Fig. 3b evidences $R_g$ increase with $L_g$ reduction, confirming above hypothesis about $R_g$ as a reason of $f_{max}(L_g)$ saturation in short devices. It is useful to note that $R_g$ values extracted in these devices are 3-4 times lower that previously reported for UTBB devices [3], thus allowing for a strong improvement of $f_{max}$.

Fig. 4 shows the dependence of extracted equivalent circuit elements on $W_f$. Firstly, $g_{me}$ slightly increases in narrow $W_f$ as a result of $R_{sd}$ improvement with $W_f$ reduction (Fig. 4a). Secondly, one can see that while $C_{gge}$ effect on total $C_{gg}$ increases with $W_f$ reduction, it is not that strong in our $W_f$ range as was previously observed in [3]. Thus, differently from [3], we do not observe strong effect of parasitic capacitance at the perimeter on $f_T$. As a result, $g_m(W_f)$ and $C_{gg}(W_f)$ trends compensate each other assuring almost $W_f$-independent $f_T$ (Fig. 2b). Thirdly, strong $R_g$ reduction with $W_f$ evidenced in Fig. 4b confirms our explanation of $f_{max}$ improvement with $W_f$ reduction. $f_T$ and $f_{max}$ reported in this work are higher than previously reported for UTBB devices [3] showing considerable process maturity ( lower $R_{sd}$, $R_g$ and $C_{gge}$). However, they are still slightly lower than ITRS requirements for LSTP logic transistors for microwave and mobile applications ($f_T = 322$ GHz and $f_{max} = 284$ GHz for $L_g = 24$ nm) [1].

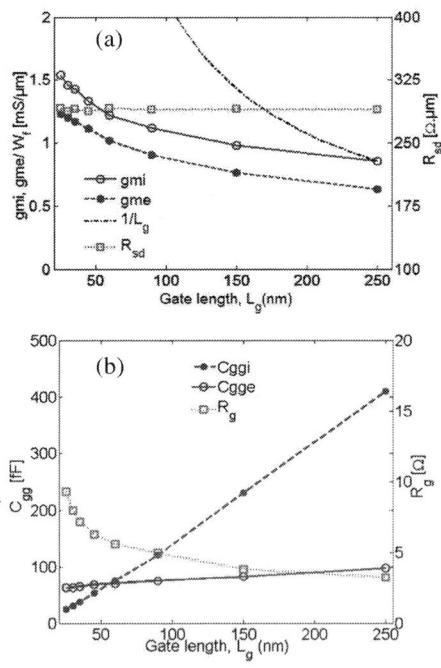

Fig. 3. (a) Normalized $g_{mi}$, $g_{me}$ and $R_{sd}$ ($R_s + R_d$) versus gate lengths ($L_g$) for $W_f = 2$μm, $N_f = 60$, (b) $C_{ggi}$, $C_{gge}$ and $R_g$ versus gate lengths ($L_g$) for for $W_f = 2$ μm, $N_f = 60$.

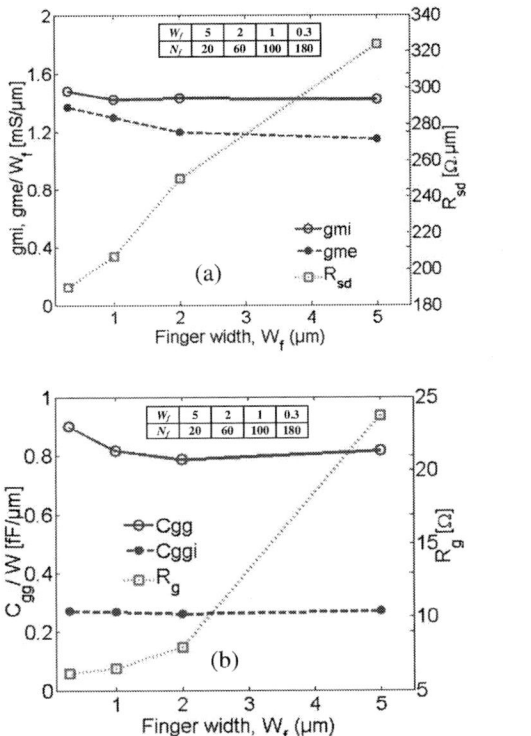

Fig. 4. (a) Normalized $g_{mi}$, $g_{me}$ and $R_{sd}$ versus gate finger widths ($W_f$), (b) normalized $C_{ggi}$, $C_{gg}$ and $R_g$ versus gate finger widths ($W_f$) for $L_g$ = 30 nm. $N_f$ is shown as inset Table in the figure.

## REFERENCES

[1] 'International Technology Roadmap for Semiconductors.', http://www.public.itrs.net.

[2] N. Planes *et al.*, '28nm FDSOI Technology Platform for High-Speed Low-Voltage Digital Applications,' *in Symposium on VLSI Technology*, pp.133-134, 2012.

[3] M. K. Md Arshad *et al.*, 'Effect of parasitic elements on UTBB FD SOI MOSFETs RF figures of merit,' *Solid-State Electronics*, (2014), in press doi: 10.1016/j.sse.2014.04.027.

[4] S. Makovejev *et al.*, 'Wide Frequency Band Assessment of 28 nm FDSOI Technology Platform for Analogue and RF Applications,' *15th International Conference on Ultimate Integration on Silicon (ULIS)*, pp. 53-56, April 2014.

[5] J.-P. Raskin *et al.*, 'Accurate SOI MOSFET Characterization at Microwave Frequencies for Device Performance Optimization and Analog Modeling,' *IEEE TED.*, vol. 45, no. 5, pp. 1017-1025, May 1998.

[6] A. Bracale *et al.*, 'A new approach for SOI devices small-signal parameters extraction,' *Analog Integrated Circuits and Signal Processing*, vol.25, no.2, pp. 157–169, 2000.

[7] H. L. Kao *et al.*, 'Limiting Factors of RF Performance Improvement as Down-scaling to 65-nm Node MOSFETs,' *in Korea-Japan MicroWave Conference (KJMW)*, April 2009 (CGU).

[8] J.-P. Raskin *et al.*, 'High-Frequency Noise Performance of 60-nm Gate-Length FinFETs,' *IEEE Transactions on Electron Devices*, vol. 55, no. 10, October 2008.

## CONCLUSIONS

Perspectives of 28 FD SOI platform for RF applications have been analyzed through cut-off frequencies dependences on $L_g$ and $W_f$. These characteristics have been further detailed based on the small-signal equivalent circuit extraction of parasitic elements. Good RF performance with $f_T$ of ~275 GHz and $f_{max}$ of ~250 GHz was demonstrated, close to the ITRS requirements. Further improvement can be achieved through the process and structure optimization in order to reduce parasitics, particularly extrinsic capacitance and gate resistance.

**ACKNOWLEDGEMENT**: This work was partially funded by FNRS (Belgium), Catrene "Reaching 22" and Eniac "Places2Be" projects.

978-1-4799-7440-5/14 $31.00 © 2014 IEEE

# An SOI Based Integrated Gate-Drivers for Automotive Application

Ken Toshiyuki, Yoshihiko Hiya, Kazuya Kinoshita, Hidemoto Tomita and Norishige Hoshikawa

Electronics Development Div. 3 Toyota Motor Corporation., Toyota, Aichi, Japan

Email: ken_toshiyuki@mail.toyota.co.jp

*Abstract*— **This paper presents our recent development of an ASIC for automotive application. With the progress of automotive electronics, there is increasing need for high temperature and high voltage operation. The ASIC for the EPS applications was integrated with a boost gate driver, three-phase inverter drivers and their components. The ASIC was designed and fabricated using SOI-BiCD process which can operate at the maximum voltage of 80V and at temperature from -40°C to 175°C.**

## I. INTRODUCTION

The development of automotive technology has been focused on the safety and environmental requirements of contemporary society while also improving comfort and usability. To achieve these, vehicles are being installed with more and more electronic devices.

Electric power steering (EPS) provides power assist even when the engine is stopped (Fig. 1). It also improves fuel economy compare to the hydraulic power steering because it is light weight and the motor consume the energy only when power assist is required. Fig. 2 shows the constitution of the EPS electric control unit (ECU). It consists of a boost converter and three-phase inverter to drive a three-phase brushless motor and to drive those. The controller of switching power supply that produce gate drive voltage and relay driver are also required.

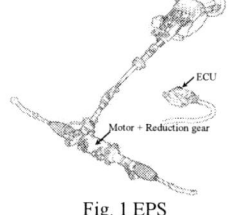

Fig. 1 EPS

In this paper, we introduced our SOI-BiCD process and discuss the driver IC that integrated these components as one chip for the EPS system.

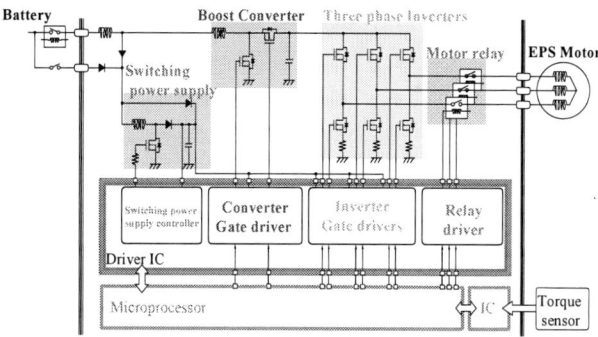

Fig. 2 EPS-ECU

## II. WIDE VOLTAGE AND HIGH TEMPERATURE BiCD PROCESS

Automotive application ASICs are operated under wide range of voltage to generate high power [1-2] and need to be installed in high temperature environments. These requirements include directly installing ASICs in the engine or transmission, achieving high performance motor systems with control circuits, and adapting ASICs to systems with simpler or no cooling devices. This ASIC is fabricated with the in-house SOI-BiCD process (Fig. 3) which can operate at the maximum voltage of 80V and at temperature from -40°C to 175°C [3].

The ASIC for the EPS application is integrated the gate drivers of a boost converter and three-phase inverter while remaining high independent by SOI-Isolation (Fig. 4). Each phase has power terminal and GND terminal. The chip size of the ASIC is 3.8 mm × 3.6 mm.

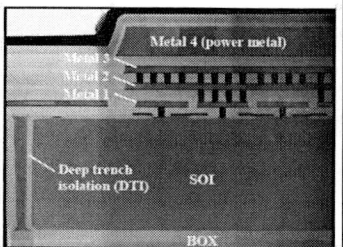 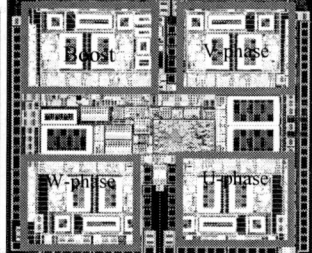

Fig. 3 Cross section image of SOI-BiCD          Fig. 4 Chip layout

## III. GATE DRIVE CIRCUIT DESIGNED

Figure 5 shows a block diagram of the gate driver. The bootstrap diode supplies the power to high-side gate drive and the level shifter coverts the incoming digital control signal from the low-side voltage level to the high side voltage level. Since the high-side source voltage swings between ground and dc bus voltage during switching operation, 80V LDMOS is chosen for both devices.

The LDMOS of the driver output stage and bootstrap diode require large space in order to flow enough current. On the other hand, pre-driver space is small by using CMOS devices even including overlap protection circuit (Fig. 6, Fig. 7).

The level shifter is surrounded by triple trench (Fig. 6) to suppress the parasitic device of the level shifter [4] and reduce coupling of the high-side source voltage.

978-1-4799-7440-5/14 $31.00 © 2014 IEEE          58

There are three protection functions, over temperature protection, under voltage detection, and over voltage detection. Fail safe function (overlap protection etc.) is also included.

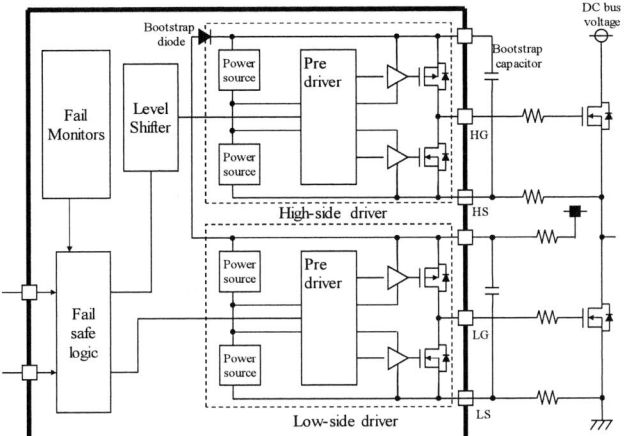

Fig. 5 Schematic of the driver

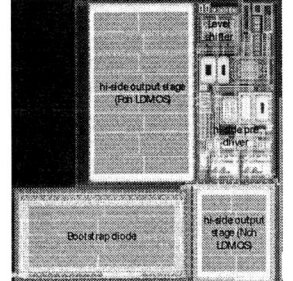

Fig. 6 The gate driver layout

Fig. 7 Area ratio of the driver

Figure 8 shows the gate drivers switching waveform. The low-side and high-side drivers operate having the dead time and the high-side source voltage swings from 0 to 32V when low-side gate switching. The structure enables to operate correctly even under switching transient without malfunctions (Fig. 8).

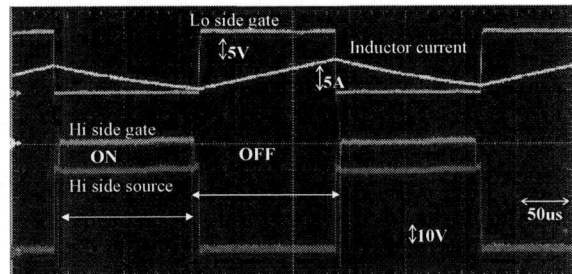

Fig. 8 Gate driver switching waveform

Figure 9 shows the temperature dependence of the gate driver output resistance from -40°C to 175°C. That is determined by the on resistance of LDMOS used as the driver output stage. The MOS resistance increases with temperature due to the decreasing the carrier mobility. That is the reason of the positive temperature coefficient. Figure 10 shows the propagation delay of the driver. Since the on resistance of the MOS is increase at the high

temperature, the propagation delay also increases at high temperature. The drivers operate successfully from -40°C to 175°C with no thermal runaway.

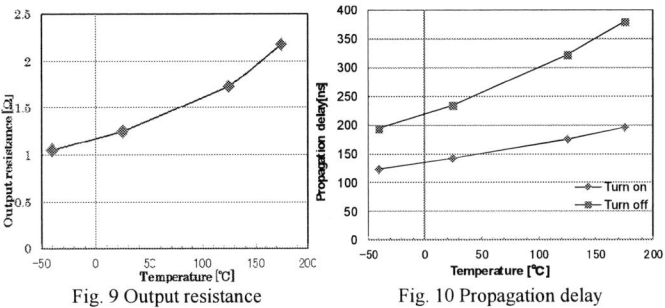

Fig. 9 Output resistance        Fig. 10 Propagation delay

To maintain gate voltage constant, the switching power supply was installed. Figure 11 shows the schematic of the switching regulator. The controller monitors output voltage and has feedback through the error amplifier. Figure 12 shows the temperature dependence of the output voltage. That shows the fluctuation of the output voltage is less than 3% between -40°C to 175°C.

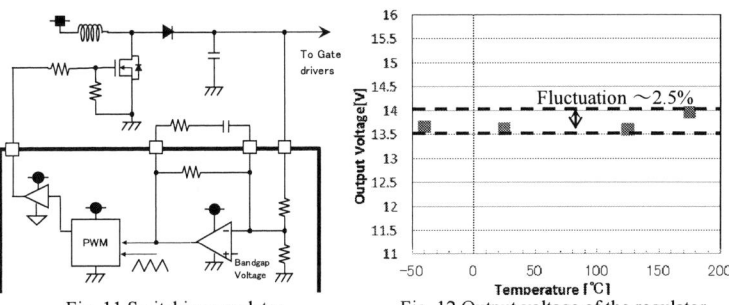

Fig. 11 Switching regulator        Fig. 12 Output voltage of the regulator

## IV. CONCLUSION

An integrated gate driver for EPS system has been presented. This ASIC is fabricated with the SOI-BiCD process which can operate at the maximum voltage of 80V and at temperature from -40°C to 175°C. The isolated gate driver IC is able to drive both the high-side and low-side devices in the power-module. The integration of the multi-phase drivers and multi-function allows for further size reduction of the gate driver as a whole.

### REFERENCES

[1] K. Hamada et al., "A 60V BiCDMOS device technology for automotive applications," Conference Record of the 1995 IEEE Industry Applications Society 30th IAS Annual Meeting, 1995, pp.986–990.
[2] F. Kawai et al., "Multi-voltage SOI-BiCDMOS for 14V & 42V automotive applications," Proc. of the ISPSD 2004, pp. 165–168.
[3] H. Tomita et al., "Wide-Voltage SOI-BiCDMOS Technology for High-Temperature Automotive Applications," Proc. of the ISPSD 2011, pp. 28–31.
[4] M.Yamaji et al., "A Novel 600V- LDMOS with HV-Interconnection for HVIC on Thick SOI," Proc. of the ISPSD 2010, p.101-104.

# Effect of Back Gate on Parasitic Bipolar Effect in FD SOI MOSFETs

Fanyu Liu[1], Irina Ionica[1], Maryline Bawedin[2] and Sorin Cristoloveanu[1]

[1]IMEP-LAHC, MINATEC, 38016 Grenoble, France
[2]Université Montpellier II, 34095 Montpellier, France
Email: fanyul@minatec.inpg.fr

*Abstract*—In short-channel fully-depleted (FD) silicon-on-insulator (SOI) MOSFETs, the drain leakage current is enhanced by the parasitic bipolar transistor. The parasitic bipolar effect is induced by band-to-band tunneling and floating-body effects. It strongly depends on film thickness and back-gate voltage. We show experimentally the possibility to reduce the parasitic bipolar effect by biasing the back gate (ground plane). Based on devices simulations, we discuss the origin of the bipolar action, its suppression and the possible applications.

*Keywords—parasitic bipolar effect; back gate; FD SOI; band-to-band tunneling*

## I. INTRODUCTION

The drain leakage current is a main concern for scaled (sub-30 nm) MOSFETs, especially due to the strong demand for low-power CMOS circuits [1]. The leakage current contains several components [2] : 1) weak inversion conduction current; 2) gate-induced drain leakage governed by band-to-band tunneling (BTBT) in off-state; 3) impact ionization (II) current when drain voltage is high enough; 4) direct gate-tunneling current. For long-channel devices, if the gate-tunneling current is neglected, the drain leakage is dominated by BTBT current [2]. In short-channel SOI devices, this BTBT current can be amplified by the parasitic bipolar transistor (PBT), where the source, drain and body work respectively as emitter, collector and base. Note that at high $V_D$, where II prevails, the II current can also be enhanced by PBT. The PBT is originated from the floating body of SOI transistors, which has been documented in relatively thick devices [3-4].

Fig. 1 shows a schematic view of the SOI MOSFET polarized in off-mode. For $V_{FG}$ < 0 V, the LDD region overlapped by the gate is depleted. A medium $V_D$ value (~ 1 V) can then cause a sufficient band bending to induce BTBT. The electrons produced by BTBT are directly recovered by the drain, while the holes will move towards the body due to the lateral electric field. The positive charge increases the body potential and hence the body-source junction can be turned on. As expected in bipolar transistors with forward-biased base-emitter junction, an amplified current of electrons can diffuse through the base and reach the collector.

In this paper, we show by both experiments and simulations that negative back-gate biasing can effectively reduce the drain leakage enhanced by parasitic bipolar effect in short-channel devices. Based on TCAD simulations, we validate that the suppression of the PBT effect is mainly driven by the increase of the energy barrier at base-emitter (body-source) junction.

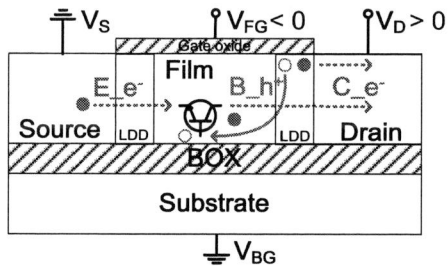

Fig. 1. Schematic cross section for an n-channel fully-depleted MOSFET. When the parasitic bipolar happens, $B\_h^+$, $C\_e^-$ and $E\_e^-$ are respectively the flows of carriers at base, collector and emitter.

## II. EXPERIMENTAL RESULTS

The structures used are advanced UTBB ($T_{si}$ = 10 nm, $T_{BOX}$ = 25 nm and ground plane) n-channel SOI MOSFETs fabricated at STMicroelectronics. High-k dielectric material and metal gate technology are adopted. Fig. 2 shows the measured drain current with various $V_{BG}$ in (a) long-channel devices ($L_G$ = 100 nm) and (b) shorter devices ($L_G$ = 30 nm). In long devices, the back-gate biasing shifts the threshold voltage but the drain leakage, dominated by BTBT [2], does not change. For shorter device (Fig. 2b), the drain leakage current is clearly higher denoting the PBT action. A negative $V_{BG}$ can drastically decrease the leakage, which is of importance for circuit applications. Beyond $V_{BG} \leq$ -2 V, the drain leakage for shorter devices reduces to the value measured on long-channel devices (symbols in Fig.2b). This trend demonstrates that negative $V_{BG}$ in short devices is effective to attenuate the drain leakage amplified by lateral PBT until it is fully suppressed. The mechanism of the suppression of the parasitic bipolar effect due to the back gate will be discussed in the next section.

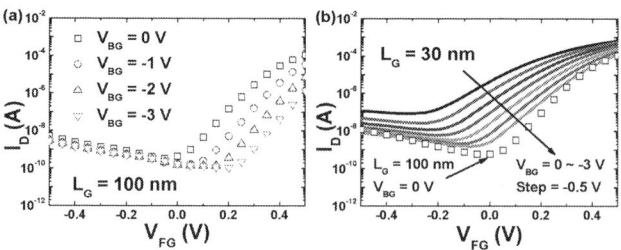

Fig. 2. Experimental drain currents *versus* front gate voltage with different $V_{BG}$ and $V_D$ = 1.5 V for (a) long- and (b) short-channel devices. EOT = 1.6 nm, $T_{si}$ = 10 nm and $T_{BOX}$ = 25 nm.

978-1-4799-7440-5/14 $31.00 © 2014 IEEE

## III. ANALYSIS AND DISCUSSION

Sentaurus TCAD simulations aimed to get further insight about the effect of back gate on the drain leakage. The dynamic non-local BTBT is used [5]. Fig. 3 shows the transient simulation drain currents with various $V_{BG}$. With BTBT on, the drain leakage for short-channel devices (solid lines in Fig. 3) is higher and decreases with negative $V_{BG}$ until it equals to the value for long-channel devices (symbol in Fig. 3). This trend is similar to the experimental results in Fig. 2b.

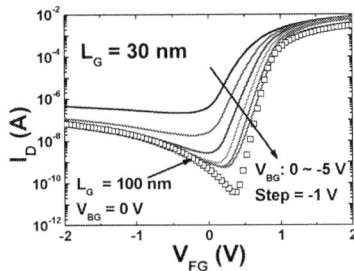

Fig. 3. Simulation drain currents for short-channel devices with different $V_{BG}$ at $V_D$ = 1.5 V. EOT = 1.6 nm, $T_{si}$ = 10 nm and $T_{BOX}$ = 25 nm.

As mentioned in [6], holes generated by BTBT act as the base current, turning on the base-emitter junction. More electrons from source can flow into the body and are finally collected by the drain. Therefore, in order to cancel the PBT, back gate must either reduce the BTBT generation or increase the barrier of base-emitter junction (or both).

### A. BTBT generation

According to [7], the BTBT current $I_{BTBT}$ can be calculated from the integration of BTBT generation rate $G_{BTBT}$:

$$I_{BTBT} = qW \iint G_{BTBT} \, dx dy \qquad (1)$$

where $q$ is the electron charge and $W$ is the device width. Fig. 4 compares the BTBT currents with $V_{BG}$ = 0 V and $V_{BG}$ = -5 V in both logarithmic (a) and linear scale (b). Note that the four curves are almost superposed in logarithmic scale (Fig. 4a). Though BTBT current with $V_{BG}$ = -5 V is a little smaller than the one with $V_{BG}$ = 0 V (Fig. 4b), the impact of back-gate bias is modest and does not account for the large difference in leakage currents of almost one order of magnitude in Fig. 3.

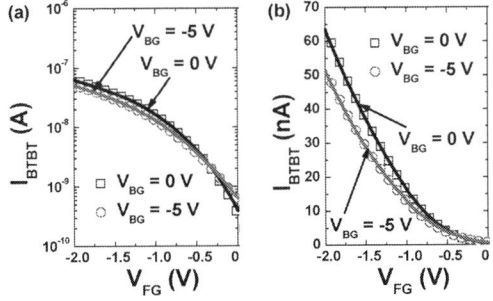

Fig. 4. BTBT current *versus* front gate voltage for $V_D$ = 1.5 V under different $V_{BG}$ in (a) logarithmic and (b) linear scale. The symbols and solid lines represent the BTBT currents for short- and long-devices, respectively.

### B. Barrier height at body-source junction

The back gate primarily affects the barrier height at body-source junction. In order to verify this argument, we compare the potential profiles along the channel (for $V_D$ = 1.5 V and $V_{FG}$ = -1 V), as shown in Fig. 5. An obvious increase of the barrier height at base-emitter junction (body-source junction) is observed when $V_{BG}$ decreases from 0 to -5 V, which helps to prevent electrons injecting from emitter. Consequently, a negative back-gate bias suppresses the parasitic bipolar effect mainly by increasing the barrier height at body-source junction.

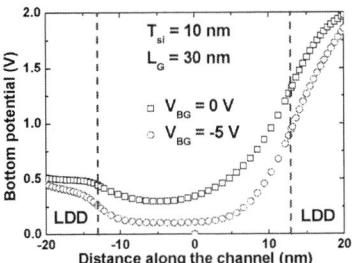

Fig. 5. Potential at the film-BOX interface *versus* position along the channel for $V_D$ = 1.5 V and $V_{FG}$ = -1 V under $V_{BG}$ = 0 V and $V_{BG}$ = -5 V. Drain and source locate on the right and left sides, respectively.

### IV. CONCLUSION

In this paper, we have studied for the first time the effect of back gate on the parasitic bipolar effect induced by band-to-band tunneling in state-of-the-art FD SOI MOSFETs. We found that a negative back-gate voltage applied to the ground-plane can effectively suppress the drain leakage amplified by the parasitic bipolar transistor. TCAD simulations show that the parasitic bipolar effect is inhibited mainly by the increase of barrier height at body-source junction. This result is of practical interest for the optimization of FD SOI circuits. Not only does the negative back bias increase the threshold voltage and reduce the off-state current, but also the bipolar action is cancelled. A comparison of leakage current measured with grounded or negatively biased back gate yields the bipolar gain which is a necessary parameter for accurate compact modeling.

### REFERENCES

[1] International Technology Roadmap for Semiconductors, 2011. http://www.itrs.net.

[2] K. Roy, S. Mukhopadhyay and H. M. Meimand, "Leakage current mechanisms and leakage reduction techniques in deep-submicrometer CMOS Circuits," in *Proc.* IEEE, vol. 91, no. 2, pp. 305-327, 2003.

[3] J. Chen, F. Assaderaghi, P. K. Ko and C. M. Hu, "The enhancement of gate-induced-drain-leakage (GIDL) current in short-channel SOI MOSFET and its application in measuring lateral bipolar current gain beta," IEEE Electron Device Letters, vol. 13, no. 11, pp. 572-574, 1992.

[4] J.-Y. Choi and J. G. Fossum, "Analysis and control of floating-body bipolar effects in fully depleted sub-micrometer SOI MOSFETs," IEEE Transaction on Electron Devices, vol. 38, no. 6, pp. 1384-1391, 1991.

[5] TCAD Sentaurus User Manual, G-2012.06., Synopsys.

[6] F. Y. Liu, I. Ionica, M. Bawedin and S. Cristoloveanu, "Parasitic bipolar effect in advanced FD SOI MOSFETs: experimental evidence and gain extraction," in *Proc.* EuroSOI 2014, Tarragona, Spain, 2014.

[7] P. C. Adell, H. J Barnaby, R. D. Schrimpf and B. Vermeire, "Band-to-band tunneling (BBT) induced leakage current enhancement in irradiated fully depleted SOI devices," IEEE Transactions on Nuclear Science, vol. 54, no. 6, pp. 2174-2180, 2007.

# High Temperature Performance of Flexible SOI FinFETs with Sub-20 nm Fins

A. Diab, G. A. Torres Sevilla, M. T. Ghoneim and M. M. Hussain

Integrated Nanotechnology Lab, King Abdullah University of Science and Technology, Saudi Arabia
Email: MuhammadMustafa.Hussain@kaust.edu.sa, Tel: (966) 544-700-072

*Abstract*—We demonstrate a flexible version of the semiconductor industry's most advanced transistor topology – FinFET on silicon-on-insulator (SOI) with sub-20 nm fins and high-κ/metal gate stacks. This is the most advanced flexible (0.5 mm bending radius) transistor on SOI ever demonstrated for exciting opportunities in high performance flexible electronics with stylish product design. For the first time, we characterize such device from room to high temperature (150 °C). And we discuss the dependence of the I-V curves with temperature.

## I. INTRODUCTION

Flexible devices have the potential for a paradigm shift from today's rigid and brittle silicon/SOI based electronics, to flexible displays, sensors, smart phones and advanced healthcare devices. Different approaches using organic materials [1], [2], pseudo-CMOS compatible techniques [3], [4] have been used to make high performance flexible electronics. Although competitive results have been shown using such methods [3], [4], often they are expensive, unorthodox, resulted films are opaque and suffer from limited bendability. In the recent past, we showed an innovative and a fully-CMOS compatible low-cost process to transform traditional devices into flexible-stretchable-transparent one while retaining high performance and integration density [5]. The aim of this work is to demonstrate and to investigate a flexible version of the SOI based FinFETs with sub-20 fins, high-κ/metal gate at high temperature.

## II. DEVICE FABRICATION

FinFETs are fabricated using 8" SOI wafers following a state-of-the-art CMOS compatible gate-first flow. After patterning the fins (down to sub-20 nm width as shown in Fig. 2(a)) using DUV followed by extreme resist trimming, we formed 10-20 nm TiN/2-4 nm $HfO_2$ gate stacks (250 nm to 1 μm $L_g$). Source and drain were formed using ion implantation followed by NiSi and activation anneal at 1000 °C for 10 sec. finally the devices had FGA ($N_2/H_2$ at 420 °C).

Fig. 1 shows the process to transform fabricated FinFETs into flexible one. In order to process each die separately, the wafer was diced into 7.5 $cm^2$ pieces. Each individual die was spin-coated with PR to protect the devices. Then, each die is bulk micro-machined in order to reduce the bulk thickness to achieve mechanical integrity and flexibility at the same time. Once the bulk achieves a thickness of 50 μm (flexibility may be controlled controlling the final bulk thickness), the devices are bendable enough without the need of any transfer process due to the support provided by the remaining substrate (see Fig. 2(b)). Finally, the PR was removed from the top (see Fig. 2(c)) and the devices are tested at different temperatures.

## III. EXPERIMENTS AND RESULTS

To study the behavior of the FinFETs at high temperature, our *I-V* measurements were performed using a Cascade probe bench where the chuck-temperature can reach 150 °C. Released and unreleased PMOS FinFET devices with $W = 3.6$ μm, $L_g = 1$ μm and $L_g = 250$ nm and 20 channels in parallel were tested. We focused on saturation region for the lowest (25 °C) and highest (150 °C) temperature curves to comprehend a full analysis.

Fig. 3 and Fig. 4 show the transfer characteristics ($I_d$-$V_g$) in linear and logarithmic scale respectively for both 'long' and 'short' FinFET. After device releasing, a very small variation of the on-current ($I_{on}$), the off-current ($I_{off}$), the threshold voltage ($V_{th}$) and the sub-threshold swing (*SS*) in the order of 0.1 % is obtained at both 'high' and 'low' temperature. The $V_{th}$ decreases and the *SS* increases in all devices as temperature is increased [6]. The decrease of $V_{th}$ with temperature tends to increase $I_d$, while the reduction of mobility due to increasing phonon scattering with temperature tends to decrease it as in a conventional MOSFET [7]. The compensation of these opposing effects around at $V_g = -0.8$ V and at $V_g = -0.9$ V for long and short devices, respectively lead to a unique points in the characteristics with zero temperature coefficient. This confirms that our devices maintain almost the same performance after the extensive fabrication process for various low and high temperatures.

The output characteristics ($I_d$-$V_d$) illustrated in Fig. 5 clearly shows that the current drive of the FinFETs decreases with temperature as expected without any significant degradation after release. The reduction of the transconductance ($g_m$) peak in Fig. 6 as a function of the temperature is explained by the mobility lowering due to the increased phonon scattering at elevated temperatures. Further confirmation is obtained by plotting the gate current ($I_g$) *versus* $V_g$ where $I_g$ increases with temperature and the variation after release sustain in the nA values range. Finally, Table I shows a comparison of the best benchmark FinFET [8] from Intel and this work.

## IV. CONCLUSION

We have demonstrated a flexible version of the industry's most advanced architecture: FinFET with sub-20 nm fins, high-κ/metal gate stacks with extended electrical characterization study at room and high temperature. The first ever measurements at high temperature of high-performance flexible electronics are reported making the originality of this work. The performance variation with temperature confirms the efficiency and the stability of our devices.

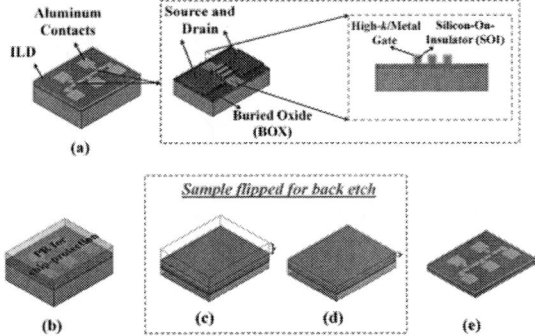

Fig. 1. Fabrication process flow: (a) fabricated FinFET devices on standard 90 nm SOI with 150 nm BOX, (b) PR coating for chip-protection during back etch process, (c) FinFET die etched back using back grinding technique, (d) FinFET devices on flexible silicon substrate (50 μm thick) and (e) PR removal and final device testing.

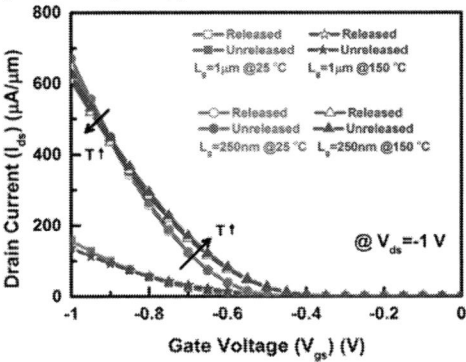

Fig. 3. Unreleased and released FinFET transfer plots in linear scale. Variation with temperature shows no performance degradation.

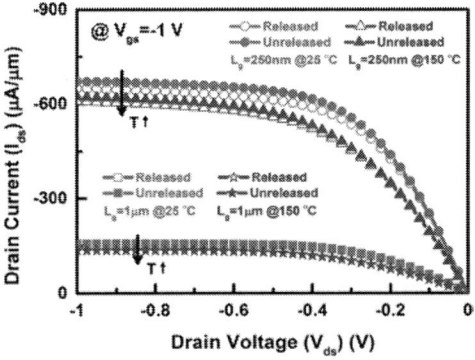

Fig. 5. Unreleased and released FinFET output plots. Variation with temperature shows no device performance degradation.

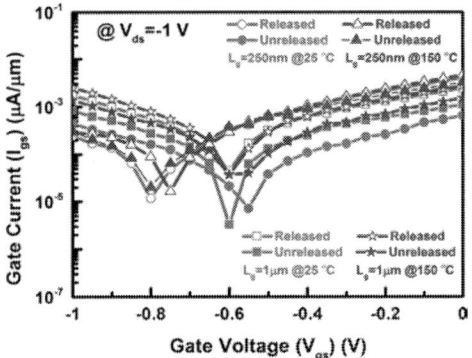

Fig. 7. $I_{gs}$ versus $V_{gs}$ curves in unreleased and released FinFETs showing the good maintain of the gate leakage values.

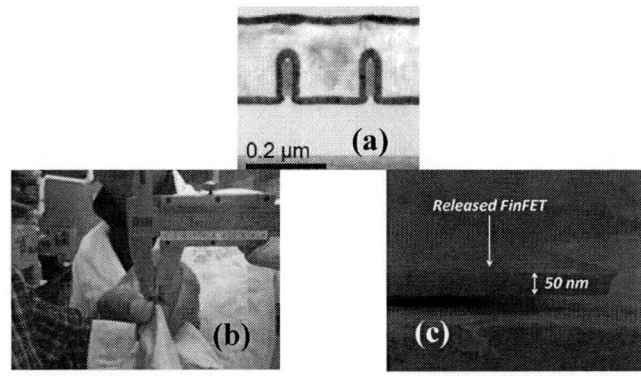

Fig. 2. (a) TEM cross-section of fabricated fins (fin width = Sub-20 nm or higher). Process optimization can result in more uniform fin patterning. (b) Record flexibility (0.5 mm bending radius) in flexible FinFET and (c) SEM for released FinFET on SOI.

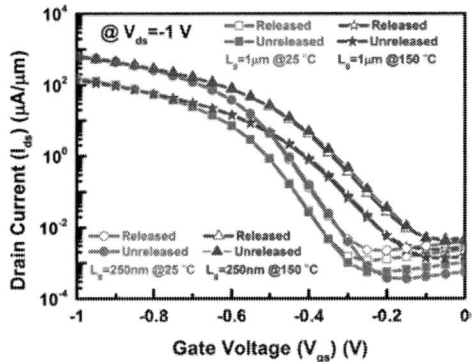

Fig. 4. Unreleased and released FinFET transfer plots in logarithmic scale. $V_{th}$ decreases and SS increases with temperature increasing. A small increase of $I_{off}$ is observed with released compared to unreleased FinFETs at 25 °C.

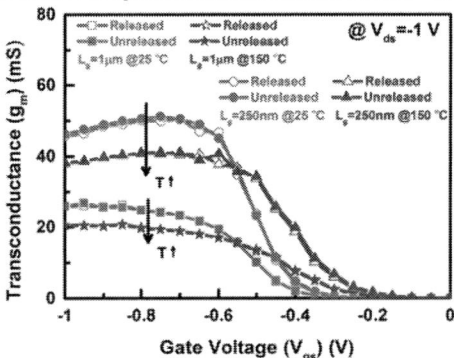

Fig. 6. Transconductance curves *versus* gate voltage for unreleased and released FinFET. The $g_m$ peak decrease with increasing temperature.

TABLE I COMPARISON OF EXTRACTED PARAMETERS

| | $L_g$ (nm) | $I_{on}$ (mA/μm) | $I_{off}$ (nA/μm) | SS (mV/dec) | $\mu_{eff}$ (cm²/Vs) |
|---|---|---|---|---|---|
| Intel Tri-gate FET (rigid) | 40 | 1.1 | 100 | 90 | 200 |
| This work DG FinFET(flexible) | 250 | 0.65 | 4 | 63 | 103 |

REFERENCES

[1] Y. Xia et al., *Appl. Phys. Lett.*, vol. 90, pp. 162106-162108, 2007.
[2] P. Lin et al., *Adv. Mater.*, vol. 24, pp. 34-51, 2012.
[3] D. Shahrjerdi et al., *IEDM Tech. Dig.*, pp. 511-514, 2012.
[4] J. N. Burghartz et al., *IEEE TED*, vol. 56, pp. 321-327, 2009.
[5] G. A. T. Sevilla et al., *Adv. Mater.*, vol. 26, pp. 2794-2799, 2014.
[6] J. Colinge et al., *IEEE EDL*, vol. 27, pp. 172-174, 2006.
[7] D. S. Jeon et al., *IEEE TED*, vol. 36, pp. 1456-1463, 1989.
[8] J. Kavalieros et al., *VLSI Tech. Dig.*, pp. 50-51, 2006.

# Effects of Back-Gate Bias on Switched-Capacitor DC-DC Converters in UTBB FD-SOI

Matthew J. Turnquist[*], Guerric de Streel[‡], David Bol[‡], Markus Hiienkari[†], Lauri Koskinen[†]

[*]Dept. of Micro- and Nanosciences, Aalto University [†]University of Turku, Technology Research Center
[‡]ICTEAM Institute, Université catholique de Louvain *matthew.turnquist@aalto.fi ‡guerric.destreel, david.bol@uclouvain.be

*Abstract*—**This paper explores the effects of back-gate bias on switched-capacitor (SC) DC-DC converters in 28 nm UTBB FD-SOI. By using back-gate bias to optimize the control circuitry and switches, the SC converter can operate with a peak efficiency of 72% in sleep mode (100 nW load) and 83% in active mode (100 $\mu$W load).**

## I. INTRODUCTION

In order to reduce energy consumption in portable internet-of-things (IoT) devices, there is strong motivation to process more information in an energy-efficient manner before transmitting it since wireless transmission does not scale as advantageously as with digital processing. Thus, there has been interest to apply power-saving techniques to ULP processors such as low voltage operation, error-detection [1], and multiple operation states (e.g. sleep/active). These power-saving techniques require that the DC-DC conversion from the battery to the ULP processor be able to operate efficiently over an increasingly large power range [2], [6]. To address these requirements, this paper explores the effects of back-gate bias on switched-capacitor (SC) DC-DC converters in UTBB FD-SOI. This work focuses on a DC-DC converter with a sleep (100 nW load) and active mode (100 $\mu$W load).

The power losses of a SC DC-DC converter (i.e. shunt and series) are highly dependent on the power level [3]. Without power scaling techniques (e.g. frequency scaling), achieving high efficiency down to a typical IoT node sleep mode power level is challenging [4], [5] as shown in Fig. 1. In sleep mode, DC-DC switch sizing (see $S_1$-$S_4$ in Fig. 2) is also important to ensure an adequate ratio of the average $I_{DS}$ on-current ($I_{ON,dc-dc}$) to the average $I_{DS}$ leakage-current ($I_{OFF,dc-dc}$) as shown in Fig. 1 (sub-figure). Subsections II-A and II-B examine back-gate bias methods in UTBB FD-SOI to improve efficiency in the switches ($S_1$-$S_4$ in Fig. 2) and control circuitry, respectively.

## II. BACK-GATE BIAS IN DC-DC CONVERTERS

Body-bias has been shown to be an effective tool to extend the power range of DC-DC converters in 130 nm bulk CMOS [5]. Body-bias in UTBB FD-SOI, or back-gate bias, is more effective than in bulk CMOS for sub-100 nm nodes. With UTBB FD-SOI, the range of bias voltages is larger and the impact on $V_t$ is higher (i.e. larger body-bias factor). Thus, this paper explores the application of back-gate bias in UTBB FD-SOI to a DC-DC switched-capacitor converter with LVT transistors. For LVT transistors, a back-gate bias voltage ($V_{BB}$) > 0 V is forward-body-biased (FBB).

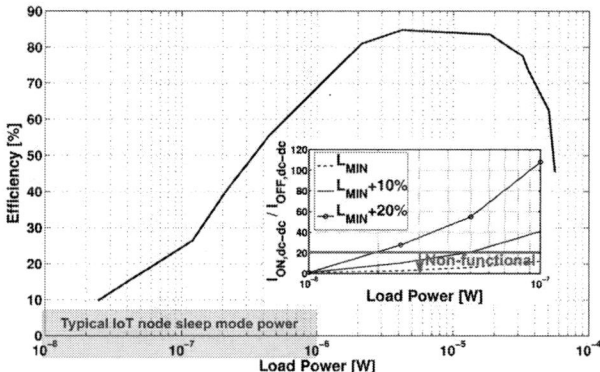

Fig. 1. Efficiency of the 2:1 DC-DC converter of Fig. 2 without any power scaling techniques (e.g., frequency scaling, body-bias, etc). The efficiency drops when the converter enters a typical IoT node sleep mode [5] due to control circuitry and switch losses. The sub-figure shows the importance of sizing the length (L) of $S_1$-$S_4$ to ensure converter functionality in sleep mode.

### A. DC-DC Core Switch Performance

A series-parallel 2:1 DC-DC converter with LVT transistors is used to test the effects of FBB (Fig. 2). As shown in Fig. 3, modulating the $V_t$ of its switches ($S_1$-$S_4$) through FBB has two main effects: 1) on-conductance ($G_{ON}$) changes 2) leakage changes. The impact of these effects depend on the DC-DC architecture, $V_{IN}/V_{OUT}$ levels, and the switch characteristics. By increasing the amount of FBB (i.e. $V_{BB}$=0 to 3 V), both the $G_{ON}$ and $I_{OFF,dc-dc}$ increase. Fig. 4 shows details of the relationships between on/off currents of LVT switches.

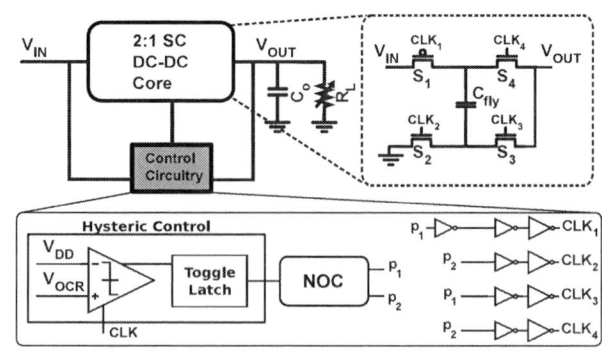

Fig. 2. A series-parallel 2:1 DC-DC converter with LVT transistors. $C_{FLY}$ is a 130 pF MIM capacitor. FBB can be applied to the DC-DC core and/or the control circuitry.

978-1-4799-7440-5/14 $31.00 © 2014 IEEE

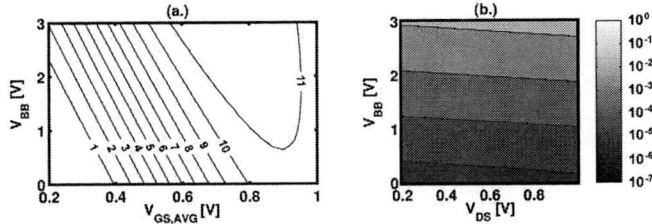

Fig. 3. Characteristics of the DC-DC core's LVT NMOS switches under different back-gate bias ($V_{BB}$)voltages. Normalized contours of (a.) $G_{ON}$ and (b.) $I_{OFF,dc-dc}$.

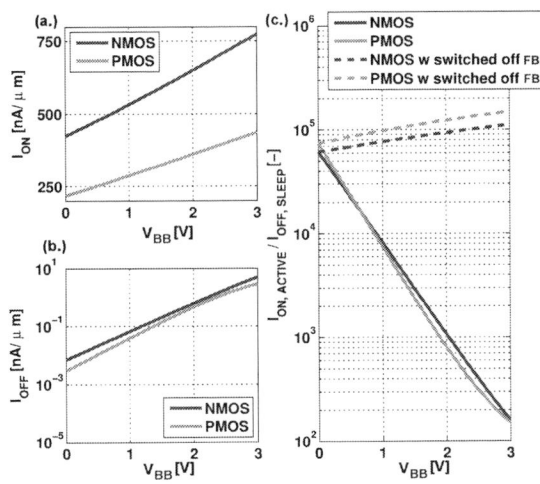

Fig. 4. (a.) $I_{ON}$ and (b.) $I_{OFF}$ of LVT switches depending on the applied $V_{BB}$. $I_{ON}$ is boosted by 2x and $I_{OFF}$ by 100x between $V_{BB}$=0V and 3 V. (c.) The ratio between $I_{ON}$ in active mode and $I_{OFF}$ in sleep mode. The solid lines show a fixed FBB in both sleep and active modes (i.e. $V_{BB}$=0 to 3 V). The dashed lines show a FBB in active mode and a switched off FBB in sleep mode (i.e. $V_{BB}$=0 V).

## B. Control Circuitry Power Control

The baselines bars in Fig. 5 (c.) shows that the relative control circuitry power compared to the load power goes from 2% to almost 350% when the load power drops from 100 $\mu$W to 100 nW. This result confirms the observations of Fig. 1. In sleep mode, we can reduce dynamic power ($E_{dyn}$) though frequency scaling without reducing the overall efficiency of the DC-DC. To reduce the leakage losses in the controller, we applied poly biasing and reduced the leakage portion of the relative efficiency from 200% to 5% of $P_{LOAD}$ (Fig. 5 (a.)). The $P_{dyn}$ is slightly increased by the driver up-sizing in order to keep the driving strength constant. In Fig. 5 (b.) we applied dynamic FBB to increase the relative efficiency of the controller. By sizing and operating the DC-DC at higher FBB in active mode and shutting down the FBB in sleep mode, we can reduce the switch and driver sizes (i.e., Switched Sized Bias or SSB). This reduces the active and leakage energy in sleep mode. In active mode, we trade dynamic energy for leakage energy, and thus, the overall penalty is negligible.

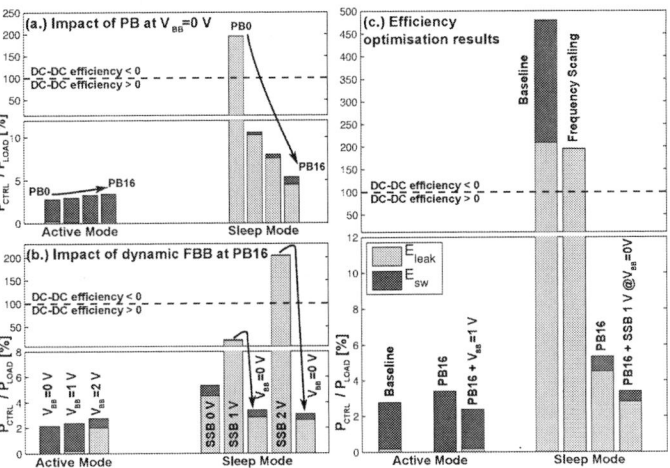

Fig. 5. (a.) Impact of 4 nm, 10 nm and 16 nm poly biasing (PB0 to PB16) and FBB on controller losses. The controller is resized for each PB value to keep the driving strength constant. (b.) For each $V_{BB}$, the DC-DC is resized using Switch Sized Bias (SSB) at 0 V, 1 V, and 2 V. In sleep mode, SSB 1V and SSB 2V use @$V_{BB}$=0V to reduce leakage power. (c.) The impact of the successive optimization on the overall controller losses.

### C. System Simulation Results

As explained in Section II, applying FBB when sizing the switches $S_1$-$S_4$, allows for reduction in control circuitry losses. In addition, it allows the converter to operate over a larger power range since leakage in $S_1$-$S_4$ does limit functionality during sleep mode (i.e. increased $I_{ON}/I_{OFF}$ for switches sized using SSB). Using previously discussed FBB techniques, the efficiency of the 2:1 DC-DC converter was found (Fig. 6).

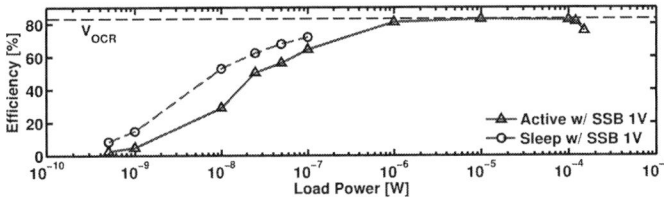

Fig. 6. Efficiency of the 2:1 DC-DC converter with SSB 1V. $V_{OCR}$ is the optimal conversion ratio voltage.

## III. CONCLUSION

This paper showed the effects of back-gate bias on SC DC-DC converters in 28 nm UTBB FD-SOI. Using FBB allowed for reductions in the control circuitry losses and improvements in switch performance. The SC converter was able to operate with a peak efficiency of 72% at 100 nW and 83% at 100 $\mu$W.

## ACKNOWLEDGMENT

AofF #124029,#13139458,GETA. National Foundation for Scientific Research (Belgium) FRIA grant.

## REFERENCES

[1] J. Mäkipää et al.,*JLPEA*, vol. 2, pp. 180–196, 2012.
[2] D. Bol et al.,*IEEE JSSC*, vol. 48, no. 1, pp. 20–32, 2013.
[3] H.-P. Le et al.,*IEEE JSSC*, vol. 46, no. 9, pp. 2120–2131, 2011.
[4] M. Wieckowski et al.,*Symposium on VLSI*, 2009, pp. 166–167.
[5] J. De Vos et al.,*IEEE FTFC*, May 2014.
[6] M.J. Turnquist et al.,*IEEE ISLPED*, Sept. 2013.

# An Optimal Probing Method of Pre-Bond TSV Fault Identification in 3D Stacked ICs

*Bei Zhang* and *Vishwani D. Agrawal*

Department of Electrical and Computer Engineering, Auburn University, Auburn, AL 36849, USA
bzz0004@auburn.edu, vagrawal@eng.auburn.edu

*Abstract*— **A fast TSV identification algorithm is proposed in this work to reduce the test time of pre-bond TSV probing. The speeding up of the algorithm comes from two aspects. First, any unnecessary session during the test is skipped. Second, the test terminates as soon as either all TSVs have been identified or a pre-specified maximum number of faulty TSVs have been identified. Experimental results demonstrate that instead of testing all sessions as stated in previous work, the algorithm always finishes the pre-bond TSV test after only a small portion of all sessions. The algorithm reduces pre-bond TSV test time and is expected to greatly reduce the pre-bond testing and the overall 3D device manufacturing costs.**

## I. INTRODUCTION

Pinpointing TSV resistive defects before die bonding is important for yield assurance of 3D ICs. Current BIST based approaches [1], [2], [3], [10] for pre-bond TSV test require dedicated test circuit to be added to each TSV and the area overhead could be huge since there can be tens of thousands of TSVs on a chip [4]. Moreover, the BIST circuits themselves are prone to the effects of process variations. References [7] and [8] propose the use of a large probe needle with active driver to short many TSVs together and conduct a parametric test. The set of TSVs contacted at once by the probe needle is referred to as a *TSV network*. The number of TSVs within a network is typically less than 20 based on the relative diameter and pitch of probe needles and TSVs [4], [9]. High measurement resolution, low hardware overhead and robustness to process variations make this technique likely to be used in practice.

Figure 1 shows a circuit model of the test set up [7], [8] for a 4-TSV network. TSV $i$ is represented by its resistance $R_i$ and capacitance $C_i$. $R_c$ represents the contact resistance between TSV and the probe. A gated scan flip-flop (GSF) is inserted between TSV $i$ and the system logic which is in accordance with the developing IEEE P1838 standard [5], [8]. All GSFs can be loaded up or read out through a boundary scan mechanism. In the normal mode, all GSFs are made transparent by opening B2 and closing B1 switches by a "bypass" signal. In the pre-bond TSV test mode, all GSFs drive respective TSVs. In Figure 1, TSVs 1 and 2 are receiving TSVs and TSVs 3 and 4 are sending TSVs. A receiving TSV receives signal from the other die and drives the on-chip logic while a sending TSV is driven by the on-chip logic and sends a signal to the other die. A GSF in scan mode drives a receiving TSV during pre-bond TSV probing

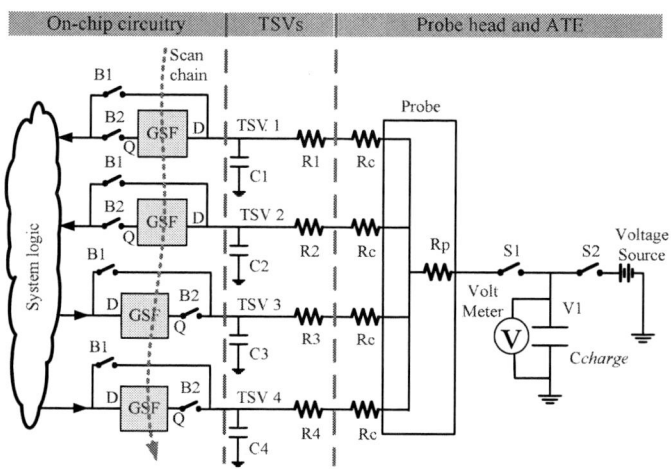

Fig. 1. Circuit model for pre-bond TSV probing.

when both B1 and B2 switches are closed. A GSF drives a sending TSV when B1 is opened and B2 is closed.

Pre-bond TSV resistance measurement starts by scanning in all GSFs with "1." $C_{charge}$ and all TSVs are then discharged through the probe needle. By configuring the switches of a GSF, a charge sharing circuit is constructed between that GSF and $C_{charge}$ through its corresponding TSV (either sending or receiving TSV). The charging rate of $C_{charge}$ is compared to a calibrated curve of a good TSV to determine the resistance of the TSV under test. Parallel TSV test can also be conducted by configuring multiple GSFs at a time. Now, $C_{charge}$ is charged faster and the measurement terminates quicker. However, the number of TSVs tested in parallel cannot exceed a constant "$r$" due to minimum measurement resolution constraint [7], [8]. This resolution of measurement refers to the minimum change in TSV resistance that can be detected by the technique and it is adversely affected by the number of TSVs tested in parallel. We call the TSVs tested in parallel within the same TSV network a *test session*. Based on this probing technique, any faulty TSV within a session will cause the session test to fail but we cannot tell which TSV(s) is (are) faulty. On the other hand, a good parallel test implies that all TSVs within the session are fault-free. The probe needle remains contacted with the corresponding TSV network, and the charging and discharging process continues until either all TSVs within the network are identified or a certain

978-1-4799-7440-5/14 $31.00 © 2014 IEEE

number of TSVs with resistive defects are pinpointed within the network. All TSV networks are tested in two groups. Networks in a group are tested simultaneously by a probe head containing a large number of needles, each making contact with a single network. Once all contacted networks are tested, the probe head is lifted and repositioned to test the remaining group [7], [8].

## II. A FAST TSV IDENTIFICATION ALGORITHM

The goal of TSV probing is to identify up to a certain number, $m$, of faulty TSVs within a $T$ TSV network where $m$ is the number of redundant TSVs in the TSV network being tested. If the number of identified faulty TSVs exceeds $m$, then not all faulty TSVs in the network can be repaired, and the chip would be discarded. Otherwise, the on-chip redundant TSVs are sufficient to replace all identified faulty ones. This goal of TSV probing can be achieved by testing each TSV individually though with unnecessarily long test time. Large time savings occur if we test TSVs in parallel without losing the capability of identifying up to $m$ faulty TSVs, and also guarantee that the size of each session does not exceed $r$. There are existing methods to generate such sessions [6], [8], [12]. However, those assume that all sessions need to be tested and they do not identify TSVs based on the sessions. This work proposes a fast TSV identification algorithm to further speed up pre-bond TSV test from two aspects. First, during the identification process, any "currently unnecessary" session is skipped. Second, TSV test is terminated as soon as either all TSVs have been identified or the number of identified faulty TSV exceeds $m$ as the chip can be discarded due to lack of redundant TSVs and further test is useless.

The pseudo-code of the algorithm is shown in Figure 2 where argument $t$ represents the test time of sessions. Test time of a session in this work only refers to the charging time of $C_{charge}$, and it is related to the session size [6], [8], [12]. The algorithm starts by initializing 3 empty lists named "Good", "Bad", and "F_C", respectively. The "Good" and "Bad" lists are used to contain the identified good and faulty TSVs, respectively. The faulty candidate list "F_C" is used to contain any failing session. The algorithm enumerates all the sessions generated by [8] or [12] and skip any "currently unnecessary" session which refers to a session where either all TSVs in the session have been identified so far or there is at least one identified bad TSV in the session. "currently unnecessary" session does not provide any information of TSV identification. We define a *fault map* $\rho$ as positions of defective TSVs within a TSV network. Although a session may be "currently unnecessary" for identifying some fault maps of a TSV network, it could be essential for identifying other fault maps of the same TSV network. So, none of the "currently unnecessary" sessions can be deleted. If a session is not skipped, it will be tested. If a session passes the test, all TSVs in the session are added to "Good", and we then use "Good" to refine "F_C." Here, the refinement

---

**Algorithm** Fast_TSV_Identification (*All_sessions, T, m, t*)

1. *Good*=[ ]; *Bad*=[ ]; *F_C*=[ ]; *test_time*=0; *tested_sessions*=0;
2. **foreach** *session* in *All_sessions*
   // Skip any currently unnecessary test session
3.     **if** (all TSVs in *session* have been identified) **or** (there is at least one bad TSV in *session*)
4.        Continue;
5.     *test_time*+=*t*(*session*); // Test time accumulation
6.     *tested_sessions*+=1; // Test session accumulation
   // Handle a passing session
7.     **if** *session* is tested as being good
8.        Add all TSVs in *session* to *Good*;
9.        **foreach** *FC_session* in *F_C*
10.           Remove any good TSV from *FC_session*;
11.           **if** length(*FC_session*)==1
12.              Add the TSV in *FC_session* to *Bad*;
13.              Remove the entire *FC_session* from *F_C*;
    // Handle a failing session
14.     **else if** *session* is tested as being bad
15.        Remove any good TSV from *session*;
16.        **if** length(*session*)==1
17.           Add the TSV in *session* to *Bad*;
18.        **else**
19.           Append *session* to *F_C*;
    // Termination conditions
20.     **if** ((length(*Good*)+length(*Bad*))==*T* **or** (length(*Bad*)>=*m*+1)
21.        Break;
22. Return *test_time, tested_sessions*;

Fig. 2. A dynamically optimized TSV identification algorithm.

refers to removing any identified good TSV from the targeted session (see line 10 of Figure 2). If after refinement any failing session in "F_C" contains only one TSV, that TSV is identified as defective and added to "Bad." If a session fails the test, "Good" is again utilized to refine this failing session (line 15 of Figure 2). If the session after refinement contains only one TSV, that TSV is added to "Bad." Otherwise, the refined failing session is appended to "F_C." The above procedure terminates as soon as any condition shown on line 20 in Figure 2 is satisfied.

## III. EXPERIMENTAL RESULTS

Table I shows the results of the proposed algorithm applied to various TSV networks. Column 1 shows parameters $T$ (network size), $m$ (redundant TSVs in network) and $r$ (resolution). Column 2 gives the number of faulty TSVs ($\phi$) within the network. Column 3 shows the total number of sessions and total test time (in $\mu$s) for exhaustive application of sessions optimized by ILP [12]. The test time calculation is detailed in references [6], [8] and [12]. For given $\phi$, we enumerate all possible fault maps and obtain the test time and number of tested sessions using the proposed algorithm of Figure 2. Column 4 shows the average number of tested sessions and average test time for identifying all fault maps containing $\phi$ faulty TSVs. Column 5 shows the relative reduction in column 4 over column 3. Column 6 shows the maximum number of sessions tested and the corresponding test time for identifying a fault map. Column 7 shows the relative reduction in column 6 over column 3.

TABLE I

EXHAUSTIVE [12] AND DYNAMICALLY OPTIMIZED (FIGURE 2) APPLICATION OF TSV TEST SESSIONS CONSTRUCTED BY ILP.

| Parameters $T, m, r$ | Number of faulty TSVs ($\phi$) | Optimum exhaustive test [12] (# sessions, time in $\mu s$) | Dynamically optimized test | | | |
|---|---|---|---|---|---|---|
| | | | Av. test sessions (# used, time in $\mu s$) | Average reduction (sessions, time) | Worst case sessions (# used, time in $\mu s$) | Worst case reduction (sessions, time) |
| 8, 2, 3 | 0 | (8, 3.36) | (5.0, 2.10) | (37.5%, 37.5%) | (5, 2.10) | (37.5%, 37.5%) |
| | 1 | | (5.3, 2.25) | (32.8%, 32.8%) | (6, 2.52) | (25.0%, 25.0%) |
| | 2 | | (6.4, 2.71) | (19.1%, 19.1%) | (8, 3.36) | (0.0%, 0.0%) |
| | 3 | | (7.5, 3.17) | (5.3%, 5.3%) | (8, 3.36) | (0.0%, 0.0%) |
| 12, 3, 3 | 0 | (16, 6.72) | (7.0, 2.94) | (56.2%, 56.2%) | (7, 2.94) | (56.2%, 56.2%) |
| | 1 | | (7.5, 3.14) | (53.1%, 53.1%) | (9, 3.78) | (43.7%, 43.7%) |
| | 2 | | (8.7, 3.65) | (45.5%, 45.5%) | (12, 5.04) | (25.0%, 25.0%) |
| | 3 | | (10.3, 4.32) | (35.5%, 35.5%) | (14, 5.88) | (12.5%, 12.4%) |
| | 4 | | (11.8, 4.97) | (25.9%, 25.9%) | (16, 6.72) | (0.0%, 0.0%) |
| 15, 4, 3 | 0 | (25, 10.50) | (8.0, 3.36) | (68.0%, 68.0%) | (8, 3.36) | (68.0%, 68.0%) |
| | 1 | | (9.6, 4.03) | (61.6%, 61.6%) | (14, 5.88) | (44.0%, 44.0%) |
| | 2 | | (11.1, 4.68) | (55.3%, 55.3%) | (17, 7.14) | (32.0%, 32.0%) |
| | 3 | | (12.6, 5.33) | (49.2%, 49.2%) | (20, 8.40) | (20.0%, 20.0%) |
| | 4 | | (14.3, 6.03) | (42.5%, 42.5%) | (23, 9.66) | (8.0%, 8.0%) |
| | 5 | | (15.8, 6.66) | (36.5%, 36.5%) | (25, 10.50) | (0.0%, 0.0%) |
| 20, 4, 4 | 0 | (25, 9.50) | (9.0, 3.42) | (64.0%, 63.9%) | (9, 3.42) | (64.0%, 63.9%) |
| | 1 | | (10.8, 4.10) | (56.8%, 56.7%) | (15, 5.69) | (40.0%, 39.9%) |
| | 2 | | (12.3, 4.68) | (50.6%, 50.6%) | (18, 6.83) | (28.0%, 27.9%) |
| | 3 | | (13.9, 5.31) | (44.0%, 44.0%) | (21, 7.97) | (16.0%, 15.9%) |
| | 4 | | (15.1, 5.76) | (39.3%, 39.3%) | (24, 9.11) | (4.0%, 3.9%) |
| | 5 | | (18.0, 6.85) | (27.8%, 27.8%) | (25, 9.49) | (0.0%, 0.0%) |

We make four observations from Table I. First, the average number of tested sessions and average test time is much less than the total number of sessions and total test time for any $\phi \leq m$ (reparable TSV network) or any $\phi > m$ (irreparable TSV network). For example, the average percentage reduction reaches 68.0% for parameters $T = 15$, $m = 4$, $r = 3$, and $\phi = 0$. On average, the proposed algorithm greatly speeds up pre-bond TSV identification process. Second, as $\phi$ increases the average percentage reduction decreases. This is expected as pinpointing larger number of faulty TSVs within a TSV network generally requires more sessions to be tested and cost more time. Third, in most cases even the maximum number of tested sessions is less than the total number of sessions. Fourth, as expected, the maximum number of tested sessions increases as $\phi$ increases for a given TSV network. From column 7, reduction in the worst case can be small for large $\phi$, requiring all sessions to identify a fault map. This scenario occurs when fault map contains $m$ or more faulty TSVs. The probability of such large number of faulty TSVs within a small localized silicon area may be negligible for a mature manufacturing process. Thus, the worst case percentage test time reduction could be quite significant.

## IV. CONCLUSION

The proposed TSV identification algorithm has two main advantages. First, the average number of tested sessions and test time are guaranteed to be small factions of total sessions and test time. Second, even for the worst fault map, for which most sessions are needed, not all sessions may be used. Reducing pre-bond TSV test time helps reduce pre-bond test cost in real silicon. A forthcoming paper combines this technique with several related optimizations [11].

***Acknowledgment*** This research is supported in part by the National Science Foundation Grant CCF-1116213.

## REFERENCES

[1] P. Chen, C. Wu, and D. Kwai, "On-Chip Testing of Blind and Open-Sleeve TSVs for 3D IC Before Bonding," in *Proc. 28th IEEE VLSI Test Symposium (VTS)*, 2010, pp. 263–268.

[2] M. Cho, C. Liu, D. H. Kim, S. K. Lim, and S. Mukhopadhyay, "Design Method and Test Structure to Characterize and Repair TSV Defect Induced Signal Degradation in 3D System," in *Proc. IEEE/ACM International Conference on Computer-Aided Design (ICCAD)*, 2010, pp. 694–697.

[3] S. Deutsch and K. Chakrabarty, "Non-Invasive Pre-Bond TSV Test using Ring Oscillators and Multiple Voltage Levels," in *Proc. Design, Automation & Test in Europe Conference & Exhibition (DATE)*, 2013, pp. 1065–1070.

[4] M. Jung, J. Mitra, D. Z. Pan, and S. K. Lim, "TSV Stress-Aware Full-Chip Mechanical Reliability Analysis and Optimization for 3D IC," in *Proc. 48th Design Automation Conference (DAC)*, 2011, pp. 188–193.

[5] E. J. Marinissen, J. Verbree, and M. Konijnenburg, "A Structured and Scalable Test Access Architecture for TSV-Based 3D Stacked ICs," in *Proc. 28th IEEE VLSI Test Symposium (VTS)*, 2010, pp. 269–274.

[6] B. Noia and K. Chakrabarty, "Identification of Defective TSVs in Pre-Bond Testing of 3D ICs," in *Proc. 20th Asian Test Symposium (ATS)*, 2011, pp. 187–194.

[7] B. Noia and K. Chakrabarty, "Pre-Bond Probing of TSVs in 3D Stacked ICs," in *Proc. International Test Conference (ITC)*, 2011, pp. 1–10.

[8] B. Noia and K. Chakrabarty, *Design-for-Test and Test Optimization Techniques for TSV-based 3D Stacked ICs*. Springer, 2014.

[9] K. Smith, P. Hanaway, M. Jolley, R. Gleason, and E. Strid, "Evaluation of TSV and Micro-Bump Probing for Wide I/O Testing," in *Proc. International Test Conference (ITC)*, 2011, pp. 1–10.

[10] J. You, S. Huang, D. Kwai, Y. Chou, and C. Wu, "Performance Characterization of TSV in 3D IC via Sensitivity Analysis," in *Proc. 19th Asian Test Symposium (ATS)*, 2010, pp. 389–394.

[11] B. Zhang and V. D. Agrawal, "SOS3: Three-Step Optimization of Pre-Bond TSV Test for 3D Stacked ICs," in *Proc. 32nd IEEE Int. Conf. on Computer Design (ICCD)*, Oct. 2014.

[12] B. Zhang and V. D. Agrawal, "Test Session Optimization for Pre-Bond TSV Probing in 3D Stacked ICs," in *Proc. 23rd Asian Test Symposium (ATS)*, 2014. submitted.

# Power Supply Voltage Detection and Clamping Circuit for 3-D Integrated Circuits

Divya Pathak and Ioannis Savidis

Department of Electrical and Computer Engineering
Drexel University
Philadelphia, Pennsylvania 19104.
divya.pathak@drexel.edu, isavidis@coe.drexel.edu

*Abstract*—**A circuit that detects and sets the power supply voltage of a given device plane in a 3-D IC is proposed. The circuit consists of 1) a ring oscillator in each voltage domain located in each device plane, and 2) a power module placed in a dedicated power plane. The power module consists of four components; a divide-by-40 clock divider, a frequency detector, a voltage ramp generator, and a voltage peak detector circuit. The power module provides a stable reference voltage equal to the power supply voltage of the targeted voltage domain in the device plane. SPICE simulations of the circuit indicate that power supply voltages of less than 1 V are successfully set and provided as a reference to an on-chip voltage regulator within 500 ns and with reference voltage variation of less than 1%. The power module is integrated with on-chip voltage regulators which require a precise voltage reference. The proposed implementation permits dynamic voltage and frequency scaling and point of load power delivery.**

*Index terms—3-D IC power delivery, voltage regulation.*

## I. INTRODUCTION

Through silicon via (TSV) based 3-D integrated circuits (IC) permit the integration of heterogeneous technologies with CMOS. The integrated system may include RF, analog, micro-/nano-electromechanical systems (MEMS/NEMS) as well as emerging technologies such as nano-FET and graphene-based device planes. The design and fabrication of these disparate device planes may take place at separate facilities. Unless technology specific information on each device plane is provided, the packaging facility carrying out the final 3-D integration of the device planes is unaware of the power supply voltage requirements of the different ICs [1].

In this paper, a universal power plane that includes circuits capable of detecting the power supply voltage requirement of each device plane in the 3-D IC stack is proposed. The universal power plane consists of multiple on-chip voltage regulators serving the power needs of each voltage domain in each device plane of the 3-D stack. This arrangement facilitates point-of-load power delivery through the use of TSVs. The shorter path between the power source and load leads to both lower *IR* drop and parasitic impedance of the power distribution network [2]. Each dedicated voltage regulator for a given voltage domain facilitates dynamic voltage and frequency scaling (DVFS). 3-D IC interface guidelines are required to successfully implement this circuit, and must include specifications for the location of ports for power, clock, and signal delivery for each device plane of the 3-D IC stack [1]. The components of the power supply voltage detection and clamping circuit are described in Section II. The simulation results for two device planes are discussed in Section III. Some conclusions and ongoing work are provided in Section IV.

## II. DEVICE PLANE POWER SUPPLY DETECTION AND CLAMPING

The block diagram of the various components of the 3-D IC supply voltage detection and clamping circuit are shown in Fig.1. Each voltage domain in the device plane consists of a ring oscillator with a current-starved output switching stage [3]. The ring oscillator operates at a frequency of 1 GHz when the applied control voltage equals the power supply voltage of the given domain. The oscillation frequency of 1 GHz is selected as it is easy to achieve with minimum current starved inverter stages in a deep sub-micron CMOS technology as well as in GaAs based RF circuits [4]. The power module that serves a voltage domain consists of a clock divider, frequency detector, voltage ramp generator, voltage peak detector, and a voltage regulator.

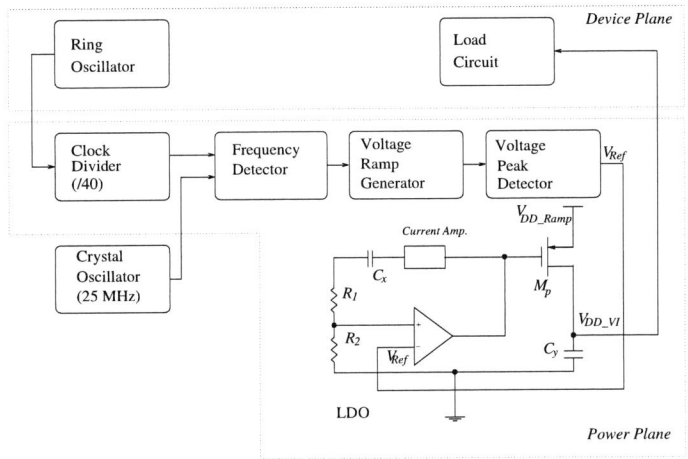

Fig. 1. Block diagram of the power supply voltage detection and clamping circuit depicting a single voltage domain in one device plane and the corresponding power module in the power plane.

The ring oscillator output frequency is down converted by a factor of 40 by a clock divider consisting of a divide-by-8 cascaded together with a divide-by-5 circuit block [5]. A frequency comparator compares the output of the clock divider with a 25 MHz clock source (off-chip crystal oscillator). The frequency comparator is implemented using four D-FFs as shown in Fig. 2(a). The frequency comparator is chosen over a phase frequency detector (PFD) as the PFD generates UP and DOWN signals (due to D-FF clock-to-Q delay) even when the inputs are phase locked [6]. In addition, the voltage detection circuit does not require that the down converted ring oscillator frequency is matched in phase with the 25 MHz reference signal.

978-1-4799-7440-5/14 $31.00 © 2014 IEEE

A voltage ramp generator circuit (RGC) includes a constant current source charging a capacitor $C_1$ and a switch $S_1$ that controls the duration of the ramp voltage, as shown in Fig. 2(b). The UP and inverted DOWN signal are logically ANDed together ($UP \bullet \overline{DOWN}$) to generate the control signal for $S_1$. When the 3-D IC is first powered on, the ramp generator provides an initial voltage to the ring oscillator, and the UP signal from the frequency detector is high as the down converted frequency from the ring oscillator is less than 25 MHz. The frequency increases as the ramp generator output voltage increases. When the ring oscillator reaches the 1 GHz frequency, the UP signal is de-asserted and any further increase in frequency asserts the DOWN signal. The switch $S_1$ is in the open state, preventing any further increase of charge (and therefore voltage) on $C_1$. The voltage $V_{RGC}$ on $C_1$ corresponds to the power supply voltage of the voltage domain ($V_{DD\_VI}$) in which the ring oscillator is placed. The discharging of the capacitor $C_1$ through the load circuit causes the ring oscillator frequency to drop below 1 GHz, which re-asserts the UP signal and places $S_1$ in the closed state. The UP signal continues to periodically toggle ensuring that the voltage $V_{RGC}$ is maintained at the desired level.

The variations in $V_{RGC}$ due to discharging of the capacitor $C_1$ are filtered using a voltage peak detector circuit (shown in Fig. 2(c)). The peak detector circuit consists of a PMOS $M_1$ which controls the current charging the capacitor $C_2$. A voltage comparator compares the output voltage $V_{Ref}$ across the capacitor $C_2$ with $V_{RGC}$. The output voltage from the comparator biases $M_1$. $V_{Ref}$ follows the positive transition of $V_{RGC}$ and at steady state equals the maximum value of $V_{RGC}$. The peak detector circuit therefore provides a steady voltage reference with less than 1% ripple voltage variation from the targeted power supply voltage of the device plane ($V_{DD\_VI}$). $V_{Ref}$ is used as the reference voltage for the voltage regulator serving the load circuit in the device plane. On-chip voltage regulator topologies like the LDO and buck converter are suitable for integration with the proposed voltage detection and clamping circuit. The stable reference voltage provided by the proposed circuit ensures superior line regulation offered by the voltage regulator.

## III. SIMULATED CIRCUIT RESULTS

Two device planes are simulated using the 45 nm low-performance ($V_{DD\_VI}$ of 1 V) and 22 nm high-performance ($V_{DD\_VI}$ of 0.8 V) predictive technology models (PTM) [7]. The designed ring oscillators for each device plane include three inverter stages in a current starved output switching configuration. The design is robust against threshold voltage variation, channel length variation, and low-field mobility variation [8].

The components of the power module are implemented with the 45 nm PTM models. The divide-by-40 clock divider and frequency detector circuits are simulated using the 45 nm low-performance PTM model ($V_{DD\_PP}$ of 1 V). The voltage ramp generator and peak detector circuits are implemented using the 45 nm thick oxide ($V_{DD\_Ramp}$ of 3 V) PTM model. A DC-DC level shifter is used to convert the UP and DOWN signals from a $V_{DD\_PP}$ of 1 V to a $V_{DD\_Ramp}$ of 3 V. The slope of the voltage ramp signal is deliberately kept low (0 V to 3 V in 2 µs) to ensure stable operation. The mimimum

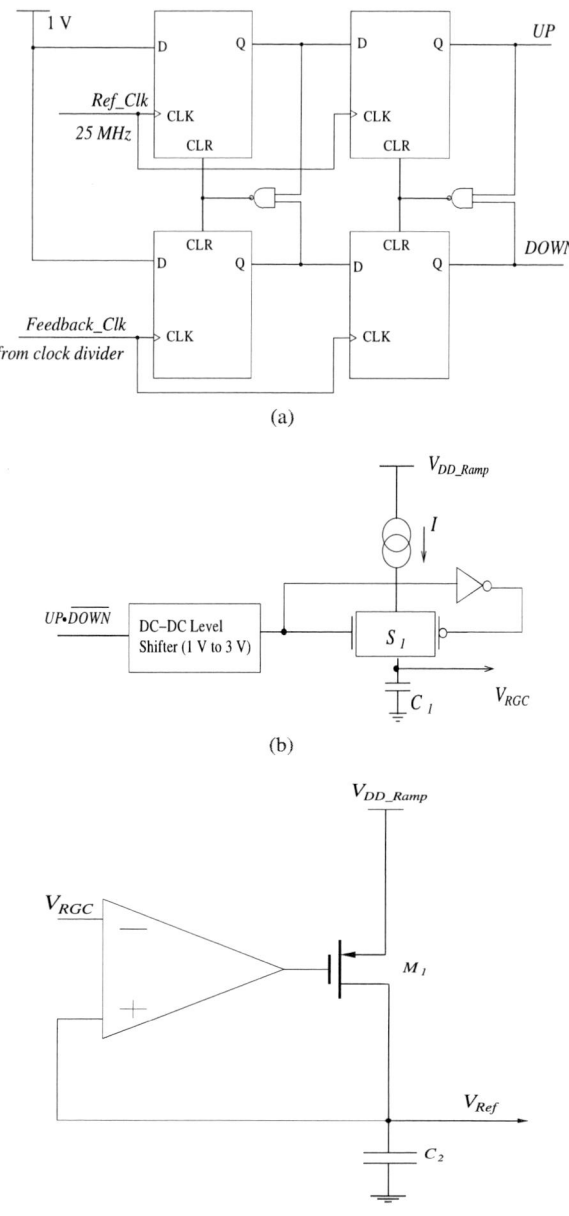

Fig. 2. Circuit schematic of (a) frequency detector, (b) voltage ramp generator, and (c) peak detector.

voltage detected reliably by the power module is 0.7 V. The maximum voltage provided by the ramp generator is 2.5 V. SPICE circuit simulations indicate a maximum variation of 1% in the reference voltage $V_{Ref}$ provided to the on-chip voltage regulator for $V_{DD\_VI}$ of less than 1 V. This is comparable to the stability of reference voltages used by off-chip buck converters, where approximately 1% variation is currently achieved [9].

The simulation results for detecting and clamping the power supply voltage for a voltage domain in a 22 nm device plane are shown in Fig. 3. $V_{Ref}$ reaches $V_{DD\_VI}$ of 0.8 V in 370 ns. The detection and clamping circuits on the 45 nm device plane require 420 ns for $V_{Ref}$ to reach the target voltage $V_{DD\_VI}$ of 1 V. The proposed circuit therefore offers a fast transition to the desired power supply voltage level at startup and is suitable for integration with on-chip voltage regulators.

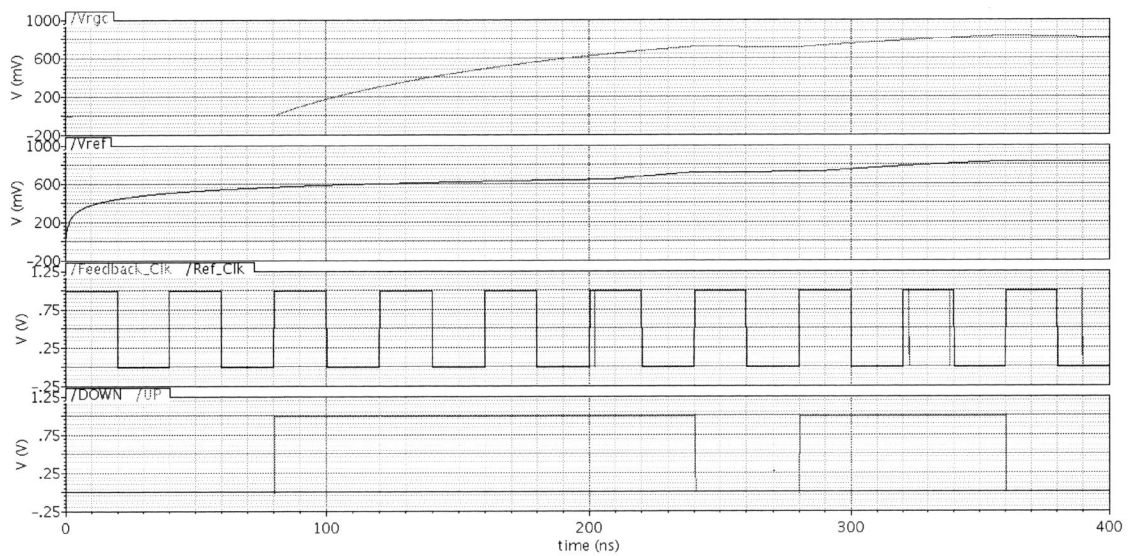

Fig. 3. Analysis of the power supply voltage detection and clamping circuit for a 0.8 V domain in a 22 nm technology.

## IV. CONCLUSIONS

A circuit to detect and reliably set the power supply voltage of a given voltage domain in a 3-D IC is described. All components except for a ring oscillator are part of the power module located in a separate and dedicated 45 nm power plane. Multiple power modules serve as point of load voltage delivery circuits to the different device planes. Correct power supply voltage detection and clamping is demonstrated through circuit simulation for two device planes, one in 22 nm and the other in 45 nm. The power module is capable of setting the power supply voltage of a device plane ranging from 0.7 V to 2.5 V. The precise voltage generated from the power module acts as a reference voltage for an on-chip LDO or buck converter. The reference voltage is within 1% of the targeted power supply voltage, as indicated by simulated results.

Further work is planned to test the precision of the generated reference voltage through statistical analysis of multiple PVT corners. The effect of the interconnect and the TSVs will be included to provide a comprehensive analysis of the 3-D IC power detection and delivery circuit.

## REFERENCES

[1] I. Savidis, *Characterization and modeling of TSV based 3-D integrated circuits*, Ph.D. Thesis, University of Rochester, 2013.

[2] I. Savidis, S. Kose, and E. G. Friedman, "Power noise in TSV-based 3-D integrated circuits," *IEEE Journal of Solid-State Circuits*, Vol. 48, No. 2, pp. 587–597, February 2013.

[3] G. Jovanovic, M. Stojcev, and Z. Stamenkovic, "A CMOS voltage controlled ring oscillator With improved frequency stability," *Scientific Publications of the State University of Novi Pazar, Series A: Applied Mathematics, Informatics and Mechanics*, Vol. 2, No. 1, pp. 1–9, December 2010.

[4] A. Teetzel and R. Walker, "A GaAs IC broadband variable ring oscillator and arbitrary integer divider," *IEEE Microwave and Millimeter-Wave Monolithic Circuits Symposium*, pp. 87–89, June 1992.

[5] J.R. Yuan and C. Svensson, "Fast CMOS nonbinary divider and counter," *Electronics Letters*, Vol. 29, No. 13, pp. 1222–1223, June 1993.

[6] R.J. Baker, *CMOS: circuit design, layout, and simulation*, Vol. 18, John Wiley & Sons, 2011.

[7] Nanoscale Integration & Modeling Group Arizona State University, "Predictive Technology Model (PTM)," http://www.eas.asu.edu/~ptm.

[8] D. Pathak and I. Savidis, "Run-time voltage detection circuit for 3-D IC power delivery," *Proceedings of the IEEE 27th International SoC Conference (SoCC)*, pp. 229–233, September 2014.

[9] Analog Devices, "ADP1821 step-down DC-to-DC controller datasheet," http://www.analog.com/static/imported-files/data_sheets/ADP1821.pdf.

# Study of Fin-Tunnel FETs with doped pocket as Capacitor-less 1T DRAM

Arnab Biswas, Adrian M. Ionescu

STI-IEL-NANOLAB, Ecole Polytechnique Fédérale de Lausanne, Switzerland

E-mail: arnab.biswas@epfl.ch

**Abstract:** In this work we propose and validate by experimentally calibrated simulations a silicon Tunnel FET(TFET) based capacitorless DRAM cell, implemented as a fully-depleted FinFET with CMOS compatible process. The devices have a conventional FinFET structure except for a $p+$ (for n-type TFET) doped pocket of length $L_{PKT}$ and doping $N_{PKT}$ between the intrinsic channel and the ($n++$) drain. This doped pocket creates a necessary condition to store holes injected from the source-to-body junction. In [1], there was a need to induce a potential well in order to store the excess charges; whereas in the present case a potential well is permanently present due to the doped pocket. The drain voltage is used as a control voltage to either fill the potential well with carriers (WRITE "1") by attracting holes from the $p++$ source or repel them to empty the well of carriers (WRITE "0"). In contrast with the SOI Z-RAM® there is no need of impact ionization to create/inject the hole charge in the device body, the holes being injected by the forward-bias $p+i$ junction, which significantly improves the device reliability. Measurements on FDSOI TFET devices as reported in [1,2] were performed at elevated temperatures and used to calibrate the non-local band-to-band (B2B) tunnelling model in Sentaurus TCAD [3]. The retention characteristics of the proposed memory cell is simulated at an elevated temperature of 85°C and is shown to be not degrading at higher temperature as is the case in conventional capacitorless DRAMs [4].

**Introduction, principle of operation:** Tunnel FETs have been recently proposed [5] as steep slope switches able to reduce the voltage operation of logic circuits below 0.2V with reduced leakage current and improved switching energy efficiency. While high-performance Tunnel FETs exploit InAs-Si heterostructures on nanowires [6], the all-Si Tunnel FET family appears primarily to have a very high $I_{ON}/I_{OFF}$ and the lowest $I_{OFF}$ [7] yet fail to achieve the requirements of $I_{ON}$. In this paper, we further build upon [1] that showed that all-Si double-gate Tunnel FET can serve for building a new class of devices: the capacitorless Tunnel FET DRAM, where the very low $I_{OFF}$ is offering a major advantage for DRAM low power consumption/low refresh rate and the zero-capacitor structure a very high potential for scalability. The proposed new structure as shown in Fig. 1. Fig. 3 shows the energy band diagram of the device in Fig. 1. The doped pocket acts as a resistance to the flow of carriers between the source and the drain and the drain bias is used to control this resistance to the current flow. Transfer curves of the device in Fig. 1 was simulated for different $L_{PKT}$ and $N_{PKT}$ as shown in Fig. 2. The values $L_{PKT}$ = 25 nm and $N_{PKT}$ = $5 \times 10^{18}$ cm$^{-3}$ were chosen to enable optimum functioning of the memory cell. Higher pocket doping results in a deeper potential well as well as a higher resistance to carrier resulting in lower currents. Lower doping on the other hand would require very high drain bias for memory operation, which is not feasible for a $L_G$=100nm device. Fig. 4 shows the simulated hole densities at the end of a WRITE "1" and WRITE "0" state. It is quite evident that the hole density in the pocket area after a WRITE "1" state is quite higher than after WRITE "0" state (also shown in Fig. 5). This is reflected in the potential profiles in the same figure as well. Fig. 6 again shows the simulated potential profiles during HOLD state (top) and READ (bottom) after both WRITE "1" and WRITE "0" states. As the device turns ON after these two states a difference in drive current is expected, which forms the basis of the memory operation.

**DRAM operation:** The programming scheme is depicted in Table 1. The effect of a back and forth sweep of the control drain bias is used to study the hysteresis effect. As shown in Fig. 7, a significant difference of 900 nA is observed at $V_D$ =1.5V between the two READ currents. Fig. 7 also shows the same curves at elevated temperatures of 55°C and 85°C. Temperature dependent simulations were carried out using the fitted parameter list calculated using [8]. A minimal temperature dependence was observed. The timing diagrams for READ, HOLD and WRITE operations are shown in Fig. 8 (top) and the corresponding simulated drain current is shown in Fig. 8(bottom). Depending on the previous stored state, a clear difference in READ current levels ($\Delta I_D$=500nA) is observed. Fig. 9 shows the retention characteristic of the proposed new memory cell. A retention time in the order of 100 µsec was observed for the simulated device. There is scope for improvement for this retention time as process and device technologies improve. The same figure also shows the retention characteristic at an elevated temperature of 85°C. It is to be note that the retention characteristic is not degrading with increasing temperature, which is related to the TFET subthreshold swing temperature stability [9-10].

**Conclusion:** In summary we have validated a new scalable 1T capacitorless DRAM cell based on DG fin Tunnel FET structure with a doped pocket, which opens a new field of applications for TFETs. A clear DRAM memory operation has been proposed and verified with simulations. Retention times of the order 100 µsec is shown to be not degrading at high temperature (85°C) which is one of the advantages for TFETs based memory cells.

**References:**

[1] A. Biswas et al. Appl. Phys. Lett. 104, 092108 (2014).

[2] F. Mayer, et al, IEDM 2008

[3] Synopsys TCAD ver. 2012.06

[4] J. Liu et al. ISCA '13, 60-71.

[5] A. M. Ionescu and H. Riel, Nature, vol. 479, Nov. 2011.

[6] K. E. Moselund et al. IEEE EDL vol. 33, 2012.

[7] E. Faraoni, et al., IEDM 2007

[8] A. Biswas et al., Microelectronic Engineering 98, 334-337.

[9] P. Agopian et al., SOI Conference 2012

[10] M. Born et al., in Proc. 15th Int. Conf. Microelectron., 2006.

978-1-4799-7440-5/14 $31.00 © 2014 IEEE

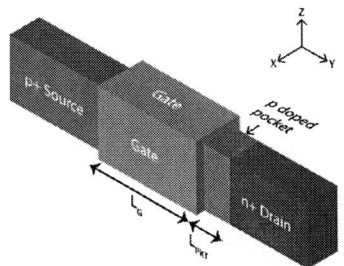

Figure 1: 3D schematic of the proposed new structure including a *p+* doped pocket between the channel and drain.

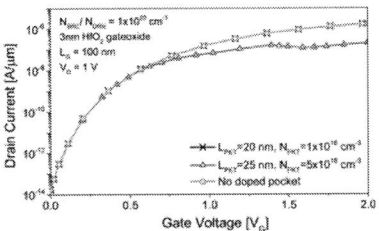

Figure 2: Simulated transfer curves for the structure in Fig. 4 for varying pocket doping and pocket length. $L_{PKT}$ = 25 nm and $N_{PKT}$ = $5\times10^{18}$ cm$^{-3}$ is chosen for the rest of the simulations.

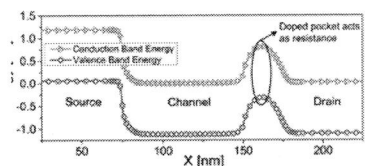

Figure 3: Energy band diagram at steady state showing the doped pocket acting as a series resistance to the flow of carriers from the source to the drain.

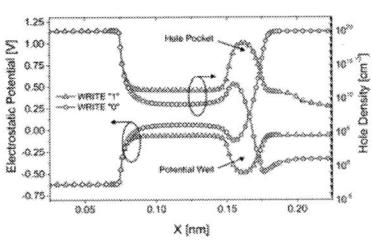

Figure 4: Simulated potential profile and hole density at 3nm below the gate oxide/channel interface after a WRITE "1" and WRITE "0" state. The higher hole density in the pocket after a WRTIE "1" state is clearly observed.

Table 1: Programming scheme for the proposed memory cell.

| State | $V_D$ (V) | $V_G$ (V) | $V_S$ (V) |
|---|---|---|---|
| WRITE "1" | -1.5 | 0.5 | 0 |
| WRITE "0" | 2.0 | 0.5 | 0 |
| HOLD | 0 | 0.75 | 0 |
| READ | 1.5 | 1.5 | 0 |

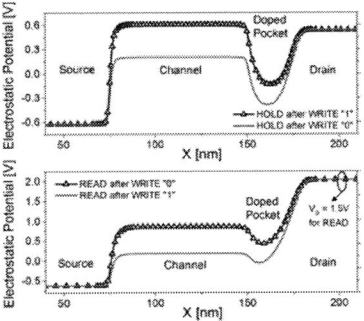

Figure 6: Simulated potential profiles at 3nm below the gate oxide/channel interface showing the HOLD (top) and READ (bottom) operations.

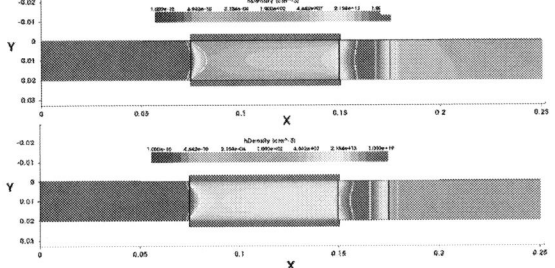

Figure 5: TCAD simulated 2D cross-section of the proposed new structure showing the hole density at a HOLD state after and WRITE "1"(top) and WRITE "0"(bottom) operation.

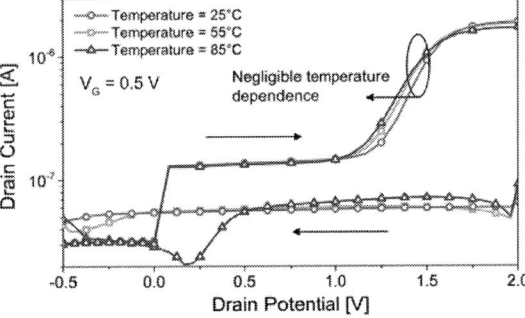

Figure 7: Simulated hysteresis curves for drain current with respect to drain potential at 25°C, 55° & 85°C. A difference of around 900 nA is observed at $V_D$ = 1.5 V between the two states, which signifies charge storage.

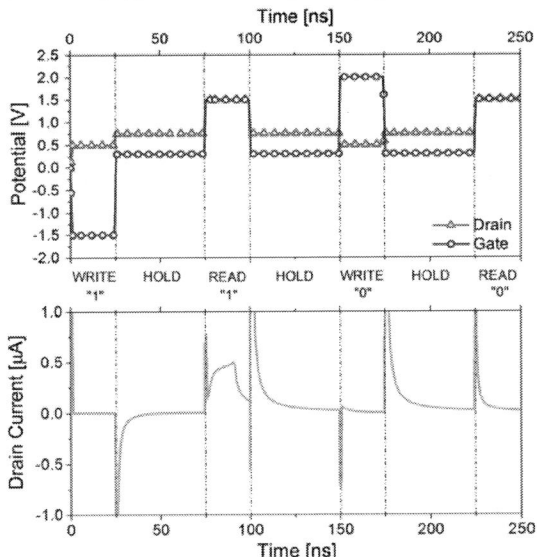

Figure 8: Continuous read/write cycle with hold time in between. A difference of 500 nA is observed between the two memory states.

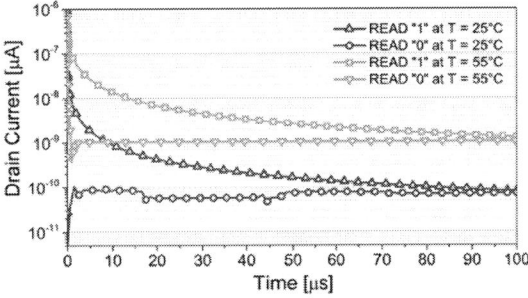

Figure 9: Retention characteristics of the proposed memory cell at 25°C & 85°C. No degradation in the retention time is observed at higher temperatures.

# UTBB/FDSOI : reasons for a success

M.Haond
STMicroelectronics, Crolles, France
Email: Michel.haond@st.com, Tel: +33 4 38 92 22 98

## Introduction

There are lots of rumours mentioning the end of Moore's Law for CMOS, since silicon is no more able to maintain the doubling of devices per unit area at an ever decreasing transistor cost every 2 years. Advanced new processes were announced some years ago by the industry leaders [1] as promising but finally appear to be very complex and expensive to follow the Law. A few years ago, we have proposed a cheaper process that allows continuing the Law thanks to the use of 2D process legacy: UTBB/FDSOI from 28 nm on. This process has entered into production and confirms its full promise of performance boost and power control. We will review the first prototype results at 28 nm and will detail the next generations allowing continuing performance increase and power reduction at 14 nm. We will also describe the ingredients ready for the 10 nm generation.

## FDSOI is a Must

It was predicted more than about 2 decades ago that MOS devices could continue shrinking and face short channel effects if the junction depth could be reduced to avoid Drain-to-source punchthrough. It was shown that the best way was to physically contain it. Hence the concept of dual or triple (and even all-around) gate control of a fully depleted films to contain the drain field penetration towards the source [2]. Two constructions finally came out: one 2D silicon film running over an oxide or a 3D vertical thin wall (or fin) wrapped around by a Gate (Fig.1).

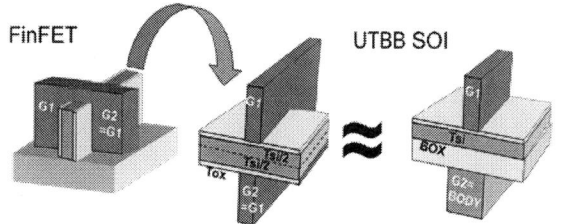

**Figure 1** : 2 FDSOI structures used for the CMOS nodes below 20 nm. 2D UTBB (right) and 3D Vertical Fins (Left).

It is quite clear that the 2D FDSOI is more straightforward as it is an evolution from the old 2D Bulk silicon CMOS technology, as soon as the silicon film thickness can be reduced and controlled to nm scale (UTBB) on 300 mm wafers. On the other hand, the 3D fin construction is very attractive for scaling purposes since drain current can flow in a vertical dimension, as soon as all 3D features can be processed and controlled in the lateral and vertical dimensions at a reasonable process complexity and die cost.

## 28 nm UTBB/FDSOI goes into Production

In 2013, 28FDSOI went into production [3]. It was confirmed that process cost was equivalent to similar Bulk process at a 30% boost in performance, since added substrate cost was balanced by the simplification of the process flow. The Power advantage was also confirmed on final industrial products. Fig.2 presents the main results obtained on an application processor heating when running in a Smartphone. The right part compares a 28 Bulk LP processor to a 28FDSOI Processor. The colouring corresponds to the heating of the respective APs.

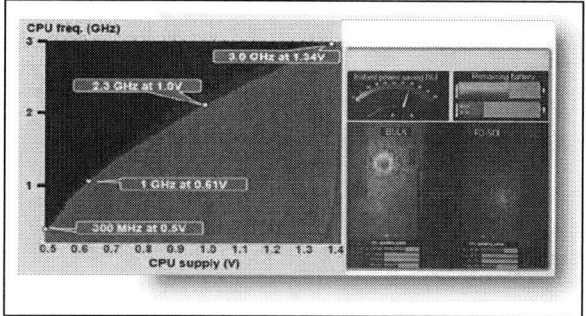

**Figure 2** : 28nm FDSOI Application Processor Product in a smartphone. Left shows high performance potential (3 GHz) and Low Voltage opportunities, whereas right part evidences the drastic heat reduction on same processor processed on FDSOI (right in green) or in Bulk (Yellow/Red on left).

The temperature contours are clearly reduced with the FDSOI Processor. On the left part of the figure, we provide a measurement of the 28FDSOI Processor Frequency vs Supply Voltage. It is shown that it can run at 3GHz for a Vdd around 1.3V. But more important is the fact that the frequency is still as high as 1GHz at 0.6V and that it is still running (300MHz) at 0.5V. This evidences one of the key application fields for this FDSOI process: Low Voltage applications for the handheld, mobile or the IOT business.

## NEXT FDSOI GENERATIONS

The FDSOI Roadmap as proposed by ST is presented in Fig.3 showing that FDSOI will continue to maintain competitive CMOS process offering at least down to the 10 nm node.

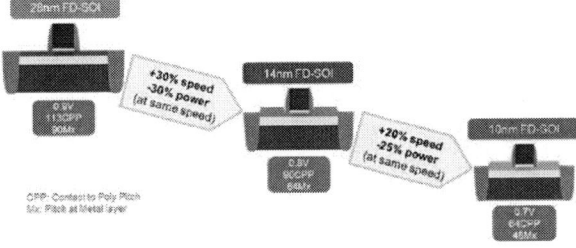

**Figure 3:** FDSOI Roadmap from 28nm FDSOI towards 10nm FDSOI

Let us describe the main features of the next FDSOI nodes.

The 14nm FDSOI is in the maturity qualification step. It follows shrinking the 28nm FDSOI and is using a 1st generation of FDSOI speed boosters [4]. The main one relies on the use of a SiGe channel fabricated by the Ge condensation process. The advantage of this process is twofold: it allows lowering the VTP and boosting the PMOS hole mobility and Ieff current by 25% [5]. Next boosters are the in-situ doped elevated S&D: eSiGe-B & eSiC-P are used for P & N resp. Ieff currents are measured at 330 and 405 µA/µm at 0.8V for P & N at 20nA/µm Ioff as extracted from plots shown in Fig. 4.

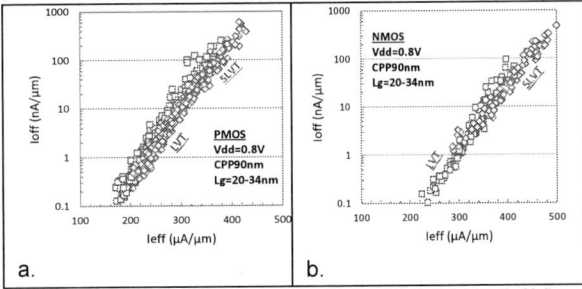

**Figure 4:** Ieff-Ioff plots of the 14 nm MOS (Vdd= 0.8V).

This drive performance leads to Ring Oscillator Frequency boost by 30 % compared to 28FDSOI devices supplied at 0.9V. This performance can be further improved by applying some Forward Biasing (FBB) to the wells acting as a back-gate

[3,4]. In this process, the SRAM bitcells can be totally embedded in a single PWell, thereby offering the opportunity to access very low Vddmin [6] and offering some area gain advantages by removing well isolation needs.

10FDSOI is the most advanced FDSOI Process under development. It is developing a new booster generation. First, it relies on the use of strained SOI (sSOI) substrates [5,7]. The tensile biaxial strain is increasing the electron mobility by as much as 90% (Fig.5a) and IonN by 20 % (Fig.5b) [8]. However a relaxation process needs to be applied in order to recover the hole mobility reduction at low field, the [Ge] needs to be optimised in the SiGe channel [8].

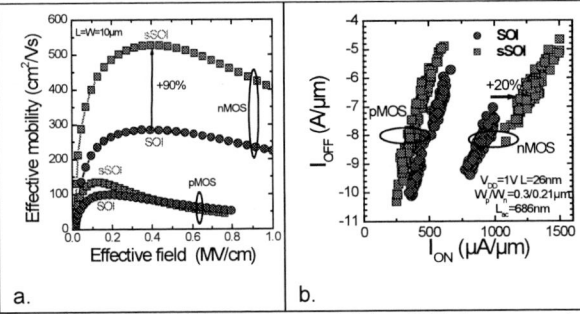

**Figure 5 :** sSOI Mobility impact (a). sSOI Ion current boost (b).

Further S&D Engineering will provide the Raccess reduction mandatory for the performance of these devices at a reduced Vdd. Moreover a further reduction of the BOX thickness allows to continue the use of efficient FBB for performance boost.

## CONCLUSIONS

We have presented the reasons of a success story for the 2D FDSOI Technology that has started at 28 nm and will continue through the 10 nm node. Competitive performance and power control is demonstrated. It is also experiencing the use of Back Biasing both for performance boost as well as for adequate Process compensation allowing reaching a tighter process control that has become key for these advanced nodes.

## REFERENCES

[1] C.Auth et al., VLSI Technology Symposium 2012.
[2] T.Skotnicki et al., IEDM 1994.
[3] N.Planes et al., VLSI Technology Symposium 2012.
[4] O.Weber et al, VLSI Technology Symposium 2014.
[5] A. Khakifirooz et al., VLSI Symposium 2012.
[6] R.Ranica et al, VLSI Technology Symposium 2013
[7] C.Fenouillet et al, VLSI Technology Symposium 2012.
[8] F.Andrieu et al, ESSDERC 2014 .

# Piezoresistivity in unstrained and strained SOI MOSFETs

R. Berthelon[1], M. Cassé[1], D. Rideau[2], O. Nier[1,2,3], F. Andrieu[1], E. Vincent[2], G. Reimbold[1]

[1]CEA-Leti, MINATEC Campus 17 rue des martyrs 38054 Grenoble Cedex 9, France
[2]STMicroelectronics, 850 rue Jean Monnet, F-38926 Crolles, France
[3]IMEP-LAHC, MINATEC 3 Parvis Louis Néel, 38016 Grenoble, France
E-mail: remy.berthelon@cea.fr

*Abstract-* **We hereby present the extraction and the study of piezoresistive (PR) coefficients in MOSFETs built on unstrained and strained SOI substrates. We have evidenced a strong dependence of these PR with the inversion charge density in particular for PMOS. These results are well explained by the Si bandstructure calculation which enlightens the effect of the strain and of the electric confinement on carrier mobility, up to high tensile strain values.**

## INTRODUCTION

Strain engineering has become necessary to keep CMOS device performance enhancement as dimensions are scaled down [1,2]. The piezoresistive (PR) coefficients are useful tools to describe the electronic transport under mechanical stress because of their simplicity [1,3]. Many works have been done to evaluate PR coefficients on bulk Si [4] or on (100) inversion surface for different device types including bulk, FDSOI, nanowires transistors and on strained channel devices [1,3,5,6,7]. However, the dependence of PR coefficients with inversion charge density has rarely been investigated. In this work, we have carefully extracted PR coefficients for both NMOS and PMOS devices fabricated on SOI and on strained-SOI (sSOI) substrates. The mobility gain from SOI to sSOI has been evaluated with the help of the piezoresistivity model and compared with experimental results and mobility calculations.

## EXPERIMENTAL SET-SUP

The devices, realized along the <110> channel direction, have been fabricated from SOI or sSOI wafers, with 7nm Si film thickness and 25nm BOX. The sSOI substrate induces a biaxial tensile stress in the channel corresponding to σ=1.35GPa. Measurements have been carried out on large (W=10μm) and long (L=10μm) devices to avoid any strain relaxation due to the technological process. PR coefficients have been extracted by measuring the linear drain current variation with an externally stress applied using a 4-point bending system (fig.1). The longitudinal ($\Pi_L$) or transversal ($\Pi_T$) piezoresistive coefficient can be extracted depending on the direction of the applied stress with respect to the current flow. From these two components, it is possible to determine $\Pi_S=\Pi_L+\Pi_T$ and $\Pi_{44}=\Pi_L-\Pi_T$ which are the biaxial and shear components, respectively. For the purpose of mobility calculation and its response to strain, a Poisson-Schrödinger solver has been used in this paper, namely UTOXPP [8]. This physically based tool uses self-consistent Poisson 6 **k.p** Schrödinger solvers and accounts for the dependence of the band structure on strain [9] and electric field. The mobility is then computed within the Kubo-Greenwood formalism [10] in a 2D electron/hole

inversion layer taking into account phonon and surface roughness.

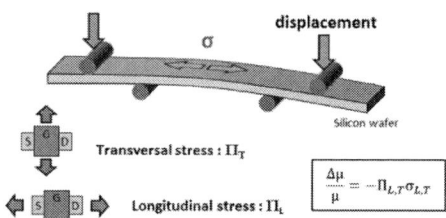

Fig. 1: Schematic of the 4-point bending setup used to extract transversal and longitudinal piezoresistive coefficients. The stress applied to the device is proportional to the displacement.

## PIEZORESISTIVE COEFFICIENTS

The piezoresistive coefficients have been extracted (fig. 2) and calculated (fig. 3) as a function of inversion charge density for both SOI and sSOI MOSFETs.

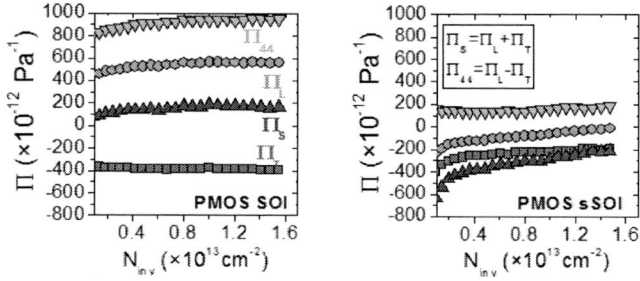

Fig. 2: Extracted piezoresistive coefficients of (left) SOI and (right) sSOI PMOS. PR coefficients sign change is observed from SOI to sSOI substrate. The dependence on inversion charge density is evidenced for sSOI in particular.

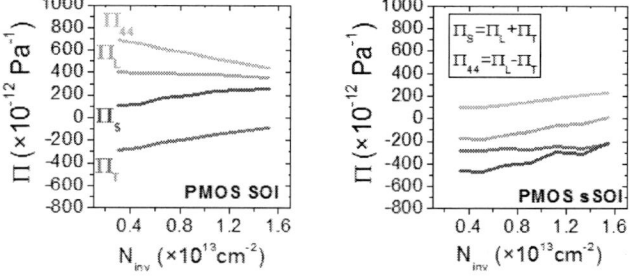

Fig. 3: Calculated piezoresistive coefficients of (left) SOI and (right) sSOI PMOS. A good qualitative agreement with experimental data is evidenced.

The PR coefficients are strongly impacted by the high level of biaxial tensile strain in sSOI devices (σ=1.35GPa). This result shows that the mobility response to an additional stress strongly depends on the initial level of stress in the channel. In particular, the biaxial components ($\Pi_S$) in PMOS switch from positive values in SOI ($\sim$200×$10^{-12}$Pa$^{-1}$) to negative values in sSOI (below -200×$10^{-12}$Pa$^{-1}$, in fig.2). This means that a biaxial tensile strain which is detrimental

978-1-4799-7440-5/14 $31.00 © 2014 IEEE          76

to hole mobility at low stress, becomes beneficial at high stress (typically above 1GPa), in agreement with some results reported in Ref.[11] on bulk Si. In addition it is found that the sSOI PR coefficients strongly depend on $N_{inv}$. In particular, $\Pi_L$ decreases from $-200\times10^{-12}Pa^{-1}$ at low $N_{inv}$ to 0 at high $N_{inv}$. This can be explained by the opposite effects of strain and electrical confinement as discussed further. The mobility gain from an initial stress-free state up to high values of stress (above 1GPa) can be calculated using the evolution of the PR coefficients with stress and integrating $d\mu/\mu=-\Pi(\sigma)d\sigma$ [3]:

$$\mu(\sigma)/\mu(\sigma = 0) = exp\left(\int_{0}^{\sigma} -\Pi(\sigma')d\sigma'\right) \quad (1)$$

The mobility gain between SOI and sSOI NMOS and PMOS can be calculated as a function of the inversion charge density. The results are in rather good agreement with the effective mobility measured by standard split-CV technique (fig. 4).

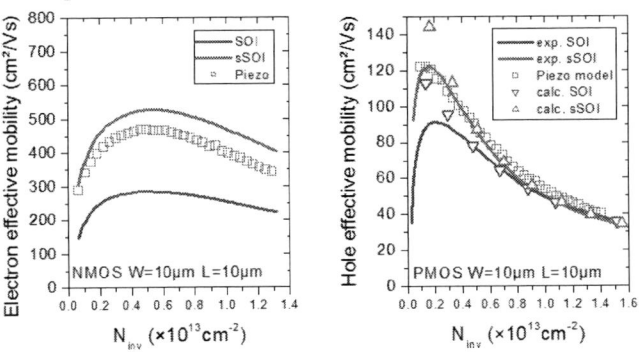

Fig. 4: (left) electron and (right) hole effective mobility $\mu_{eff}$ of SOI and sSOI. For PMOS, the mobility gain strongly decreases with inversion charge density. The trend is well reproduced by the piezoresistivity model (bold line) using the dependence on $N_{inv}$ of PR coefficients, and by the mobility calculation with Kubo-Greenwood formalism.

The effective electron mobility is strongly enhanced by the biaxial tensile strain in agreement with the extracted PR coefficients ($\Pi_S=-600\times10^{-12}Pa^{-1}$ for SOI and $\Pi_S=-200\times10^{-12}Pa^{-1}$ for sSOI, not shown). On the contrary the hole mobility in sSOI is higher than SOI up to low-medium $N_{inv}$ as previously observed in literature [11-14], but this gain disappears at high inversion charge density (typically above $1\times10^{13}cm^{-2}$) . This trend is actually reproduced by our piezoresistivity model, taking into account the dependence of PR with inversion charge density and by the mobility calculated using Kubo-Greenwood formalism.

BANDSTRUCTURE

Strain and electrical or quantum confinement both alter the Si bandstructure leading to band splitting and warping. The carrier mobility is consequently impacted, through the change of transport effective mass and the interactions with intra-valleys phonon mainly [15]. For NMOS, a biaxial tensile strain splits the energy levels of $\Delta_2$ and $\Delta_4$ valleys in the same way as the electrical confinement does. As a consequence, the electron mobility is strongly enhanced even at high $N_{inv}$. For PMOS the strain and electrical confinement have opposite effects. The valence band structure calculated by the k.p method is shown for different inversion charge with and without strain in fig.5.

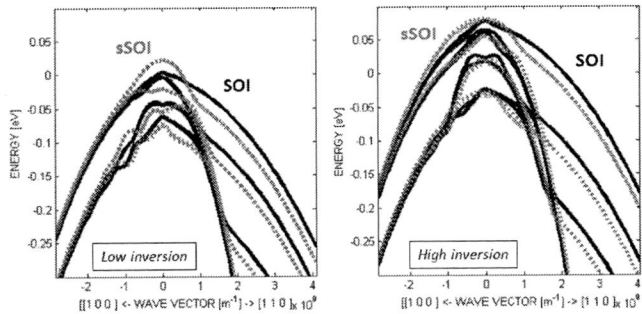

Fig. 5: Valence band structure in (solid) SOI and (dashed) sSOI device for (left) low and (right) high inversion density. 6 level k.p model using L=-6.69, M=-4.62 and N=-8.56 together with strain parameters of l=-1.2eV, m=3.6eV, and n=-8.2446eV. One ought to mention that the 2 former parameters correspond to b=-1.6eV as proposed in Ref [16] which is a bit lower than the value ($\approx$-2eV) determined e.g. by ab-initio simulation [17] or by direct measurements [18].

Under biaxial tensile strain, the light hole (LH) and heavy hole (HH) subbands degeneracy is lifted. The LH subband with lower effective mass is preferably populated (as seen on fig. 5 left) and phonon scattering is reduced [16,19-21], leading to a higher hole effective mobility. The split between LH and HH is reduced at high inversion. The relative band position and curvature lead to the reduction of the hole mobility enhancement consistently with experimental results.

CONCLUSION

We have extracted and analyzed the piezoresistive coefficients for devices built on SOI and sSOI substrates. We demonstrated their capability to describe the stress-induced mobility enhancement as a function of inversion carrier density. The mobility behavior has been explained by the strong coupling between biaxial tensile strain and quantum-confinement and has been highlighted through band structure and mobility calculations. In particular, a strong biaxial tensile strain can enhance PMOS mobility up to moderate inversion carrier density.

ACKNOWLEDGEMENTS

This work was partially carried out in the frame of the LETI/ST/IBM joint program and in the DYNAMICULP and PLACES2BE projects.

REFERENCES

[1] Y. Sun, S. Thompson, and T. Nishida,"Strain effect in semiconductors", Springer (2010)
[2] S. Thompson et al., IEDM Tech. Dig., p.221 (2004)
[3] O. Weber et al., IEDM Tech. Dig.,p.719 (2007)
[4] C.S. Smith, Physical Review 94 (1), pp.42–49 (1954)
[5] F. Rochette et al., Solid-State Electron. 53, p.392 (2009)
[6] C. Gallon et al., IEEE TED 51 (8), pp.1254–61 (2004).
[7] M. Cassé et al., IEDM Tech. Dig., p.637 (2012)
[8] D. Garetto et al., Proc. 14th Intern. Nanotech Conf. (2011).
[9] A. Pham et al., IEEE TED 54 (9), p.2174 (2007)
[10] M. Fischetti et al., J. Appl. Phys. 94 (2), p.1079 (2003).
[11] M. Liao et al, J. Appl. Phys. 98 (6), p.066104 (2005)
[12] K. Rim et al., Proc. Symp. VLSI Tech., p.98 (2002)
[13] K. Rim et al., IEDM Tech. Dig., pp. 49–52 (2003)
[14] M. Lee et al., Proc. ISTDM, p.87 (2004)
[15] S. Takagi et al., J. Appl. Phys. 80 (3), p.1567 (1996)
[16] E. Wang et al., IEEE TED 53 (8), p.1840 (2006)
[17] D. Rideau et al., Phys. Rev. B 74, p.195208 (2006)
[18] C. Van de Walle et al., Phys. Rev. B 34, p.5621 (1986)
[19] A. Khakifirooz and D. Antoniadis, IEEE EDL 27, p.402 (2006)
[20] S. Thompson et al., IEEE TED 51 (11), p.1790 (2004)
[21] Y. Sun et al., J. Appl. Phys. 101 (11), p.104503 (2007)

# nFET FDSOI Activated By Low Temperature Solid Phase Epitaxial Regrowth: Optimization Guidelines

L. Pasini[1,2,3], P. Batude[1], M. Cassé[1], L. Brunet[1], P. Rivallin[1], B. Mathieu[1], J. Lacord[1], S. Martinie[1], C. Fenouillet-Beranger[1], B. Previtali[1], N. Rambal[1], M. Haond[2], G. Ghibaudo [3] and M. Vinet[1]

[1]CEA, Leti, MINATEC Campus, 17 rue des Martyrs, F38054 GRENOBLE, Cedex 9, France
[2] STMicroelectronics, 850 Rue Jean Monnet, F-38926 Crolles, France
[3]IMEP-LAHC, Minatec/INPG, 3 parvis Louis Néel, F38016 Grenoble, France
Tel:+33438782044 e-mail:luca.pasini@cea.fr

**Abstract** – Low temperature (LT) activation on Fully Depleted Silicon On Insulator by SPER is needed for 3D sequential integration and also provides interest to obtain highly doped abrupt junctions in the standard planar technology. In this work, through the confrontation of electrical data and KMC process simulation we identify the efficient lever to optimize the low temperature device performance. This work evidences that the most suitable integration for LT FET implies a LDD implantation before the first spacer and the raised source drain epitaxy.

## I. INTRODUCTION

In 3D sequential integration scheme, the use of Low Temperature (LT) processes is mandatory for the top FETs fabrication in order to preserve bottom FETs performance [1]. The most critical step in term of thermal budget is dopant activation, usually carried out by spike annealing (>1000°C). LT (<600°C) Solid Phase Epitaxial Regrowth (SPER) dopant activation is promising to reach higher activation level and junction abruptness than the conventional High Temperature (HT) activation [2]. In this work, LT SPER and HT activation are compared in terms of their impact on NFET performance. The electrical characteristics will be analyzed in the light of KMC simulations in order to offer process optimization guidelines for LT FDSOI devices

## II. DEVICE FABRICATION

The process flows of Process Of Reference (POR) device (so called HT device) and of devices with LT activation (so called LT devices) are described in Fig.1. For HT device, a first ion implantation (LDD) is performed after Raised Source/Drain (RSD) epitaxial deposition. A second ion implantation (HDD) is performed after the second spacer formation. Finally, spike activation at 1050°C is carried out. For LT devices, only one implantation step is performed of either phosphorous or arsenic (two different devices). Then, two minutes annealing at 600°C is applied.

For HT device, the total implanted dose is $5x10^{15}$at/cm$^2$ with As and P co-doping. For LT devices a dose of $8x10^{14}$ at/cm$^2$ (P device) or $6x10^{14}$at/cm$^2$ (As device) has been chosen in order to avoid dopant clustering as shown in [3].

## III. ELECTRICAL RESULTS

Both carrier mobility in the channel at low electric field (Fig.2) and DIBL (Fig.3) show the same behavior for HT and LT devices.

In addition, similar Equivalent Oxide Thickness (EOT) is extracted for HT and LT device (EOT$_{HT}$=11Å EOT$_{LT}$=11.2Å). EOT is also found to be stable with gate length variation for both HT and LT devices (not shown).

However, the comparison of access resistance, defined as $R_D$= $R_{SD}$/2, extracted by Y function method [4] shows a strong degradation between HT and LT devices accentuated in case of As ($\Delta R_D^{HT/LT\_P}$=68Ω·µm; $\Delta R_D^{HT/LT\_As}$=131Ω·µm) (Fig.4).

$R_{SD}$ increase is translated in Ion-Ioff degradation between HT and LT device (Fig.5). At constant $I_{OFF}$, 20% degradation in LT best case (P) is observed. Higher performance degradation is observed

for As device (44%). In order to get full comprehension of performance degradation a deeper analysis of $R_{SD}$ is proposed.

## IV. SIMULATIONS AND ANALYSIS

By means of SPROCESS KMC simulation the dopant concentration under Spacer 1 (SP1) has been extracted. The results are depicted in Fig.6 (HT), Fig.7a (LT P), Fig.7b (LT As). As expected in LT devices, due to the quasi-negligible diffusion, the dopant concentration below the first spacer (SP1) is small (below $1x10^{19}$ at/cm$^3$).

### A. DIBL analysis

While dopant concentrations below the spacer are clearly different in the case of LT and HT splits, slight DIBL variation is observed (<20mV/V for $L_{mask}$=26nm on Fig.3). This is explained by the fact that a dopant concentration below $1x10^{19}$ at/cm$^3$ has relatively low impact on Short Channel Effects (SCE), and thus all three devices are in an underlapped configuration (Fig.6 and 7). In this case the electrical gate length is roughly defined by the physical gate length.

### B. $R_{SD}$ analysis

The dopant concentration profile under SP1 is different for the three devices: in HT device the $1x10^{20}$ at/cm$^3$ concentration is reached below SP1 (Fig.6) while for the LT P &As devices dopant concentration is smaller (Fig.7). The situation is illustrated in Fig.8 in which the evolution of iso-concentration of each dopant at $5x10^{19}$at/cm$^3$ is shown for the tested devices.

The resistance value corresponding to the region under the SP1 ($R_{SP1}$, see Fig.6) has been extracted by SDEVICE simulations and reported in Fig.9. The results show that the $R_{SD}$ degradation between HT and LT is mainly explained by $R_{SP1}$ increase.

It is worth mentioning that to overcome this $R_{SP1}$ increase, the use of tilted implant cannot be implemented. Indeed with this integration scheme it is not possible to tune an implantation dose that allows to have sufficient dopant concentration under the SP1 without full amorphization of the film.

An alternative solution could be to implant before the spacer and epitaxy deposition as described in Fig.10 (Extension First scheme). This allows to place dopants under SP1 and then improve $R_{SP1}$.

Despite the challenge of growing an epitaxy on an implanted film, in [5] it has been shown how choosing low energy implantation (~1KeV) good RSD morphology can be obtained.

## V. CONCLUSION

We have demonstrated that the region under the spacer, with lower dopant concentration, is the main contribution responsible for performance degradation of LT devices w.r.t. HT devices. On the other hand, the other electrical parameters, i.e. mobility, DIBL and EOT are left unchanged. An alternative architecture such as Extension First may allow optimizing the LT activated FDSOI devices and reach HT performance.

### Acknowledgments

This work is partly funded by the ST-IBM-LETI Alliance program and by Qualcomm.

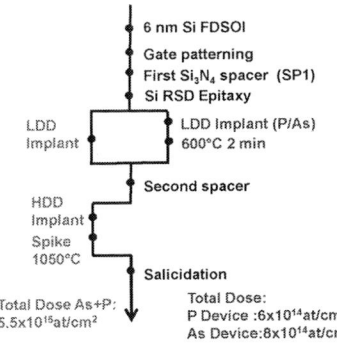

**Fig.1:** Process Flow for HT (red) and LT (blue) activated devices.

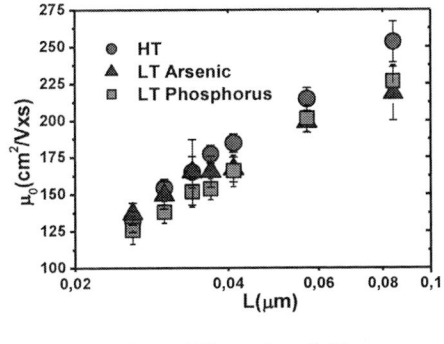

**Fig.2:** Carrier mobility at low field versus gate length. HT and LT devices show similar behavior.

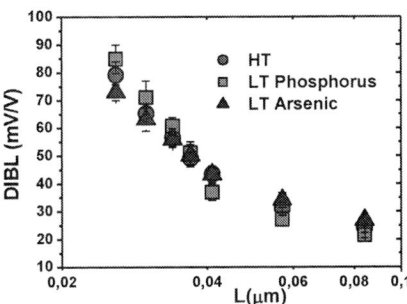

**Fig3:** DIBL versus gate length. HT and LT devices show similar behavior.

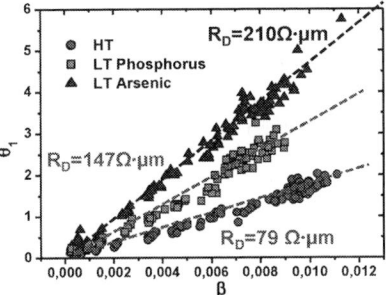

**Fig.4:** Access resistance $R_D = R_{SD}/2$ extracted by Y-function for the tested devices. A degradation is shown for LT devices compared to HT.

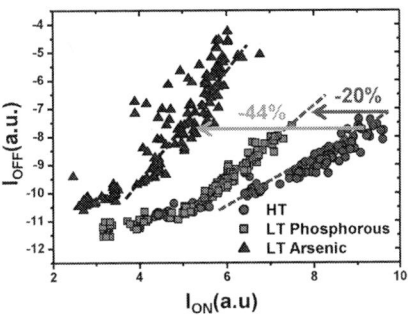

**Fig.5:** $I_{ON}$-$I_{OFF}$ trade-off of HT/LT devices with gate length between 26nm and 1μm. At constant $I_{OFF}$ a degradation of 20% is showed for best LT device (P implant) compared to HT.

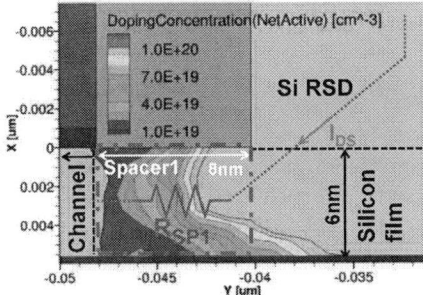

**Fig.6:** Dopant concentration under Spacer 1 extracted from KMC simulation for HT device. $R_{SP1}$ is also defined.

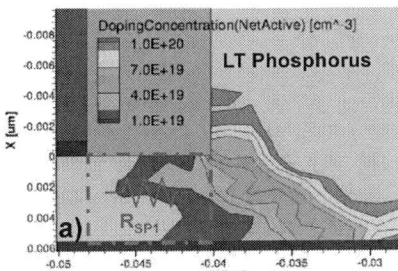

**Fig.7a:** Dopant concentration under SP1 extracted from KMC simulation for P LT device

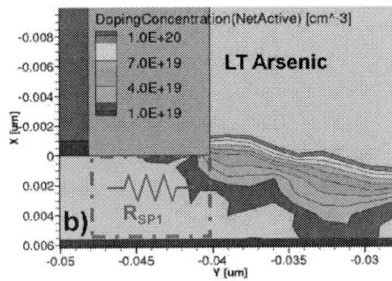

**Fig.7b:** Dopant concentration under Spacer 1 for As LT device

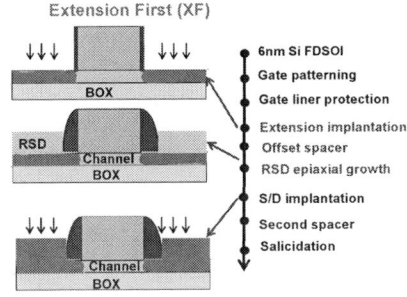

**Fig.10:** Extension First (XF) process flow. The dopants are implanted before Spacer 1 deposition within the very thin silicon film and then the RSD is grown.

**References**

[1]. P. Batude *et al.*, VLSI Technology 2009.
[2]. L Lindsay et al., J. Vac. Sci. Technol., Vol.22, No. 1, January 2004, p.306.
[3]. L. Pasini *et al.* IWJT 2014
[4]. G. Ghibaudo, Electron.Lett., vol.24, no9, pp.543-545, Apr. 1988
[5]. S. Ponoth *et al.*, SOI Conference 2011

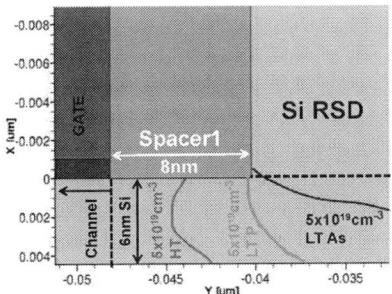

**Fig.8:** Evolution of Iso-concentration at $5 \times 10^{19}$at/cm$^3$ for each dopant of the tested devices.

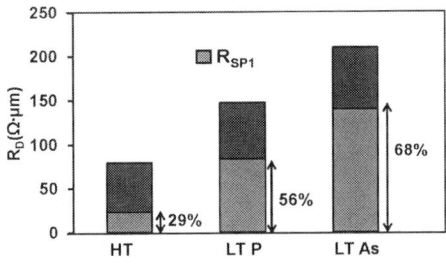

**Fig.9:** $R_{SP1}$ contribution extracted by SDEVICE simulation. The percentage of $R_{SP1}$ on total $R_D$ is also shown.

978-1-4799-7440-5/14 $31.00 © 2014 IEEE

# In depth characterization of hole transport in 14nm FD-SOI pMOS devices

M. Shin[1,3], M. Shi[1], M. Mouis[1], A. Cros[2], E. Josse[2], G.T. Kim[3] and G. Ghibaudo[1]

[1]IMEP-LAHC, Grenoble INP, Minatec, BP 257, 38016 Grenoble, France; [2]STMicroelectronics, 850, rue J. Monnet, BP 16, 38921 Crolles, France; [3]School of Electrical Engineering, Korea University, 136-701, Seoul, South Korea

## Abstract

In this paper, we studied hole transport in highly scaled (down to 14nm-node) FDSOI devices, from 77K to 300K in the coupling condition. We studied mobility enhancement by Ge% and back biasing. Then, mobility degradation in short channel devices was intensively analysed and additional scattering mechanisms were revealed in terms of their origin and location.

## Introduction

To continue downscaling of CMOS technology to 14nm node and beyond, SiGe channel has been exploited in PMOS devices thanks to superior hole transport properties and intrinsically lower threshold voltage [1]. In this paper, hole transport characterization is performed with varying Ge concentration in FDSOI devices. The scattering mechanisms were extracted from the temperature dependence of mobility and their evolution with gate length was also analysed.

## Devices under test

The devices under test were fabricated by STMicroelectronics as the 14nm node, UTBB FDSOI devices with high-k and metal gate. The body and BOX thickness were thinned down to 8nm and 25nm. For the PMOS devices, SiGe (10%, 13%, 15% and 20% of Ge) layers were fabricated by condensation method [2]. In addition, we have two different types of gate-oxide stacks (GO1/GO2). EOT were 1.8 nm/ 4.5nm for GO1/ GO2 devices because of a thicker SiON interfacial layer. On-mask gate length, $L_M$, ranged from 10 μm down to 30 nm for GO1 and from 10 μm down to 100 nm for GO2.

## Results and discussion

Fig.1 shows transfer curves for long channel device with different Ge% at 300K ($V_d$=20mV), which clearly demonstrates increase of $I_d$ depending on Ge%. In Fig. 2, we also confirmed the identical value of EOT for different devices and shift of $V_T$ according to the Ge% due to discrepancy of valence band offset by using the $C_{gc}$ and its derivative curves [3]. In the cryogenic experiments, performance enhancement by increasing Ge% is still effective at 77K (Fig. 3). To see the influence of back bias engineering, we applied voltage from 8V to -8V on the backplane (Fig. 4-5). When $V_b$ was negative, $I_d$ and $g_m$ are increased as in NMOS devices with $V_b$>0 [4]. This feature can be caused by channel centroid getting far from the high-k layer. In order to have further insight of transport properties, we extracted $\mu_{FE}$ (Eq. 1) at the max $g_m$ values and plotted them in Fig.6. Back channel exhibited the superiority of transport properties, especially for $V_b$=-6V/-8V, regardless of Ge%. In order to assess the transport properties induced by high-k layer, we used the additional mobility approach (Eq. 2) [5]. Good agreement between experiments and empirical model (Eq. 3) was obtained in Figs 7-8. We found that surface optical phonon (SOP) is predominant, which can be reduced by carrier centroid moving away from front interface proximity and that SOP is less prevailing as increasing the Ge% (Fig. 8).

In order to study the evolution of mobility as a function of channel length, we used low field mobility ($\mu_0$) with $V_b$=0 [6]. At 300K, the mobility gain with Ge% in long devices is lost in sub-100nm devices (Fig. 10). In order to have deeper insight of transport mechanisms, we characterized our devices for different temperature and gate length as comparing with empirical model (Eq. 4) for investigating dominant scattering contributions (Fig. 11) [7,8]. In Fig. 12, neutral defects (ND) mobility ($\mu_{nd}$) varies linearly as decreasing $L_{eff}$ below 1μm, which can be originated from S/D processes [4]. Unlike NMOS case, $\mu_{nd}$ depends on the gate stack layers nature and thickness [4]. Also, Phonon scattering increases in short channel devices, which might be due to modification of phonon interaction close to S/D regions. In terms of $\mu_{nd}$ benchmarking with different Ge% (Fig. 13), $\mu_{nd}$ is inversely proportional to Ge% and low as compared to NMOS case.

Finally, we applied Eq. 4 to short channel devices ($L_{eff}$=15nm) with coupling conditions and extracted $\mu_{nd}$ (Fig. 14). Unlike NMOS devices [4], it is more sensitive to $V_b$, so it shows surface-like behavior, which means that ND centers could be located at front interface proximity.

## Conclusions

The hole transport of SiGe FDSOI devices were investigated under low temperature in the coupling condition. In long channel devices, we demonstrated that back bias engineering and Ge incorporation influence transport properties. Also, SOP from high-k layer is prevailing for $\mu_{add}$. Mobility degradation below 100nm was clearly attributed to NDs induced by S/D formation. We also confirmed that ND density is proportional to Ge% and that ND could be denser close to front interface.

## Acknowledgements

This work was partly supported by REACHING 22 CATRENE European project, Places2Be ENIAC project and by the National Research Foundation of Korea (NRF) funded by the Ministry of Education, Science and Technology (Global Frontier Research Program, No. 2011-0031638).

978-1-4799-7440-5/14 $31.00 © 2014 IEEE

## Equations

$$(1)\ \mu_{FE} = \frac{g_m L}{W C_{ox} V_{ds}}$$

$$(2)\ \frac{1}{\mu_{add}} = \frac{1}{\mu_{GO1}} - \frac{1}{\mu_{GO2}}$$

$$(3)\ \frac{1}{\mu} = \frac{T}{300\,\mu_{ph}} + \frac{300}{T\,\mu_C}$$

$$(4)\ \frac{1}{\mu} = \frac{T}{300\,\mu_{ph}} + \frac{1}{\mu_{nd}} + \frac{1}{\mu_{bal}}$$

$$\mu_{bal} = \frac{q \cdot v_T \cdot L}{2kT}$$

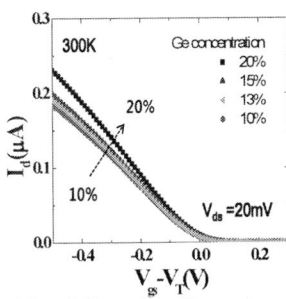

**Fig. 1** $I_d$ vs $V_{gs}$ for various Ge%. ($V_d$=20mV, 300K, GO1, L=10µm, W=1µm).

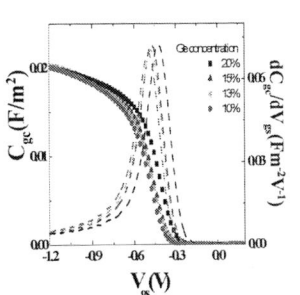

**Fig. 2** $g_m$ and $dg_m$ vs $V_{gs}$ for various Ge%. ($V_d$=20mV, 300K, GO1, L=10µm, W=1µm).

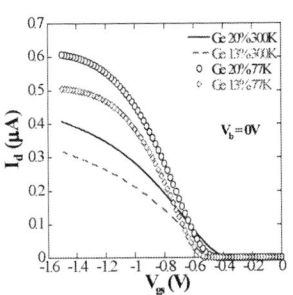

**Fig. 1** $I_d$ vs $V_{gs}$ for various Ge% and temperature. ($V_d$=20mV, GO1, L=10µm, W=1µm).

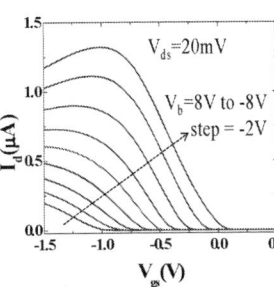

**Fig. 4** $I_d$ vs $V_{gs}$ with coupling condition from 8V to -8V ($V_d$=20mV, 77K, GO1, L=10µm, W=1µm).

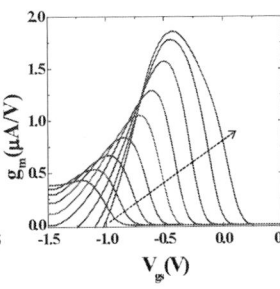

**Fig. 5** $g_m$ vs $V_{gs}$ with coupling condition from 8V to -8V ($V_d$=20mV, 77K, GO1, L=10µm, W=1µm).

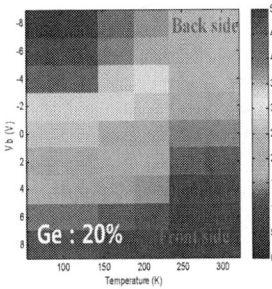

**Fig. 6** 2D-maping of $\mu_{FE}$ with different temperatures and back gates (Ge: 20%).

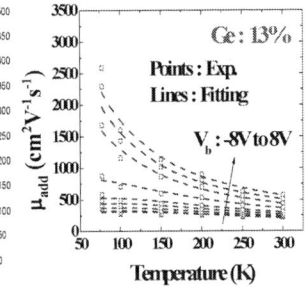

**Fig. 7** $\mu_{add}$ vs temperature with coupling condition ($V_d$=20mV, GO1, L=10µm, W=1µm).

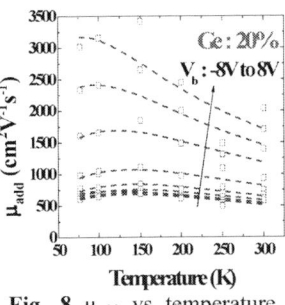

**Fig. 8** $\mu_{add}$ vs temperature with coupling condition ($V_d$=20mV, GO1, L=10µm, W=1µm).

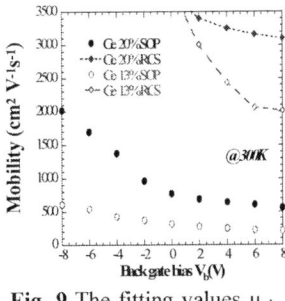

**Fig. 9** The fitting values $\mu_{ph}$ and $\mu_C$ represent the extracted values of SOP and RCS at 300K.

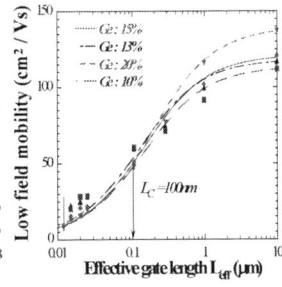

**Fig. 10** Evolution of $\mu_0$ as a function of $L_{eff}$ for devices with various Ge%.

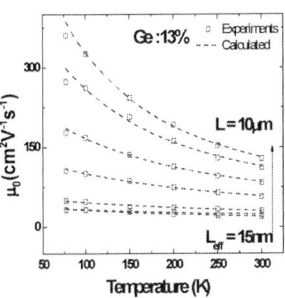

**Fig. 11** Evolution of $\mu_0$ as a function of temperature with various $L_{eff}$.

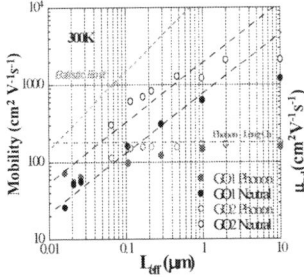

**Fig. 12** Extracted $\mu$ with different dominant scattering contributions vs. $L_{eff}$ at 300K.

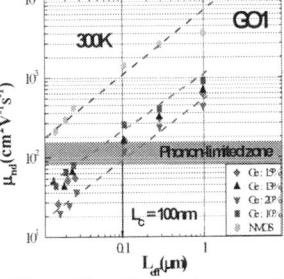

**Fig. 13** Comparison of extracted $\mu_{nd}$ for different Ge% and NMOS as a function of $L_{eff}$.

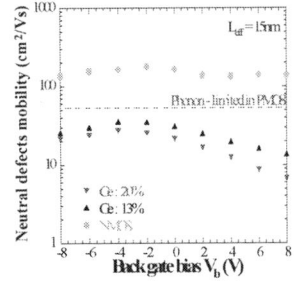

**Fig. 14** Comparison of extracted $\mu_{nd}$ for different Ge% and NMOS as a function of $V_b$ (W=1µm and $L_{eff}$=15nm).

## References

**[1]** Q. Liu, IEEE Int. Electron Devices Meet., 9.2.1–9.2.4, (2013). **[2]** T. Tezuka, Jpn. J. Appl. Phys., 40, 2866 (2001) **[3]** M. Rieger Phys. Rev. B, 48, 19, 14276 (1993). **[4]** M. Shin, Ultimate Integration On Silicon (ULIS), Stockholm (2014). **[5]** M. Casse, IEEE Trans. Electron Devices, 53, 759 (2006). **[6]** G. Ghibaudo, Electron. Lett., 2, 543–545 (1988). **[7]** A. Cros, IEDM Tech. Dig.,1 (2006). **[8]** G. Ghibaudo, Semiconductor-On-Insulator Materials for Nanoelectronics Applications, Springer (2011).

# Influence of Underlap on UTBB SOI MOSFETs in Dynamic Threshold Mode

K. R. A. Sasaki[1], M. Aoulaiche[2], E. Simoen[2], C. Claeys[2,3] and J. A. Martino[1]

[1]LSI/PSI/USP, University of Sao Paulo, Sao Paulo, Brazil,

[2]imec, Leuven, Belgium,

[3]E.E. Dept., KU Leuven, Leuven, Belgium

email: katia@lsi.usp.br

*Abstract* - **This paper discusses the influence of extensionless lengths (0nm-self aligned, 15nm and 20nm) on UTBB (Ultra-Thin-Body-and-Buried oxide) SOI (Silicon-On-Insulator) devices operating in conventional ($V_B$=0V), Dynamic Threshold (DT2, where $V_B$=$V_G$) and enhanced Dynamic Threshold (eDT, where $V_B$=$kV_G$) modes. The extensionless device of 20nm (underlap between gate and source/drain) presents better SS (Subthreshold Swing), DIBL (Drain Induced Barrier Lowering), GIDL (Gate Induced Drain Leakage), transistor efficiency ($gm/I_D$), $V_{EA}$ (Early Voltage) and $A_V$ (Intrinsic Voltage Gain). A large improvement was also observed experimentally when these devices operate under DT2 and eDT modes thanks to better coupling between the front and back gates, except for the GIDL that degrades due to a higher tunneling current near the drain caused by the higher transversal electric field.**

## I. Introduction

UTBB SOI devices have been studied as a promising candidate for sub 28nm technology node, since the better electrostatic coupling between the front and back gate bias provides a better control of short channel effects (SCE), while preserving the planar structure [1-4]. It also presents a better threshold voltage ($V_T$) control by the back gate bias ($V_B$) [1, 4-6], which can be improved by adding the ground plane (GP) implantation below the buried insulator. Moreover, extensionless devices, also called underlap devices, have demonstrated a better subthreshold behavior and SCE for low voltage applications [4] as well as for analog and memory applications [7-10].

Another possible improvement of UTBB SOI devices for low voltage applications can be obtained using the new generation of dynamic threshold voltage configuration (DT2), where the back-gate is tied to the gate ($V_B$=$V_G$) [3, 13]. The effect is similar to what is observed for the original dynamic threshold voltage technique (DTMOS) proposal on PD SOI MOSFET [11], where the body (channel) is tied to the front gate terminal. This makes the threshold voltage a function of the front gate voltage ($V_G$), which reduces $V_T$ while $V_G$ increases and improves the device performance [11].

In this paper, the impact of the length of the underlap (including self-aligned devices) on UTBB SOI MOSFETs is analyzed when submitted to a conventional and a dynamic threshold voltage (DT2) configuration. In order to enhance the back gate influence on $V_T$, a back gate bias is also used as a multiple value of the front gate voltage $V_B$=$k*V_G$ [3, 13], for $k$>1, that will be called in this paper the eDT mode.

## II. Device Characteristics

Figure 1 represents an UTBB SOI in DT2 operation mode. The measured devices have channel dimensions (W, $L_{eff}$, $t_{Si}$) of 1μm, 70nm and 14nm respectively. The gate stack consists of a TiN metal gate deposited on 5nm $SiO_2$ and the buried oxide thickness ($t_{BOX}$) is 18nm. The channel doping level ($N_A$) is $1\times10^{15}$ cm$^{-3}$. For the standard devices, extensions were fabricated by an As-implantation and 30nm-wide nitride spacers. For the extensionless ones, 15nm and 20nm-wide nitride spacers were used in order to form the underlap regions. Below the buried oxide there is a p-type ground plane, which is used as a back gate. More process information can be found in [12].

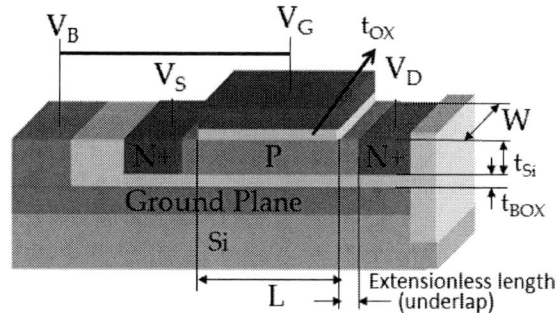

**Fig. 1.** DT2-UTBB SOI operation mode scheme.

## III. Results and Analysis

### A. Digital Performance

Figure 2 shows the SS as a function of k values for self-aligned and extensionless (15nm and 20nm) devices. The lower SS for extensionless devices is due to the longer effective channel length in weak inversion [9,10]. Together with this improvement, there is also an SS reduction when the k-factor is increased due to the stronger DT effect (the dynamic reduction of the $V_T$ due to the $V_B$ increasing, during the $V_G$ sweep).

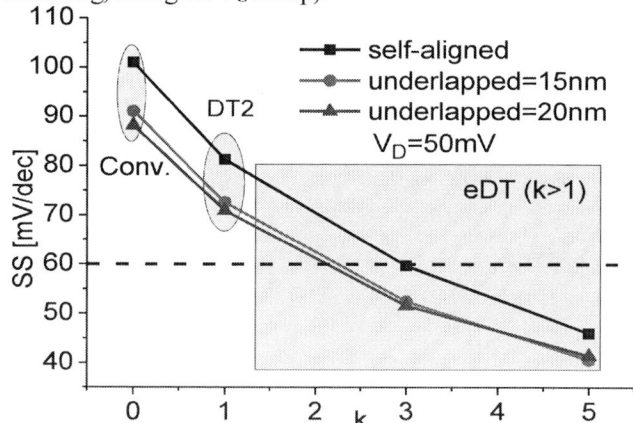

**Fig. 2.** Subthreshold swing as a function of k value for self-aligned and extensionless (15nm and 20nm) devices.

The maximum transconductance ($gm_{max}$) is shown as a function of the k-factor for self-aligned and for both extensionless devices in figure 3. A higher total resistance ($R_T$) for extensionless devices (figure 4) can explain the lower transconductance obtained in these devices. Regarding the k values, the improvement (higher $gm_{max}$ and lower total resistance) observed can be explained by the DT effect, which means a higher current flowing for the same $V_G$ and drain bias ($V_D$).

978-1-4799-7440-5/14 $31.00 © 2014 IEEE

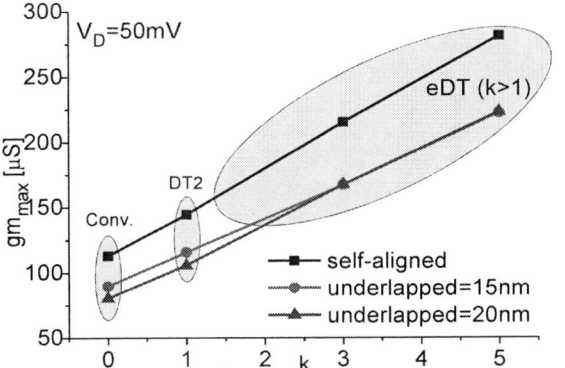

**Fig. 3.** Maximum transconductance as a function of k value for self-aligned and extensionless (15nm and 20nm) devices.

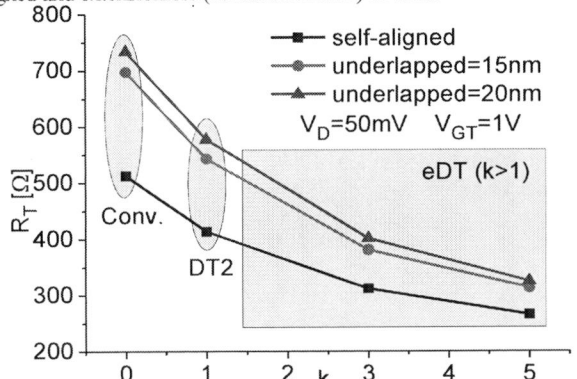

**Fig. 4.** Total resistance as a function of k value for self-aligned and extensionless (15nm and 20nm) devices.

In figure 5, one can observe the threshold voltage reduction when the k factor is increased for self-aligned and extensionless devices. A lower $V_T$ was obtained for higher k, since the $V_T$ reduces with higher $V_B$. Furthermore, the lower lateral electric field of the extensionless devices allows a better front/back gate coupling, which results in a stronger DT effect and, therefore, a further reduced $V_T$. It is important to emphasize a $V_T$ lower than 0.35V for the eDT mode, which implies that it allows application for a supply voltage of 0.7V, considered as a requirement for the future technology node [4].

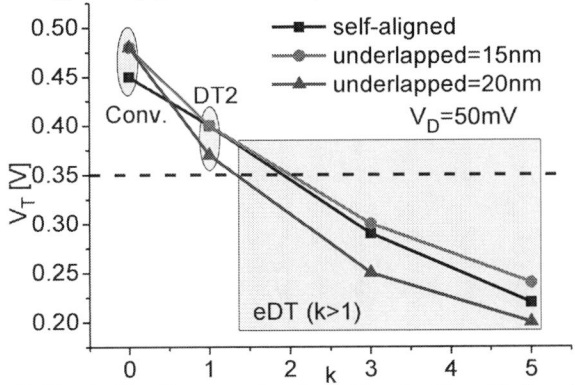

**Fig. 5.** Threshold voltage as a function of k value for self-aligned and extensionless (15nm and 20nm) devices.

The DIBL is shown in figure 6 for various k-factors and different devices. DIBL values were lower than 100mV/V for all k values in extensionless devices and the DT effect is stronger for a longer underlap region (20nm). Thus, thanks to their lower lateral electric field and superior front/back gate control, the extensionless devices

operating in DT, and even more in eDT mode, present better DIBL.

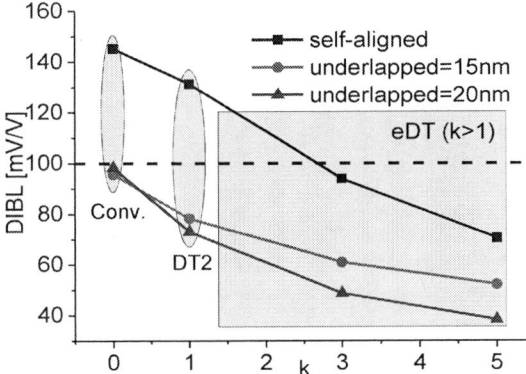

**Fig. 6.** Drain induced barrier lowering as a function of k value for self-aligned and extensionless (15nm and 20nm) devices.

Figure 7 presents the GIDL as a function of the k values and for standard and extensionless junctions. In spite of the GIDL leakage current increase for DT2 and eDT operation mode, due to the higher tunneling near the drain caused by the higher transversal electric field, the extensionless devices present better values thanks to a lower drain electric field coupling into the channel.

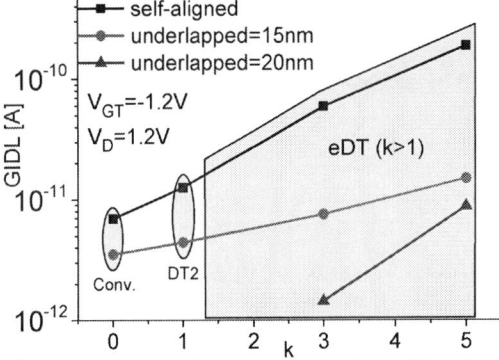

**Fig. 7.** Gate induced drain leakage as a function of k value for self-aligned and extensionless (15nm and 20nm) devices.

### B. Analog Performance

Figure 8 shows the efficiency of the transistor, i.e., the $gm/I_D$ for various k values and drain configurations. Since the stronger DT effect in extensionless devices resulted in lower SS (figure 2), these devices also present a greater $gm/I_D$ ratio in weak inversion since it is inversely proportional to the SS value.

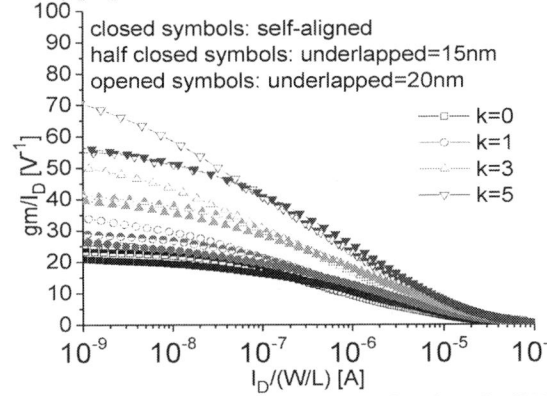

**Fig. 8.** Efficiency of the transistor ($gm/I_D$) as a function of $I_D/(W/L)$ for various k value and for self-aligned and extensionless (15nm and 20nm) devices.

The Early voltage ($V_{EA}$) can be observed in figure 9 for various k values and drain structures. Due to the underlap regions, the lateral electric field in these devices is lower and, therefore, a higher $V_{EA}$ can be achieved. Besides, the stronger DT effect for higher k values improves the front/back gate coupling, weakening the influence of the lateral field. Thus, a higher $V_{EA}$ can be obtained for eDT mode.

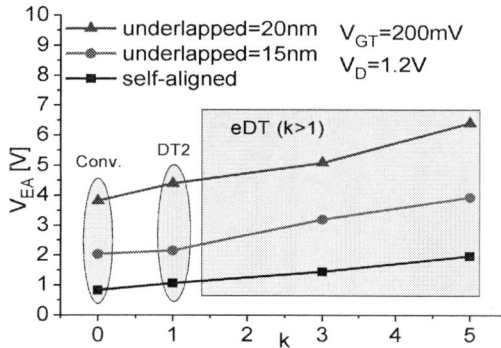

**Fig. 9.** Early voltage as a function of k value for self-aligned and extensionless (15nm and 20nm) devices.

Figure 10 shows the output conductance (left axis) and the transconductance in saturation regime (right axis) as a function of the k values for self-aligned and extensionless devices (15nm and 20nm). These parameters were used to obtain the intrinsic voltage gain ($|A_V| = gm/g_D$), which can be seen in figure 11, in V/V (left axis) and dB (right axis). From figure 11 one can see that in spite of the lower gm for extensionless devices, the improvement of $g_D$ thanks to a better front to back gate coupling and, consequently, lower drain electric field influence, results in a higher $|A_V|$. The behavior is further improved using DT2 and eDT operation mode mainly on 20nm underlapped devices.

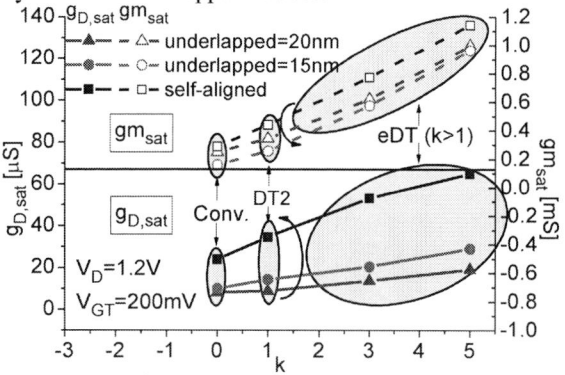

**Fig. 10.** Output conductance as a function of k value for self-aligned and extensionless (15nm and 20nm) devices.

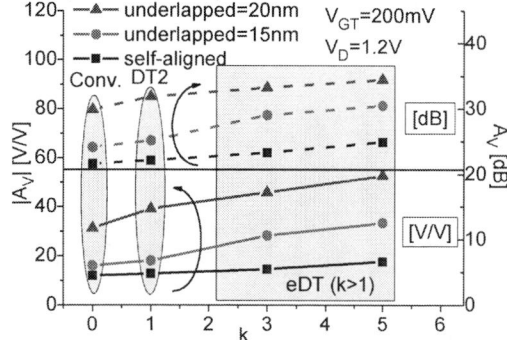

**Fig. 11.** Intrinsic voltage gain as a function of k value for self-aligned and extensionless (15nm and 20nm) devices.

## IV. Conclusions

This work shows a comparison of the digital and analog behavior of self-aligned and extensionless UTBB SOI in the conventional ($V_B$=0V), dynamic threshold (DT2) and enhanced (eDT) operation mode. The best results were obtained for 20nm extensionless eDT-UTBB devices where SS≅41mV/dec, $V_T$=0.2V, DIBL≅38mV/V, gm/$I_D$≅70V$^{-1}$ (in the weak inversion), $V_{EA}$≅6.4V, $|A_V|$≅52V/V or 34dB for a transistor channel length of 70 nm. The performance enhancement for 20nm underlap devices is also shown to be more susceptible with the increase of the k-factor. Therefore, the extensionless device in eDT operation presented the best subthreshold behavior and analog performance for low voltage applications. The only drawback for devices operating in DT2 and eDT modes was observed for the GIDL parameter that degrades due to the higher tunneling current near the drain. However, the extensionless devices presented better GIDL compared to the self-aligned one thanks to the lower drain electric field, which compensates this drawback.

## Acknowledgments

The authors acknowledge FAPESP, CAPES, CNPq and FWO for the financial support.

## References

[1] C. Fenouillet-Beranger et al., "FDSOI devices with thin BOX and ground plane integration for 32nm node and below", *Solid-State Electronics*, 53, p.730, 2009

[2] M. Fujiwara et al., "Impact of BOX scaling on 30nm gate length FD SOI MOSFETs", *Proc. IEEE Int. SOI Conf.*, p.180, 2005

[3] V. Kilchytska et al., "On the UTBB SOI MOSFET performance improvement in quasi-double-gate regime", *ESSDERC.*, p.246, 2012

[4] B.-Y. Nguyen and C. Maleville, *Advanced Substrate News*, 2014, in: www.advancedsubstratenews.com/2014/03/fd-soi-back-to-basics-for-best-cost-energy-efficiency-and-performance.

[5] R. Yan et al., "LDD depletion effects om thin-BOX FDSOI devices with a ground plane", *Proc. IEEE Int. SOI Conf.*, p.1, 2009. DOI: 10.1109/SOI.2009.5318771

[6] T. Ohtou, et al., "Variable-body-factor SOI MOSFET with ultrathin buried oxide for adaptive threshold voltage and leakage control", *IEEE Electron Device Lett.*, vol. 55, p. 40, 2008

[7] V. Trivedi et al., "Nanoscale FinFETs with gate-source/drain underlap", *IEEE Trans. Electron Devices*, v.52, p.56, 2005

[8] K.-W. Song et al., "55nm capacitorless IT-DRAM cell transistor with non-overlap structure", *IEDM Tech. Digest*, 1, 2008

[9] T. Nicoletti et al., "The impact of gate length scaling on UTBOX FDSOI devices: the digital/analog performance of extension-less structures", *Proc. of 13th Ultimate Integration on Silicon*, 121, 2012

[10] J.G. Fossum et al., "Physical insights on design and modeling of nanoscale FinFETs", *IEDM Tech. Digest*, 29.1.1, 2003

[11] J. P. Colinge, *IEEE Trans. Electron Devices*, 44, 845, 1987

[12] N. Collaert et al., "Analysis of sense margin and reliability of 1T-DRAM fabricated on thin-film UTBOS substrates", *Proc. IEEE Int. SOI Conf.*, p.1, 2009. DOI: 10.1109/SOI.2009.5318781

[13] K.R.A.Sasaki et al., "Ground plane influence on Enhanced Dynamic Threshold UTBB SOI nMOSFETs", *Proc. IX ICCDCS Conf.*, p.57, 2014

# Near-0.1V Ultra-low Voltage Operation of SOTB 1M Logic Gates

Yasuhiro Ogasahara, Masakazu Hioki, Tadashi Nakagawa, Toshihiro Sekigawa,
Toshiyuki Tsutsumi, and Hanpei Koike,
Nanoelectronics Research Institute, AIST, Tsukuba 305-8568, Japan
E-mail:ys.ogasahara@aist.go.jp

*Abstract*—**This paper demonstrates the near-0.1V minimum operation voltage of SOTB process for one million logic gates on silicon. In a circuit simulation, the lowest energy/cycle is estimated to be obtained at near 0.1V when near- or sub-60mV/dec. SS transistors are introduced, and lowering minimum operation voltage of logic circuit will be more important from the viewpoint of obtaining high energy efficiency. The variability is the main obstacle of low voltage operation, and lowering operation voltage to near 0.1V is difficult. Comparison of measurement results between 65nm SOTB and bulk processes indicate that the low-variability SOTB process notably lowers the minimum operation voltage to near 0.1V where high energy efficiency is obtained with near- or sub-60mV/dec. SS transistors in simulation.**

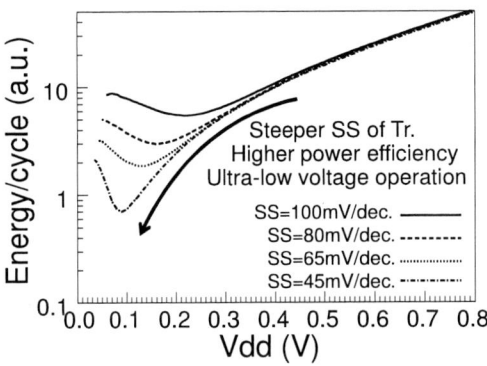

Fig. 1.   Energy efficiency simulation results of inverter chains for SS=100, 80, 65, 45mV/dec..

## I. INTRODUCTION

The near- or subthreshold designs bring the high energy efficiency by lowering operation voltage of the circuit, and can be applied to low power sensor nodes and so on [1]. On the other hand, several new structure FETs like FinFET, GAA(Gate All Around), Tunnel FET, and so on, [2], [3], are discussed for further advancement of CMOS processes. These transistors have near- or sub-60mV/dec. SS (subthreshold slope), and reduce leakage current at a low voltage. Figure 1 [4] shows the energy/cycle of a 54-stages inverter chain in several SS conditions which we simulated based on ref. [5]. SS of planer CMOS process is about 100mV/dec., and the minimum energy/cycle is obtained at about 0.3V [5]. On the other hand, 3X and 8.5X energy efficiency improvement is obtained at near 0.1V in 65mV/dec. and 45mV/dec conditions because of reduction of the leakage current. When SS of transistors is improved, the minimum energy point is lowered, and lowering operation voltage will be more effective for obtaining high energy efficiency in future processes.

However, a low-voltage operation is restricted by variability in transistor performances, and an impact of a local random variation is serious in large-scale circuits [5]. Reference [6] reported the minimum operation voltage of ring oscillators (RO) in a 90nm bulk process. Though minimum operation voltage of 11 stages RO was 90mV, that of 1M stages RO reached 343mV. As for SRAM, it has relatively high minimum operation voltage, and methods for lowering operation voltage were frequently discussed. Conventional 6T-SRAM has about 0.8V minimum operation voltage [7], and 7T [8], 8T [9], 10T [10], and improved-6T [11] SRAMs operated at 0.44V, 0.25V, 0.16V, and 0.208V respectively.

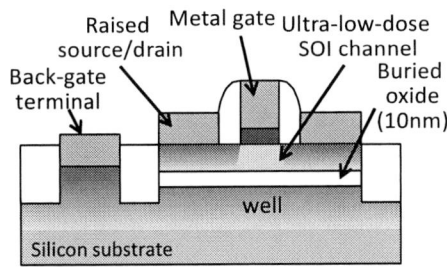

Fig. 2.   Schematic cross section of SOTB transistor.

This paper is a primary report of minimum operation voltage of the million-gate logic circuit in SOTB (Silicon on Thin Buried Oxide) [12] process. SOTB (Fig. 2) notably reduces random dopant variability because of an ultra-low dose channel, and lowering operation voltage is expected. Though 0.37V minimum operation voltage of 6T-SRAM in SOTB process is reported [7], we focus on a million-gate logic circuit. We measure 1011 ROs with 1001-stages on test chips fabricated in 65nm SOTB and bulk processes, and the notable impact of the low-variability SOTB process on low-voltage operation is demonstrated.

## II. TEG IMPLEMENTATION

We designed 337 patterns of ROs and 3 ROs for each patterns were located on a chip. Each RO consists of 1000 stage measurement target gates and 1 NAND gate for controlling

978-1-4799-7440-5/14 $31.00 © 2014 IEEE

TABLE I
NUM. OF DESIGNED RO PATTERNS. 3 ROs FOR EACH RO DESIGN WERE
PLACED ON A CHIP.

| Included | # of RO designs containing | | |
|---|---|---|---|
| gate(s) | 1 gate | 2 gates | 3 gates |
| INV | 14 | 43 | 20 |
| NAND | 21 | 40 | - |
| NOR | 18 | 40 | - |
| INV&NAND | - | 45 | - |
| INV&NOR | - | 45 | - |
| NAND&NOR | - | 51 | - |

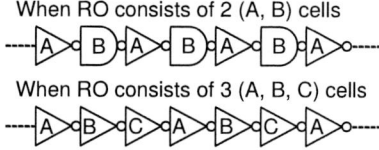

Fig. 3. The gate connection in RO when RO includes two or three cell patterns.

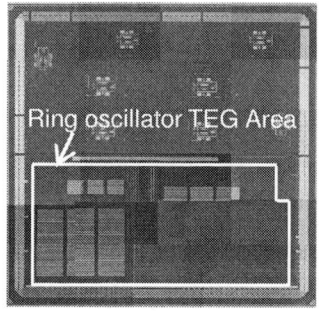

Fig. 4. A micrograph of the test chip.

oscillation. The RO pulse is counted by the counter on the chip. To measure ultra-low voltage operation of ROs, we implemented three power supply and ground lines pairs for ROs, the counter, and the buffers. The buffers are located between ROs and the counter. Table I describes the implemented ROs. Nominal design of INV, NAND, and NOR cells has 1:1, 1:1, and 2:1 P:N ratios respectively.

The RO containing only 1 cell design may results in too optimistic low voltage performance. 53 patterns of ROs include single cells and 284 patterns are composed of two or three cells. Figure 3 depicts connection of gates when RO include two or three patterns of gates.

Complex gates in standard cells frequently include serially stacked transistors, transmission gates, or unbalanced P:N ratio. Impacts of these designs are different from those of MOS W change. This RO TEG is combined with TEG for evaluation of layout dependence, and INV, NAND, and NOR cells, whose nMOS or pMOS width, width/finger, or STI conditions were modified, were designed for evaluation of narrow channel and STI stress effects. These effects can cause $V_{th}$ or mobility variation, and will be beneficial for estimating low voltage behavior of logic circuits.

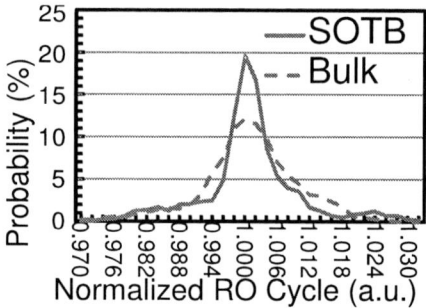

Fig. 5. Measured RO cycles on SOTB and bulk chips at 0.4V and 0.55V respectively.

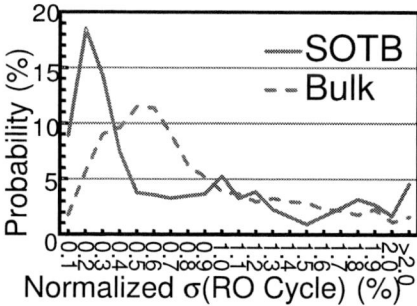

Fig. 6. Standard deviations of cycle of each RO pattern on each chip.

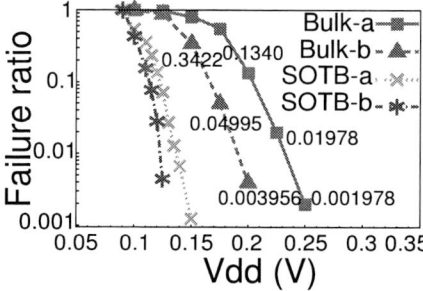

Fig. 7. Measurement results of the RO operation. Y-axis is the ratio of the number of ROs which failed to operate.

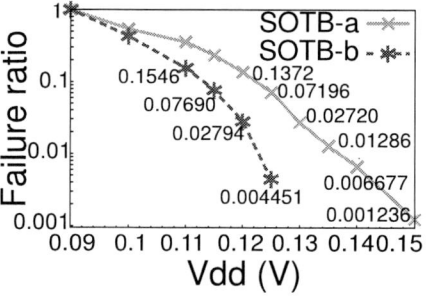

Fig. 8. Measurement results of SOTB process at $V_{dd}$=0.09V-0.15V.

The test chips were fabricated both in 65nm SOTB and bulk processes. Figure 4 shows the micrograph of the fabricated test chip. The size of the test chips is 5.8x5.8mm, and about 5.0mmx2.5mm area is allocated for the RO TEG.

## III. VARIABILITY IN SOTB AND BULK PROCESSES

Figure 5 shows a histogram of measured RO cycles at 0.4V(SOTB) and 0.55V(bulk) where average of RO cycles are about 500ns. RO cycles were normalized by average of RO cycle of each pattern. Distribution of RO cycles was more concentrated on the average cycle in SOTB process than that in bulk process. We also calculated standard deviations ($\sigma$) of RO cycle for each of 337 ROs patterns on each chip, and Fig. 6 is the histogram of calculated $\sigma$. The peaks of $\sigma$(RO cycle) distribution of SOTB and bulk are 0.2% and 0.5% respectively. Both of Figs. 5 and 6 indicated lower variability of SOTB process than bulk process, and the random variability was effectively suppressed in SOTB process.

## IV. LOW VOLTAGE OPERATION

1011 ROs with 1001-stages on the test chips fabricated both in bulk and SOTB processes were measured. We measured 4 SOTB chips and 4 bulk chips. At near $V_{dd}$ =0.1V, the margin of $V_{th}$ mismatch is very small, and even the inevitable slight $V_{th}$ mismatch seriously affects. We applied backgate biasing to compensate $V_{th}$ mismatch, and PMOS and nMOS backgate bias voltages, VBP, VBN, are set to be following conditions.

Bulk-a: $VBP = V_{dd}, VBN = 0V$
Bulk-b: $VBP = V_{dd} + 1.5V, VBN = 0V$
SOTB-a: $VBP = V_{dd} + 0.8V, VBN = -0.8V$
SOTB-b: $VBP = V_{dd} + 1.0V, VBN = -0.6V$

Figures 7 and 8 show the measurement result of the RO operation. X-axis is a supply voltage and y-axis is ratios of the number of ROs which failed to operate. The yield of 10k-gate and 100k-gate circuit can be calculated as $(1 - Y)^{10}$ and $(1 - Y)^{100}$ where Y is failure ratio. When the backgate bias voltages were equally applied for pMOS and nMOS, the first failure were observed at 0.25V and 0.15V for bulk and SOTB respectively. The failure rates at this voltage enables practical yield of a 100k-gate circuit. 0.072 failure ratio of SOTB at 0.125V corresponds to about 47.3% yield of 10k-gate circuit, and 0.225V is required for the bulk process to obtain this yield. Adjusting the backgate bias lowered the first failure voltage to 0.2V and 0.125V for bulk and SOTB respectively. The over 40% yield for a 10k-gate circuit is achieved at 0.115V in SOTB process while 0.175V is required for bulk process.

These measurement results validated that the low-variability of SOTB notably lowers the minimum operation voltage of the large scale logic circuit, and the low-variability of SOTB technology enables near-0.1V operation of logic circuits. At near 0.1V, margin against $V_{th}$ mismatch is seriously small in comparison with 0.2V- to 0.4V-class low voltage operation. Even the $V_{th}$ control by the foundries for low-voltage operation will not sufficiently eliminate the $V_{th}$ mismatch. Measurement results also indicated that backgate biasing is effective for obtaining operation margin in these extremely low-voltage and small-$V_{th}$-margin operation.

## V. CONCLUSION

We implemented the RO TEG chips both in 65nm SOTB and bulk processes, and measured the minimum operation voltages of the logic gates. The SOTB process achieved fine yield down to 0.15V-0.11V depending on circuit size and backgate bias conditions while bulk process required 0.25-0.175V supply voltages. The low variability of the SOTB process significantly contributed to lowering the minimum operation voltage of the logic circuit. The minimum operation voltage achieved by SOTB process is close enough to the voltage where minimum energy/cycle is obtained with near- or sub-60mV/dec. SS transistors in simulation. Measurement results also indicated that applying backgate bias voltage is effective for the small $V_{th}$ mismatch margin at ultra-low voltage, which is beyond the general process control.

## ACKNOWLEDGMENT

A part of this work was performed as "Ultra-Low Voltage Device Project" funded and supported by the Ministry of Economy, Trade and Industry (METI) and the New Energy and Industrial Technology Development Organization (NEDO). A part of the device processing was operated by the SCR Operation Office, the National Institute of Advanced Industrial Science and Technology (AIST), Japan.

## REFERENCES

[1] H. Fuketa, T. Yasufuku, S. Iida, M. Takamiya, M. Nomura, H. Shinohara, and T. Sakurai, "Device-Circuit Interactions in Extremely Low Voltage CMOS Designs," in *IEEE IEDM Tech. Dig*, pp 559–562, Dec. 2011.

[2] M. Bohr, "The New Era of Scaling in an SoC World," *Dig. of Tech. Papers of IEEE ISSCC*, pp. 23–28, Feb. 2009.

[3] M. Bohr, "The Evolution of Scaling from the Homogeneous Era to the Heterogeneous Era," in *Proc. IEEE IEDM*, pp. 1–6, Dec. 2011.

[4] Y. Ogasahara, C. Ma, M. Hioki, T. Nakagawa, T. Sekigawa, T. Tsutsumi, H. Koike, M. Tada, and T. Sakamoto, "Utility of High On-off ratio, High Off Resistance Rewritable Device to EEPROM for Ultra-low Voltage Operation of Steep Subthreshold Slope FETs," in *Proc. IEEE IMW*, pp 111–114, May 2014.

[5] T. Sakurai, "Designing Ultra-Low Voltage Logic," in *Proc. IEEE/ACM ISLPED*, pp. 57–58, Aug. 2011.

[6] T. Niiyama, Z. Piao, K. Ishida, M. Murakata, M. Takamiya, and Takayasu Sakurai, "Increasing Minimum Operating Voltage (VDDmin) with Number of CMOS Logic Gates and Experimental Verification with up to 1Mega-Stage Ring Oscillators," in *Proc. IEEE/ACM ISLPED*, pp. 117–122, Aug. 2008.

[7] Y. Yamamoto, H. Makiyama, H. Shinohara, T. Iwamatsu, H. Oda, S. Kamohara, N. Sugii, Y. Yamaguchi, T. Mizutani, and T. Hiramoto, "Ultralow-Voltage Operation of Silicon-on-Thin-BOX (SOTB) 2Mbit SRAM Down to 0.37 V Utilizing Adaptive Back Bias," *Dig. of Tech. Papers of Symp. on VLSI Tech.*, pp. 212–213, June 2013.

[8] K. Takeda, Y. Hagihara, Y. Aimoto, M. Nomura, Y. Nakazawa, T. Ishii, and H. Kobatake, "A Read-Static-Noise-Margin-Free SRAM Cell for Low-VDD and High-Speed Applications," *IEEE JSSC*, vol. 41, no. 1, pp. 113–121, Oct. 2008.

[9] M. E. Sinangil, N. Verma, and A. P. Chandrakasan, "A Reconfigurable 8T Ultra-Dynamic Voltage Scalable (U-DVS) SRAM in 65 nm CMOS," *IEEE JSSC*, vol. 44, no. 11, pp. 3163–3173, Nov. 2009.

[10] I. J. Chang, J.-J. Kim, S. P. Park, and K. Roy, "A 32kb 10T Sub-Threshold SRAM Array With Bit-Interleaving and Differential Read Scheme in 90nm CMOS," *IEEE JSSC*, vol. 44, no. 2, pp. 650–658, Feb. 2009.

[11] B. Zhai, S. Hanson, D. Blaauw, and D. Sylvester, "A Variation-Tolerant Sub-200mV 6-T Subthreshold SRAM," *IEEE JSSC*, vol. 43, no. 10, pp. 2338–2348, Oct. 2008.

[12] N. Sugii, R. Tsuchiya, T. Ishigaki, Y. Morita, H. Yoshimoto, and S. Kimura, "Local $V_{th}$ Variability and Scalability in Silicon-on-Thin-BOX (SOTB) CMOS with Small Random-Dopant Fluctuation," *IEEE Trans. on Electron Devices*, vol. 57, no. 4, pp. 835–845, Apr. 2010.

# Performance Prediction for Multiple-Threshold 7nm-FinFET-Based Circuits Operating in Multiple Voltage Regimes Using a Cross-Layer Simulation Framework

Shuang Chen, Yanzhi Wang, Xue Lin, Qing Xie, Massoud Pedram

University of Southern California, California, USA

{shuangc, yanzhiwa, xuelin, xqing, pedram}@usc.edu

## Abstract

Because of their many attractive attributes, FinFETs are emerging as the device of choice for CMOS process technology nodes below 20nm. This paper is the first work that investigates the effectiveness of building CMOS circuits operating in the near-threshold regime and above with 7nm FinFET technology through a cross-layer design and simulation framework. Three types of FinFET devices with different threshold voltages are designed using Sentaurus TCAD to accommodate the need for constructing both high-speed cells and low-power cells in the same library. Compact and SPICE-compatible device models are extracted with high accuracy using current source modeling techniques. Standard cell libraries with two different (near- and super-threshold) supply voltages are generated. Circuit syntheses are performed on extensive benchmarks to compare the performance with the state-of-the-art planar CMOS counterparts. Simulation results demonstrate the benefit of 7nm FinFET-based circuits from both aspects of speed and energy efficiency.

**Keywords:** FinFET, near-threshold computing, multiple-threshold devices, cross-layer simulation, current source modeling, energy consumption

## Introduction

As the geometric dimension of transistors scales down, FinFET devices are proved to better address the challenges facing conventional planar CMOS devices, e.g. high leakage power dissipation, significant short channel effect, etc [1]. Due to manufacturing limitations, deeply scaled FinFET devices below 10nm feature size have not been manufactured. Nevertheless, it is crucial to investigate the performance of such devices with lower feature sizes in order to shed some light on further studies on novel process techniques and circuit structures. A well-known predictive FinFET model, PTM-MG [2], which is based on the BSIM-MG model [3], have been proposed. In this paper, we present an alternative method based on TCAD device simulation and the *current source modeling* (CSM) technique to design and characterize FinFET devices with different threshold voltages, which is believed to be more accurate. Besides, this paper for the first time presents a cross-layer design and simulation framework in which the performance of 7nm-FinFET-based circuits is evaluated in both the near- and super-threshold regimes.

## Framework Overview

As shown in Fig. 1, the proposed design framework is mainly comprised of three parts, i.e. (1) FinFET device modeling and design, (2) compact and SPICE-compatible device model extraction, and (3) standard cell library construction and circuit synthesis results.

### A. FinFET Device Modeling and Design

We model and simulate 7nm FinFET devices using Sentaurus TCAD tools [4]. The device model is shown in Fig. 2. To mitigate the direct source-to-drain tunneling (DSDT) current [5], the gate underlap is introduced. It is worth noting that the designs of both nfets and pfets have three versions with different threshold voltages that vary in the range between 0.20V and 0.45V. The design with the lowest $V_{th}$ can be used on the critical path(s) of a circuit for high speed operation, while the designs with higher $V_{th}$'s can be used on the non-critical paths in order to reduce power dissipation. The three different $V_{th}$'s are achieved by tuning the gate workfunction of the devices. Table I shows the design parameters.

The Sentaurus device simulations apply the hydrodynamic carrier transport model, Oldslotboom bandgap narrowing model, and the density gradient quantization correction model. The carrier mobility degradation resulting from high doping, high field saturation, and scattering at silicon-insulator interfaces is also taken into account.

### B. Compact Model Extraction

We extract the compact models of FinFET devices using the CSM technique, in which an nfet or pfet is modeled as a set of current sources and parasitic capacitances. In this paper, we only consider the case in which the front gate and the back gate of a fin are tied together, with the corresponding current source model shown in Fig. 3. The terminal voltage dependent output currents and capacitance values are characterized by performing proper DC and/or transient terminal voltage sweepings similar to the method described in [6]. And finally, similar to [7], we import the extracted parameters into lookup-table-based Verilog-A models that are SPICE-compatible, for the convenience of further circuit level simulations.

### C. Standard Cell Library Construction and Circuit Synthesis

We build up the standard cell libraries in an industrial standard format, the Liberty library format (.lib), in which logic cells, either combinational or sequential, are modeled by recording the timing and power parameters under a specific supply voltage and process technology. The standard cell library can be used to synthesize arbitrary digital circuits using Synopsys Design Compiler. In this paper, we consider both the near-threshold and super-threshold computing scenarios for the multi-threshold FinFET devices, which have the supply voltage of 0.3V and 0.45V, respectively. The timing and power parameters are obtained through HSPICE simulations based on the Verilog-A models under different input and output conditions.

## Results and Discussion

Fig. 4 and Fig. 5 show the $I_d$-$V_g$ curves and the $I_d$-$V_d$ curves of three types of designs of nfets. Nfet 1 has the lowest threshold voltage, Nfet 3 has the highest threshold voltage, and Nfet 2 has a threshold voltage between the other two. For all three types of designs, the subthreshold slope is ~80mV/dec. When $V_d = V_g = 0.5V$, the drain current of Nfet 1 is 1.8x larger than Nfet 2 and 6.7x larger than Nfet 3.

Fig. 6 shows the minimum energy point (MEP) and minimum energy-delay product point (MEDP) of the FinFET designs with three different threshold voltages. We test a 20-stage inverter chain with varying activity factors. As shown in Fig. 6, the MEP lies around 0.2V, which is in the subthreshold regime, whereas the MEDP begins to occur in the near-threshold regime.

Table II shows the delay and power consumption comparison in a number of benchmark circuits using the synthesis results of our 7nm FinFET-based cell library with OSU 45nm CMOS standard cell library and NANGATE 45nm CMOS cell library as baselines. Comparing the low-threshold 7nm FinFET cell library at 0.3V with the OSU 45nm CMOS standard cell library (operating at 1.1V), we achieve a maximum of 8.7x reduction in circuit delay/clock period and maximum of 1400x reduction in energy per operation.

## Conclusion

This paper provided a 7nm FinFET standard cell library for HSPICE simulations based on device models built in Sentaurus TCAD. The simulations run on some benchmarks show significant improvement in terms of delay and energy consumption when using FinFETs instead of planar CMOS devices.

## References

[1] L. Chang, *et al*, *Proc. of the IEEE*, 2003. [2] S. Sinha, *et al*, *Proc. DAC*, 2012. [3] M. V. Dunga, *et al*, *VLSIT*, 2003. [4] [Online] http://www.synopsys.com/tools/tcad/. [5] R. A. Vega, *et al*, *IEEE TED*, 2009. [6] H. Fatemi, *et al*, *Proc. DAC*, 2006. [7] A. A. Goud, *et al*, *DRC*, 2013.

---

This research is sponsored in part by grants from the Defense Advanced Research Projects Agency and the National Science Foundation.

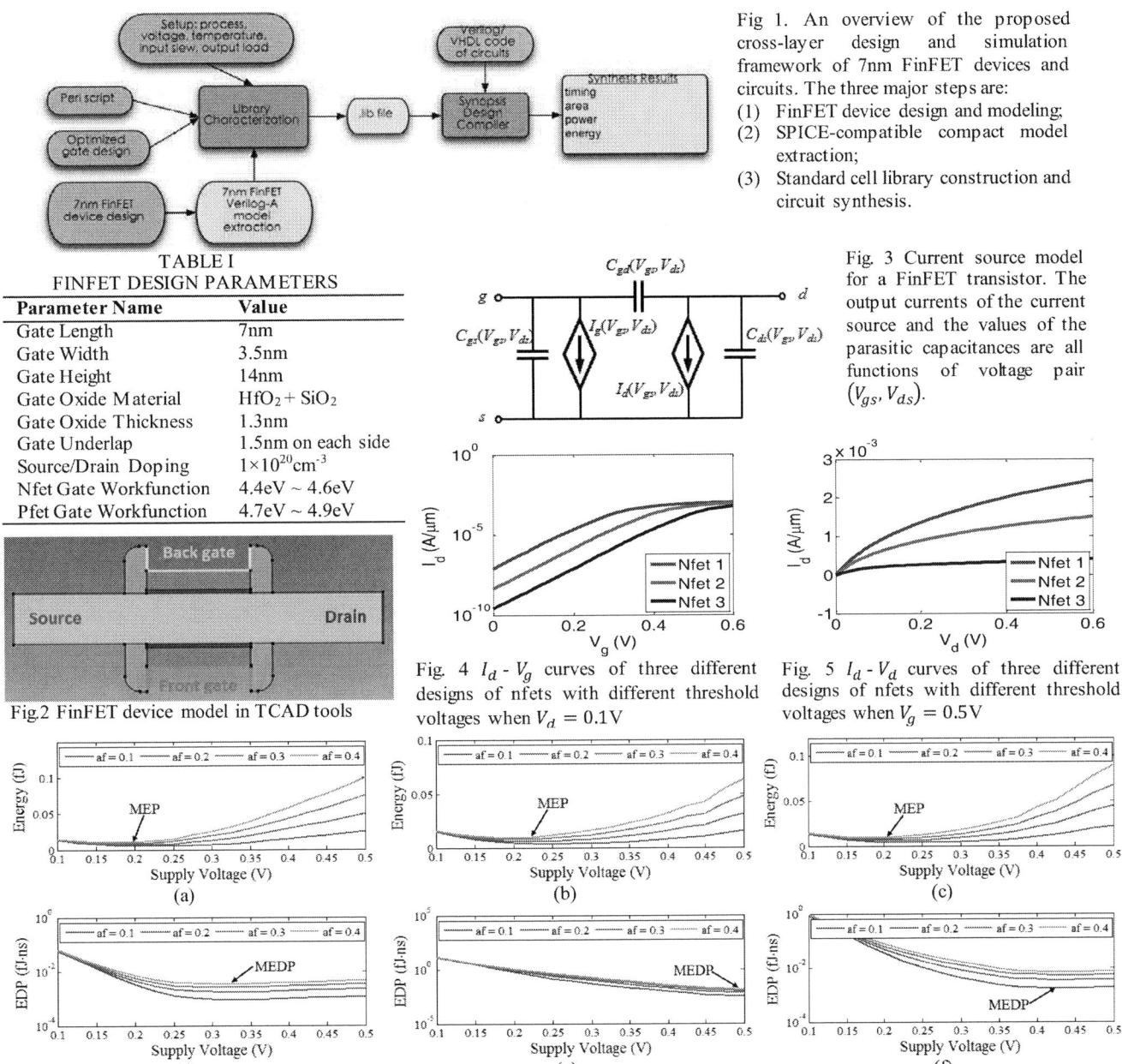

Fig 1. An overview of the proposed cross-layer design and simulation framework of 7nm FinFET devices and circuits. The three major steps are:
(1) FinFET device design and modeling;
(2) SPICE-compatible compact model extraction;
(3) Standard cell library construction and circuit synthesis.

TABLE I
FINFET DESIGN PARAMETERS

| Parameter Name | Value |
|---|---|
| Gate Length | 7nm |
| Gate Width | 3.5nm |
| Gate Height | 14nm |
| Gate Oxide Material | $HfO_2 + SiO_2$ |
| Gate Oxide Thickness | 1.3nm |
| Gate Underlap | 1.5nm on each side |
| Source/Drain Doping | $1 \times 10^{20} cm^{-3}$ |
| Nfet Gate Workfunction | 4.4eV ~ 4.6eV |
| Pfet Gate Workfunction | 4.7eV ~ 4.9eV |

Fig.2 FinFET device model in TCAD tools

Fig. 3 Current source model for a FinFET transistor. The output currents of the current source and the values of the parasitic capacitances are all functions of voltage pair $(V_{gs}, V_{ds})$.

Fig. 4 $I_d$ - $V_g$ curves of three different designs of nfets with different threshold voltages when $V_d = 0.1V$

Fig. 5 $I_d$ - $V_d$ curves of three different designs of nfets with different threshold voltages when $V_g = 0.5V$

Fig. 6 (a) ~ (c) are the Energy-Supply voltage curves under different activity factors, ranging from 0.1 to 0.4, obtained from a 20-stage inverter chain built from the low-, high-, and medium-$V_{th}$ FinFETs. (d) ~ (f) are the Energy delay product-Supply voltage curves under different activity factors of the same set of FinFET devices. The MEPs and MEDPs are marked in the figure. All MEPs are in the subthreshold region. MEDPs begin to occur in the near-threshold regime.

TABLE II BENCHMARK CIRCUIT PERFORMANCE ON OUR CELL LIBRARY AND 45NM CMOS STANDARD CELL LIBRARIES
(a) ISCAS benchmark circuit C499; (b) ISCAS benchmark circuit C3540; (c) 16-bit adder; (d) 16-bit multiplier

| Cell Library | Delay (ns) | Energy per Operation (J) | Cell Library | Delay (ns) | Energy per Operation (J) | Cell Library | Delay (ns) | Energy per Operation (J) | Cell Library | Delay (ns) | Energy per Operation (J) |
|---|---|---|---|---|---|---|---|---|---|---|---|
| Lib1 | 0.100 | 1.7507e-15 | Lib1 | 0.130 | 3.44e-15 | Lib1 | 0.150 | 407.86e-18 | Lib1 | 0.350 | 6.67e-15 |
| Lib2 | 7.500 | 1.2500e-15 | Lib2 | 11.20 | 1.96e-15 | Lib2 | 10.00 | 246.62e-18 | Lib2 | 25.00 | 4.57e-15 |
| Lib3 | 0.800 | 741.51e-18 | Lib3 | 1.000 | 2.00e-15 | Lib3 | 0.700 | 349.40e-18 | Lib3 | 2.000 | 5.74e-15 |
| Lib4 | 0.080 | 3.1732e-15 | Lib4 | 0.100 | 6.44e-15 | Lib4 | 0.070 | 1.3700e-15 | Lib4 | 0.200 | 22.6e-15 |
| Lib5 | 0.300 | 2.6187e-15 | Lib5 | 0.400 | 5.77e-15 | Lib5 | 0.300 | 895.73e-18 | Lib5 | 0.800 | 14.7e-15 |
| Lib6 | 0.100 | 3.0896e-15 | Lib6 | 0.130 | 7.12e-15 | Lib6 | 0.100 | 1.3819e-15 | Lib6 | 0.350 | 19.4e-15 |
| Base1 | 0.600 | 997.28e-15 | Base1 | 1.500 | 2.39e-12 | Base1 | 1.310 | 219.79e-15 | Base1 | 2.710 | 6.30e-12 |
| Base2 | 0.600 | 590.36e-15 | Base2 | 1.500 | 1.18e-12 | Base2 | 1.450 | 102.48e-15 | Base2 | 2.890 | 3.30e-12 |
| (a) | | | (b) | | | (c) | | | (d) | | |

Lib1 ~ Lib3 are cell libraries obtained under supply voltage of 0.3V with low-, high-, and medium-$V_{th}$ FinFETs, respectively. Lib4 ~ Lib6 are cell libraries obtained under supply voltage of 0.45V with low-, high-, and medium-$V_{th}$ FinFETs, respectively. Base1 is OSU 45nm CMOS standard cell library, and Base2 is NANGATE 45nm CMOS standard cell library. Base1 and Base2 operate at 1.1V supply voltage.

978-1-4799-7440-5/14 $31.00 © 2014 IEEE

# A Cross-Layer Design Framework and Comparative Analysis of SRAM Cells and Cache Memories using 7nm FinFET Devices

Alireza Shafaei, Shuang Chen, Yanzhi Wang, and Massoud Pedram

Department of Electrical Engineering, University of Southern California, Los Angeles CA 90089, USA

Email: {shafaeib, shuangc, yanzhiwa, pedram}@usc.edu

## I. INTRODUCTION

FinFET devices are currently viewed as the technology-of-choice beyond the 10nm regime [1]. This is mainly due to the improved gate control over the channel which makes FinFETs more immune to short channel effects. On the other hand, SRAM caches, because of occupying a large portion of the chip area, and high sensitivity to device mismatches, are considered as the major bottleneck of the $V_{dd}$ scaling [2]. Hence, FinFET-based SRAMs have emerged as a solution to a more robust and energy efficient memory design [3]. This paper thus adopts a cross-layer design framework (Fig. 1) in order to study the effect of different deeply-scaled (7nm) FinFET devices on memory designs: (1) at device-level, different FinFET devices for 7nm process are designed using TCAD tools [4], (2) at circuit-level, Verilog-A models are extracted from the device simulator for performing fast SPICE-based simulations, (3) and finally at architecture-level, the overall characteristics of an on-chip cache is assessed using a modified version of CACTI tool with FinFET support.

## II. CROSS-LAYER DESIGN FRAMEWORK

### A. Device-level Design

Since no industrial data for deeply-scaled FinFET devices exist, 7nm FinFET devices are modeled (Fig. 2) and simulated using Sentaurus TCAD tools [4]. Gate underlap is introduced to mitigate the *direct source-to-drain tunneling* (DSDT) current [5]. We develop seven different designs of 7nm FinFET devices with different parameters such as gate length $L_{FIN}$, oxide thickness $t_{ox}$, fin width $W$, and underlap length $ul$. Table 1 reports the design parameters of the baseline (standard) FinFET device, whereas Table 2 shows the design parameters of other devices with only one parameter changed per device. Based on device simulations, we also extract SPICE-compatible Verilog-A models for fast circuit-level simulations, e.g. deriving ON/OFF currents of FinFET devices, *static noise margin* (SNM), as well as other parameters for integration into architecture-level simulators. According to Fig. 3(a), the highest ON current is achieved by the *high_w* device which has a larger fin width (which means a larger effective channel width) compared with the baseline device. On the other hand, as a result of the $V_{th}$ roll-off effect, the lowest OFF current (Fig. 3(b)) and the highest ON/OFF current ratio (Fig. 3(c)) are obtained by using the *high_l* device (with a longer gate length).

### B. Circuit-level Design

FinFET devices are next incorporated into 6T and 8T [6] SRAMs in order to find a robust and functional cell under this 7nm FinFET process. Since the P-type fin is (1.6x) weaker than the N-type counterpart, we only need to increase the number of fins of pull-down transistors for the 6T cell to ensure proper operation. Therefore, 6T-*n* is used to refer to a 6T cell whose pull-down transistors have *n* fins each, where *n>1* since 6T-1 cell does not work properly in our 7nm FinFET process (because of weak pull-downs). On the other hand, 8T cell, by dedicating separate paths to read and write operations, does not need stronger pull-down transistors, and hence all transistors can be single-fin. Area (for memory density) and SNM (for robustness) of SRAMs are calculated based on cell layouts (Fig. 4, 5) and butterfly curves (Fig. 6), respectively. In general, the SNM is higher if the corresponding FinFET device has higher ON/OFF current ratio. The highest SNM is achieved by 8T cell using *high_l* devices (Fig. 7(b)) at the cost of 21% larger area compared with the smallest working 6T cell (Fig. 7(b)). The reason is higher SNM of 8T cell compared with 6T cell, and the highest ON/OFF current ratio in *high_l* devices.

### C. Architecture-level Design

In order to evaluate SRAM cells at the architecture-level, deeply-scaled FinFET devices along with FinFET models are integrated into CACTI [7], which is a widely-used cache modeling tool. For this purpose, a 4MB cache (Table 3) is assumed. Cache area and access energies do not change significantly when using different FinFET devices, and are thus omitted. Access latency is mainly determined by the ON current of the underlying device, and hence, the shortest access latency is achieved by using *high_w* devices (Fig. 8(a)). On the other hand, the OFF current of the SRAM cell is the major component of the cache leakage power. Accordingly, the *high_l* device achieves lowest cache leakage power (Fig. 8(b)). Meanwhile, due to the usage of all single-fin transistors, 8T cell experiences less power consumption compared with working 6T cells. In summary, 8T cell using *high_l* devices has the lowest leakage power, with 18% latency penalty compared with the fastest 6T cell.

## III. CONCLUSION

Seven FinFET devices optimized for 7nm technology along with three SRAM cells were evaluated and compared. The *high_l* device has the lowest OFF current and the highest ON/OFF current ratio. Moreover, 8T SRAM cell achieves the highest SNM which guarantees its robust operation. Hence, 8T SRAM cell using *high_l* devices is suggested as the choice of memory cell for the discussed 7nm FinFET process.

### ACKNOWLEDGMENT

This work is funded by the DARPA PERFECT program.

### REFERENCES

[1] E. Nowak et al., *IEEE Circuits and Devices Magazine*, 20(1), 2004. [2] Baravelli et al., *Solid-State Electronics*, 54(9), 2010. [3] Guo et al., *ISLPED*, 2005. [4] http://www.synopsys.com/tools/tcad. [5] A. Goud et al., *DRC*, 2013. [6] L. Chang et al., *Symposium on VLSI Technology*, 2005. [7] http://www.hpl.hp.com/research/cacti/.

978-1-4799-7440-5/14 $31.00 © 2014 IEEE

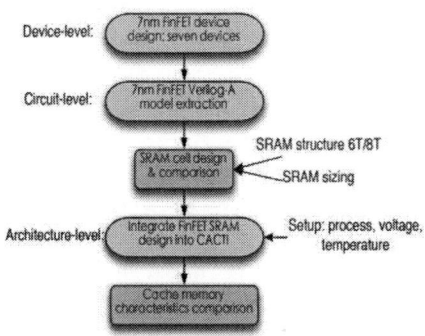

Fig. 1. Cross-layer design framework.

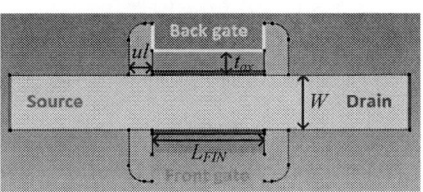

Fig. 2. 2-D model for 7nm FinFET in TCAD device simulator [4].

Table 1. Design parameters of the baseline 7nm FinFET device.

| Parameter name | Value |
|---|---|
| Gate length ($L_{FIN}$) | 7nm |
| Fin width ($W$, or $T_{SI}$) | 3.5nm |
| Fin height | 14nm |
| Gate oxide material | $SiO_2+HfO_2$ |
| Gate oxide thickness ($t_{ox}$) | 1.3nm |
| Gate underlap ($ul$) | 1.5nm |
| Source/Drain doping | $1\times10^{20}cm^{-3}$ |
| Gate work function (NFET) | 4.4eV |
| Gate work function (PFET) | 4.9eV |

Table 2. Design parameters of other 7nm FinFET devices. For each device, only one parameter is changed.

| Device | Parameter | Value |
|---|---|---|
| low_w | $W$ | 3.2nm |
| high_w | $W$ | 3.8nm |
| low_tox | $t_{ox}$ | 1.1nm |
| high_tox | $t_{ox}$ | 1.5nm |
| high_ul | $ul$ | 2.25nm |
| high_l | $L_{FIN}$ | 8nm |

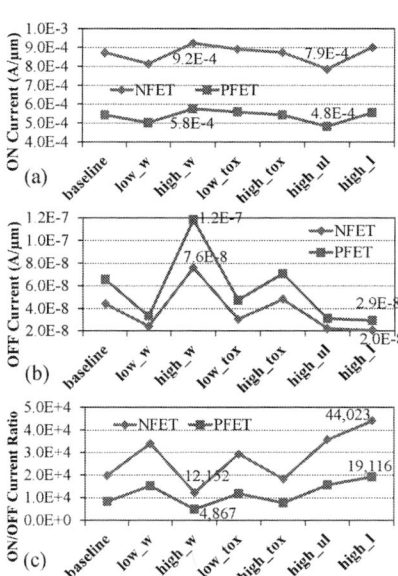

Fig. 3. (a) ON currents, (b) OFF currents, and (c) ON/OFF current ratios of N- and P-type FinFET devices. (†)

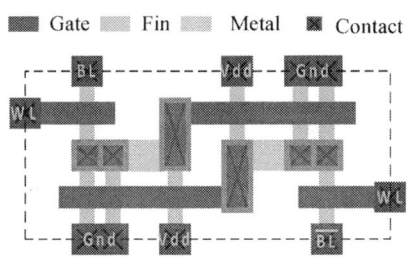

Fig 4. Layout of 6T SRAM cell. Pull-down transistors have 2 fins each. Other transistors have one fin.

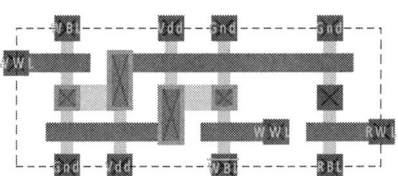

Fig. 5. Layout of 8T SRAM cell with single-fin devices. Separate read path increases the SNM.

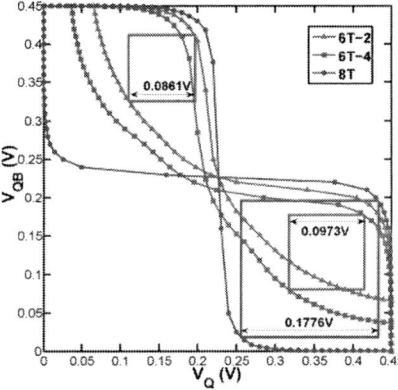

Fig. 6. Butterfly curves of SRAM cells during read access using the baseline 7nm FinFET device. The butterfly curve is derived by combining the *voltage transfer curves* (VTCs) of the two inverters with one VTC inverted. *Static noise margin* (SNM) values are also shown. SNM of 8T cell is 1.8x higher than that of the best 6T. (*)

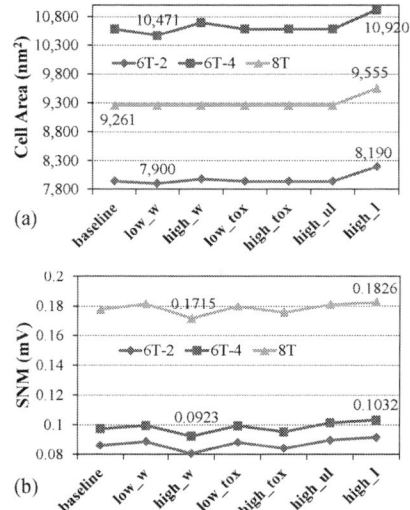

Fig. 7. (a) Layout areas, and (b) SNM values of SRAM cells using different 7nm FinFET devices. (†) (*)

Table 3. Cache configuration.

| Parameter | Value |
|---|---|
| Cache size | 4MB |
| Cache level | L3 |
| Block size | 64B |
| Associativity | 8 |
| Number of read/write ports | 1 |
| Cache model | UCA |
| Number of banks | 4 |
| Output/input bus width | 512 |
| Temperature | 300K |

UCA: Uniform Cache Access

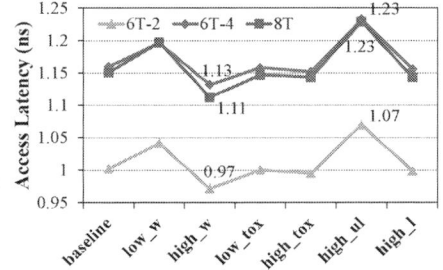
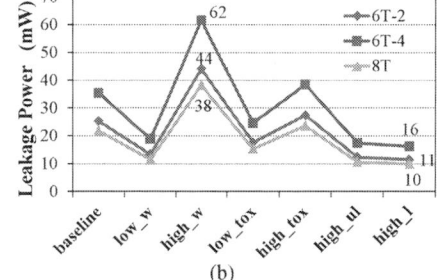

Fig. 8. (a) Access latency, and (b) leakage power of the cache for various combinations of SRAM cells and FinFET devices. Higher ON current leads to shorter access latency (a), whereas higher OFF current causes larger leakage power dissipation (b). (†) (*)

(†) Numbers on each plot show maximum and minimum values.

(*) 6T-*n* denotes a 6T SRAM cell whose pull-down transistors have *n* fins each.

978-1-4799-7440-5/14 $31.00 © 2014 IEEE          91

# Efficient ultra low power rectification at 13.56 MHz for a 10 μA load current

P.-A. Haddad, G. Gosset, J.-P. Raskin and D. Flandre

Institute of Information and Communication Technologies, Electronics and Applied Mathematics (ICTEAM)
Université catholique de Louvain
Place du Levant 3, Louvain-la-Neuve, B-1348, Belgium
pierre-antoine.haddad@uclouvain.be

*Abstract*— A 3-stage Greinacher rectifier designed at 13.56 MHz for a 1 Vpp sinusoidal input and a 10 μA load current in 250 nm CMOS bulk technology is characterized. The measured output voltage is 1.923 V DC with an estimated 72% power conversion efficiency. This ultra low power and high efficiency AC/DC power converter with 0.13 mm² chip area can be used with RF energy harvesters to power implantable or wearable biomedical devices in body sensor networks.

*Keywords*— *rectifier; AC/DC; energy harvesting; ultra low leakage; ultra low power; voltage multiplier; Gradient's method; high efficiency.*

## I. INTRODUCTION

Progress in sensors, energy harvesters and high-efficiency power converters miniaturization has made applications such as industrial and implanted biomedical wireless autonomous sensor networks possible. These can be used to track environmental changes, monitor infrastructure, as well as in health care (Body Area Networks) and security [1].

We consider an ultra low power (ULP) biomedical sensor equipped with a miniature RFID coil antenna operated at 13.56 MHz (RF source), with respect to ISO15693 standard for biomedical applications. This wireless link is used to provide power as well as to retrieve the sensor measurements. With about 30 μW harvested power and limited energy storage capabilities a high efficiency AC/DC converter consisting of cascaded diodes and capacitors can be used to rectify the input AC voltage signal and elevate the output DC supply value. Various novel but complex CMOS diode architectures have been proposed to improve the power conversion efficiency (PCE) of such converters.

In this paper, we present measurements of a 3-stage Greinacher AC/DC converter design based on the composite ULP CMOS diode concept [2] and fabricated in 250 nm bulk CMOS technology. We first present an overview of the circuit and the design methodology that was used and published in [2]. The measurement setup

is described in Section III and the results are presented and discussed in Section IV.

## II. AUTOMATED DESIGN METHODOLOGY

The circuit schematic is shown in Fig. 1. Standard MOS diodes (i.e. MOSFET with gate connected to drain) were replaced by innovative ULP diodes depicted in Fig. 1 [2]. Circuit operation uses a sinusoidal voltage source $V_{in}$ to charge the capacitors when diodes are forward biased, adding up the DC voltage from one capacitor to the next along the ladder. With ideal diodes (i.e. zero-threshold voltage, Vth=0), the output DC voltage $V_{out}$ would reach $N \times V_{in}$ for N stages if $V_{in}$ is taken as the peak-to-peak input voltage. In practice non-zero Vth and reverse bias leakage degrade performance. The ULP diodes address those non-idealities by implementing a quasi-zero-Vth diode with ultra low leakage thanks to a unique subthreshold circuit operation [2].

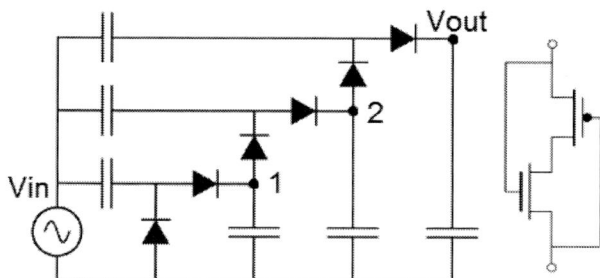

**Fig. 1.** Greinacher 3-stage AC/DC converter (left) and schematic of the ULP diode (right)

The portable and automated design methodology used to determine the sizes of each transistor of the diodes described in [2] is based on a gradient's method. This convex optimization method was used to optimize the efficiency stage by stage on a numerical model of the circuit. The predicted model values were next confirmed by SPICE simulations. A variation-tolerant circuit design obtained with this method is also discussed in [2]. The resulting chip size is 0.13 mm².

978-1-4799-7440-5/14 $31.00 © 2014 IEEE

## III. MEASUREMENT SETUP

The measurement setup shown in Fig. 2 highlights the on-chip ESD protections with the 2.5V power supply rail that were present on the I/O ports of the silicon chip under test (DUT). The output load of the MSO8104A Agilent infiniium oscilloscope used to measure $V_{out}$ is modelled by an impedance of 1 M$\Omega$ // 13 pF.

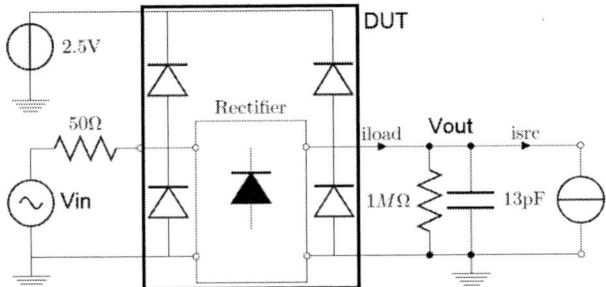

**Fig. 2.** Measurement setup with diodes illustrating the ESD protection circuits at the rectifier I/O.

The 13.56 MHz $V_{in}$ source signal was delivered by an Agilent 33250A waveform generator with 50 $\Omega$ source impedance. Two separate Keithley 2400 source meters were used to impose the 2.5 V DC supply of the ESD protections and the isrc DC current load.

## IV. RESULTS AND DISCUSSION

The output characteristic curve of the rectifier is presented in Fig. 3. Electrical simulations using the Spice foundry model show good agreement with measurements. The $V_{out}$ saturation observed around 3.2 V for input voltages above 1.3 Vpp are consistent with an ESD diode voltage drop above the 2.5 V supply rail.

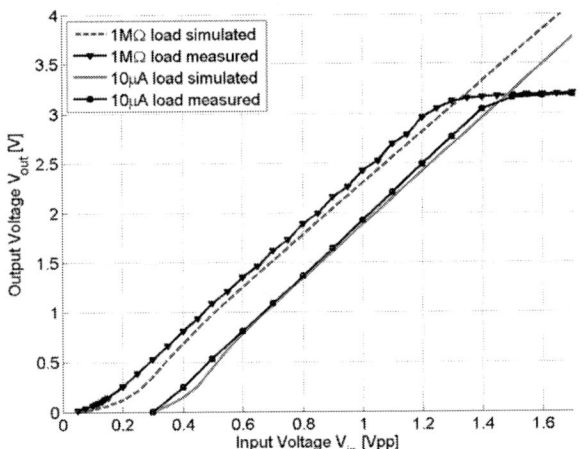

**Fig. 3.** Rectifier output at 13.56 MHz and 25°C.

To apply a 10 µA iload current (see Fig. 2) while still probing the output voltage, the isrc source was adjusted iteratively to isrc=10 µA-$V_{out}$/1 M$\Omega$.

Additional measurements for PCE extraction are shown in Fig. 4. A 10 $\Omega$ resistance was placed in series

with the generator and the voltage drop was measured with a 200 k$\Omega$ differential voltage probe. The DUT input voltage was measured with a 1 M$\Omega$ // 10 pF probe. These signals were multiplied by the probing TDS7104 Tektronix oscilloscope and averaged to calculate the input power. The ESD circuit input power loss was estimated by applying the input signal to the DUT output port and subtracting the simulated rectifier leak power. This estimated power loss is then subtracted from the measured input power in efficiency calculations. The various measurement uncertainties are drawn as confidence intervals.

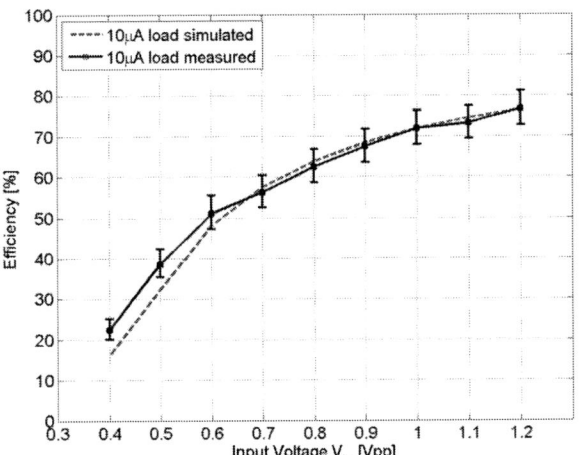

**Fig. 4.** Rectifier efficiency for a 10 µA load.

## V. CONCLUSION

A 3-stage Greinacher AC/DC converter design using ULP diodes was fabricated in 250 nm bulk CMOS technology. The predicted performance values for a 1 Vpp input voltage at 13.56 MHz with a 10 µA load current at 37°C are 1.985 V DC output with 72% PCE in the typical process corner. The measured values obtained on our first silicon under the same input conditions at 25°C are 1.923V DC output with an estimated 72±4% PCE validating the simulations and design methodology presented in [2]. Our PCE is higher than in other comparable but more complex works, e.g. a 66% rectifier PCE for a 21.6 µW load [3].

## REFERENCES

[1] R.J.M. Vullers, R. van Schaijk, I. Doms, C. Van Hoof, R. Mertens, "Micropower energy harvesting", *Solid-State Electronics*, vol. 53, no. 7, 2009, pp. 684-693.

[2] P.-A. Haddad, G. Gosset and D. Flandre, "Design of an Ultra-Low-Power Multi-Stage AC/DC Voltage Rectifier and Multiplier Using a Fully-Automated and Portable Design Methodology", *Journal of Low Power Electronics*, vol. 8, no. 2, 2012, pp. 197-206.

[3] J. Hao Cheong *et al.*, "An Inductively Powered Implantable Blood Flow Sensor Microsystem for Vascular Grafts", *IEEE Transactions on Biomedical Engineering*, vol. 59, no. 9, 2012, pp. 2466-2475.

# Experimental Model of Adaptive Body Biasing for Energy Efficiency in 28nm UTBB FD-SOI

Martin Cochet[12], Bertrand Pelloux-Prayer[1], Mehdi Saligane[123], Sylvain Clerc[1],
Philippe Roche[1], Jean-Luc Autran[2], Dennis Sylvester[3]

[1]STMicroelectronics, Crolles, France, [2]IM2NP – Aix-Marseille University, France, [3]University of Michigan, Ann Arbor, USA

*Abstract*— **In Ultra-Thin Body and BOX Fully Depleted Silicon-On-Insulator (UTBB FD-SOI) technology, body biasing can be used to achieve better energy efficiency. We propose a simple Time, Energy, Power (TEP) model based on Ring Oscillators (RO) measurements to predict optimal (Vdd;Vbb) point for complex circuits and validate it against direct experimental measurements. The model predicted Adaptive Forward Body Biasing (A-FBB) energy gain is compared to Vdd only methods: supply scaling, Vdd hopping and power gating.**

## I. INTRODUCTION

FD-SOI technology allows full scale body biasing which can be used to increase operating speed for a given Vdd at the expense of higher leakage. So, contrary to bulk, a target operating frequency can be obtained with several (Vdd;Vbb) couples [1] and it is important to estimate which yields to the lowest energy. This will highly depend on design and operating conditions such as gate length, activity rate or target frequency. Moreover the practical on-chip generation of these voltages must be considered, as continuous tuning is much easier for biasing than for supply voltage.

## II. PROPOSED TEP MODEL

ROs are simple test structures providing access to timing and energy characteristics of a process. However because of the ROs maximum activity rate, results like optimal (Vdd;Vbb) point cannot be directly applied to complex circuits.

The TEP model is based on three key device level parameters measured form the ROs: gate delay Tp, dynamic energy per transition Edyn and static power Pstat.

Based on these data, the TEP model predicts the energy consumption of a full circuit, knowing its equivalent critical path duration N (i.e. its minimum period expressed in FO3 units) and activity rate $a$ (defined as the number of transitions par clock cycle). For a target period T:

$$Etot = a \times Edyn + Pstat \times T \qquad (1)$$

And the optimization method finds the (Vdd;Vbb) couple to minimize Etot while $N \times Tp < T$.

Measurements of these parameters have been performed on 100-inverter ROs, at 25°C, on 30 dice, for 6 Vdd (0.3V to 1V), 8Vbb (0V to 2V) and two gate lengths (24nm and 40nm).

Last, a fractional coefficient f24 is introduced to reflect the gate length mix in a full circuit:

$$Etot = f24 \times E24 + (1 - f24) \times E40 \qquad (2)$$

Where E24 and E40 are computed from (1) using data from L=24nm and L=40nm ROs respectively.

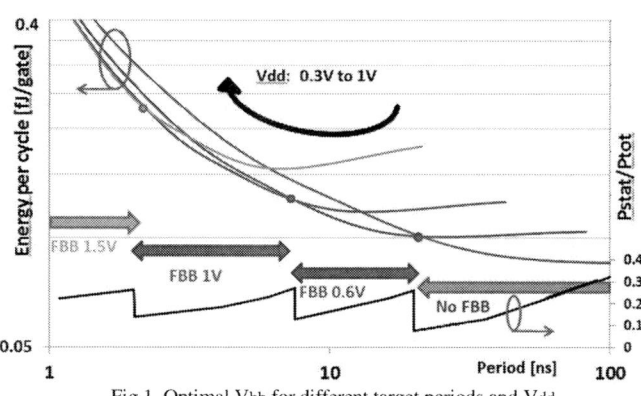

Fig 1. Optimal Vbb for different target periods and Vdd

First, we can use this model on a theoretical case with N=100, $a$=25% and f24=30%. Fig. 1 shows that optimal bias is highly dependent on the target period. In the low speed (T>20ns) leakage prone case, FBB effects a too small reduction in Vdd to make up for the extra leakage consumption. On the contrary, for T<2ns, up to 1.5V bias proves optimal.

## III. TEP MODEL VS. DIRECT MEASUREMENTS

First, the model is compared to direct silicon measurements of two circuits: a high performance custom DSP at 1.5GHz operation [2] and a low power 150MHz 32bit ARM M4® processor running a Dhrystone benchmark. The f24 parameter was obtained from netlist cell count (N24…N40) and (3), N and $a$ parameters from measured timing and leakage (Table I).

$$f24 = (N24 + \tfrac{10}{16}N28 + \tfrac{6}{16}N34)/(N24 + N28 + N34 + N40) \qquad (3)$$

TABLE I
f24, N AND $a$ FITTING PARAMETERS FOR PROCESSOR AND DSP

| Gate length | 24nm | 28nm & 34nm | 40nm | f24 value |
|---|---|---|---|---|
| ARM M4® | 50% | 27% | 23% | **0.63** |
| Custom DSP | 38% | 9% | 53% | **0.43** |

| Vdd/Vbb: | | 0.8V/1V | 0.6V/0V |
|---|---|---|---|
| **Equivalent critical path duration** | N | Fmax [MHz] (measured/*predicted*) | |
| ARM M4® | 135 | **652** / *612* | **186** / *192* |
| Custom DSP | 60 | **1442** / *1389* | **460** / *432* |
| **Activity rate** | $a$ | Pstat/Ptot at Fmax (measured/*predicted*) | |
| ARM M4® | 12% | **29%** / *35%* | **10%** / *9%* |
| Custom DSP | 20% | **2.8%** / *2.9%* @1V no bias (CAD) | |

978-1-4799-7440-5/14 $31.00 © 2014 IEEE

For different bias values the $V_{dd}$ needed to meet the target period and the resulting energy were computed, and compared to direct measurements. The results are plotted in Fig. 2. On both cases the optimal 500mV-step $V_{bb}$ point is correctly predicted (1.5V and 0V) and the corresponding $V_{dd}$ matches within 30mV. In addition to finding minimal energy configurations, the model offers a satisfying prediction of the energy loss when moving away from that point.

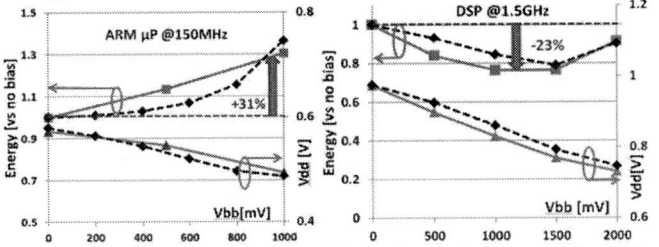

Fig 2. Direct measurements of $V_{dd}$ (blue) and enrgy (red) compared to TEP model predictions (black dashes) for different $V_{bb}$.

## IV. FINE VOLTAGE STEPS

Now, the TEP model can be readily used to estimate the benefit of adaptive body biasing while varying the input parameters. First we can compare the predicted $(V_{dd};V_{bb})$ couples for activity rates of 3% and 20%. Previously, $V_{bb}$ was fixed to discrete values, and $V_{dd}$ tuned continuously. We will now consider a more realistic case where both $V_{dd}$ and $V_{bb}$ are set in small steps (resp. 100mV and 200mV).

For low activity rates (Fig. 3a), the body biasing only serves as an extra lever to smooth $V_{dd}$ transitions: before going a full 100mV step, the bias is increased. This results in an average gain of 6.6% compared to $V_{dd}$ scaling alone. For the high 20% activity rate (Fig. 3b), the phenomenon is similar for T>50ns, but for shorter periods the bias provides continuous energy gain, leading to an average gain of 17.9%.

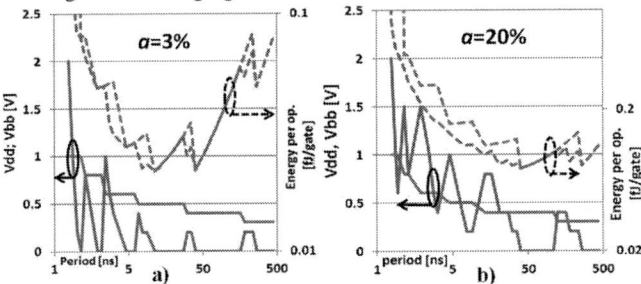

Fig 3. Optimal ($V_{dd}$ – blue; $V_{bb}$ – red) couples from TEP model at low and high activity rates. Resulting energy (dashed red) compared with $V_{dd}$ scaling only (dashed blue).

## V. DISCRETE $V_{dd}$ VALUES

Last, even this 100mV supply step hypothesis can prove unrealistic for on-chip voltage regulation. LC-DC/DCs are complicated to implement on-chip, and Switched Capacitor DC/DCs (SC-DC/DC) only operate efficiently at a few fixed conversion ratios [3]. $V_{dd}$ hopping [4] has been proposed to emulate continuous $V_{dd}$ operation. On the contrary, any standard SC-DC/DC can generate continuous bias values: no current is drained from the wells, so the actual DC/DC energy yield is not important. Hence, we will consider three $V_{dd}$ values

(0.4V, 0.6V and 1V) and continuous $V_{bb}$. We can compare the energy efficiency from our TEP model (N=100, $a$=10%, f24=30%) for different power management methods leveraging $V_{dd}$ only (a,b,c) or both $V_{bb}$ and $V_{dd}$ (d,e):

a) *Standard operation:* constant $V_{dd}$

b) *Power gating:* the circuit operates just as long as needed. This corresponds to hopping between $V_{dd}$ and 0V.

c) *$V_{dd}$ hopping:* alternates between two voltages.

d) *ABB:* $V_{dd}$ is kept to one of the three values and $V_{bb}$ is set to match the target frequency.

e) *ABB hopping:* hopping between two $(V_{dd};V_{bb})$ points.

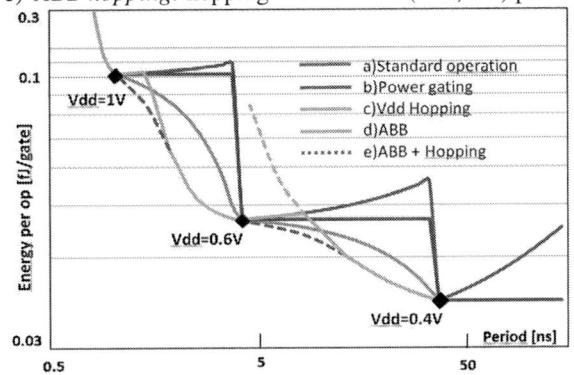

Fig 4. TEP model energy plots for different power optimization schemes

TABLE II
COMPARED ENERGY PERFORMANCES

| Method | Power gating | Vdd Hop | ABB | ABB + Hop |
|---|---|---|---|---|
| Avg. gain | 11% | 24% | 30% | 32% |

As seen on Fig. 4 and Table II, *ABB* proves more efficient than *Vdd hopping*, even without including the extra cost of hopping (~5% according to [4]) or power gating. The added benefit of *ABB hopping* is only visible around the 5-6ns and 1.5-1ns period where $V_{bb}$>1V bias is less efficient than switching or hopping to the next $V_{dd}$ step.

## ACKNOWLEDGMENTS

The authors wish to thank Damien Croain and Sébastien Haendler for diligently conducting the circuits' measurements.

## CONCLUSION

In this paper we have presented a simple modeling of energy consumption of complex circuits to predict optimal supply/bias voltages. The model results were successfully compared to measured data from a DSP and a microprocessor, predicting same optimal bias and corresponding $V_{dd}$ value correct to 30mV. When used with realistic supply constraints of discrete 3 points $V_{dd}$ and continuous $V_{bb}$, TEP data proved the benefit of body biasing for energy efficiency with 30% energy gain, compared to 11% for standard power gating.

## REFERENCES

[1] F. Arnaud et al., *IEDM,* **2012** 3.2.1-3.2.4

[2] R. Wilson et al., *ISSCC Dig. Tech. Papers,* **2014**, pp. 452-453

[3] J. De Vos et al., *IEEE Trans. CaS,* **2013**

[4] E. Beigné et al., *ACM/IEEE NoC,* **2008**, pp 129-138

# UTBB FD-SOI Front- and Back-Gate Coupling Aware Random Telegraph Signal Impact Analysis on a 6T SRAM

K.C Akyel[1,3], L.Ciampolini[1], O. Thomas[2], D.Turgis[1] and G.Ghibaudo[3]

[1] STMicroelectronics, Crolles, France [2] CEA,LETI Campus Minatec, Grenoble, France
[3] IMEP-LAHC, Minatec, Grenoble, France

e-mail: kaya-can.akyel@st.com, Tel: (033) 476-92-51-81

*Abstract*- This work investigates the impact of Random Telegraph Signal (RTS) noise on a 6-Transistor Single-P-Well Static Random Access Memory (6T-SRAM) in 28nm Ultra-Thin Body and Buried Oxide Fully-Depleted Silicon-On-Insulator (UTBB FD-SOI) technology. RTS noise impact is observed through Write-Ability measurements on a 143Kb SRAM macro. A SPICE-level bias- and time-dependent RTS model peculiar to UTBB FD-SOI, introducing the back-gate as a second RTS noise source and considering the front- and back-gate coupling, is used for simulations to confirm silicon observations. It is shown that the body-biasing feature of UTBB FD-SOI does not introduce critical RTS noise compared to the one originated from the device front gate.

## INTRODUCTION

Random Telegraph Signal (RTS) Noise, which is caused by capture/emission of carriers in oxide traps, is seen as a main dynamic variability concern in highly-scaled digital circuits, in particular for Static Random Access Random Memories (SRAM) [1-2]. The Ultra-Thin Body and Buried Oxide Fully-Depleted Silicon-on-Insulator (UTBB FD-SOI) technology offers new design opportunities thanks to its very efficient body-biasing capabilities [3], introducing the body of the device as a second gate. As a consequence, both front-and back-gate (FG and BG) of a device could act as RTS noise sources (Fig. 1).

The RTS noise impact in UTBB FD-SOI is investigated on a STMicroelectronics 6-Transistor Single-P-Well Single-Port High-Density SRAM Bitcell (6T SPW SPHD) [4] through the Write-Ability (WA) [5] failure criterion. Silicon measurements evidence the impact of RTS on the WA Bit Error Rate (BER). A SPICE-level RTS modeling peculiar to UTBB FD-SOI is used for simulations and shows good agreement with silicon results.

## UTBB FD-SOI RTS MODEL

The capture and emission of an electron causes a two-state $V_{th}$ fluctuation between $V_{th}$ and $V_{th}+\delta V_{th}$, where $\delta V_{th}$ is dependent on the dielectric geometry ($W_{eff}$, $L_{eff}$) and on the trap to dielectric/silicon interface distance ($x_{tox/tbox}$). Since BG dielectric (BOX) is much thicker ($t_{box}$=25nm) compared to the FG dielectric ($t_{ox}$=1.7nm), BG traps have significantly larger amplitudes (A) (up to 60mV) but the final impact on device $V_{th}$ is attenuated by the FG-BG coupling (body factor) [3]. The physical characteristics for a particular trap are summarized in Fig.2. The average number of traps in a dielectric is fixed for a given technology and related to the dielectric geometry and to the oxide quality ($N_t$). The average capture and emission times, $<\tau_c>$ and $<\tau_e>$ are related to the applied bias and satisfy Shockley-Read-Hall recombination model [1]. The inversion charge $Q_i$ at each interface placed at $x_i$ is modeled as a Dirac's delta function with its amplitude depending on the surface potential, $V_s$ (Fig.3). The UTBB FD-SOI FG and BG $V_s$ coupling is modeled by a non-linear system of equations solving $V_s$ from the applied FG and BG voltages ($V_g$ and $V_b$) (Fig.3). The proposed analytical model is validated through TCAD simulations (Fig.4). The resulting $V_{th}$ fluctuation caused by a particular trap can be modeled in the SPICE-level netlist through the use of voltage sources. For each trap in a given device, a voltage source R is connected to the device gate [2]. The Fig. 5 presents the RTS-aware bitcell netlist. The $V_{th}$ fluctuation waveform (Fig.6) for each R is generated by a MATLAB program. The bias-and time-dependency of traps are extracted through a nominal simulation and used as input of the generator program, in which they serve as look-up tables for bias-and time-dependent $<\tau_c(t)>$ and $<\tau_e(t)>$ calculation. The stochastic nature in the evolution of a state change for a particular trap at a given bias condition is implemented through exponentially distributed $\tau_c(t)$ and $\tau_e(t)$. The RTS-aware simulation flow is shown in Fig. 7.

## MEASUREMENT AND SIMULATION SETUP

Measurements are performed on a 143Kb SPW SPHD macro using a dynamic characterization module implemented on test chip. Silicon-calibrated surface-potential based SPICE models [6] are used for Monte Carlo simulations. The bitcells are initialized with a "1" in left internal node L. In measurements, an overdrive phase with 1.9V Vddm (Fig.5) is performed between initialization and the SRAM operations to accelerate trapping. With this initial condition, the PMOS PU1 and the NMOS PD2 are in strong inversion during the overdrive, and have higher probability to be weakened by RTS noise. The analysis is performed at very low Vddm to push the bitcell in the failure zone where RTS impact can be exposed.

## RESULTS AND DISCUSSION

Fig. 8 presents WA measurement BER vs. $V_b$ at 0.34V Vddm and 80ns world-line pulse-width for a Single-Write test; the results are normalized w.r.t those obtained at $V_b$ =-0.1V. The BER is estimated under 4 different conditions: no overdrive, only-FG-overdrive, only-BG-overdrive and both FG-and BG-overdrive. The BER increases w.r.t negative $V_b$, since errors are generated by the write discharge failure mechanism [5]. Indeed, PU1 becomes stronger with increasingly negative $V_b$ (forward body-bias) while PG1 becomes weaker (reverse body-bias) preventing the discharge of node "L". After overdrive, PU1 is weakened by the accelerated trapping due to RTS noise, which favorably reduces the BER. The difference becomes more significant at high-BER zone since more bitcells are close to fail and thus more sensitive to small $V_{th}$ fluctuations caused by RTS noise. Compared to the BER estimated after only-FG-overdrive and FG-and BG-overdrive, only-BG-overdrive does not lead to a significant BER reduction: The RTS noise is dominated by FG traps, in other words BG traps do not increase the noise impact. The same trend is reproduced with the proposed RTS-aware netlist. The simulation BER, normalized as for the measurements, is shown on Fig.9.

## CONCLUSION

The impact of RTS noise on UTBB FD-SOI 6T SPW SPHD SRAM is investigated through write-ability measurements in which an overdrive is first applied to accelerate trapping. RTS-aware netlist simulations are performed using a SPICE-level bias-and time-dependent RTS model peculiar to UTBB FD-SOI, considering both front-and back-gate (FG and BG) as RTS noise source. Simulations in good agreement with measurements evidence RTS noise in very aggressive conditions as a canceling out of some strong PMOS-related write failures. It is shown that the RTS noise impact is dominated by FG traps and the body-biasing does not introduce a critical RTS noise.

References: [1] L. Brusamerello et al. *Microelec. Reliability* 49, 2009, [2] M. Tanizawa et al. *VLSI Tech* 2010, [3] J.P. Noel et al. *TED* Aug. 2011, [4] O. Thomas et al *SOI Conf.* 2012, [5] K.C Akyel et al. *ISCAS* 2013, [6] O. Rozeau et al. *SOI Conf.* 2011

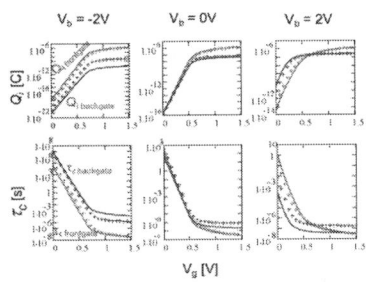

Fig. 1: Interface trapping in UTBB FD-SOI device. Both front and back dielectric traps are considered.

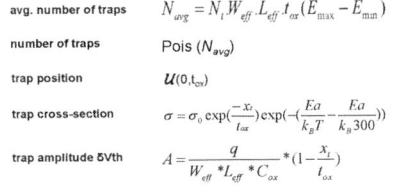

Fig. 2: Physical characteristics of a particular trap. It is assumed that traps are uniformly distributed.

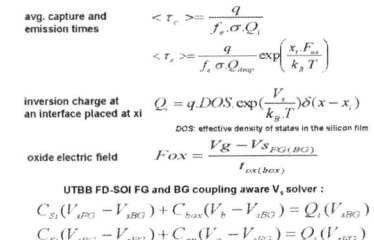

Fig. 3: Bias-and time-dependent capture and emission time analytical model considering UTBB FD-SOI FG and BG coupling.

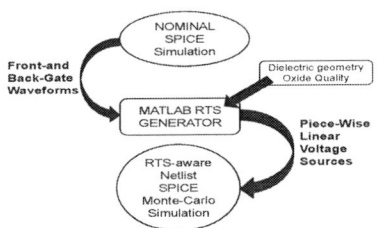

Fig. 4: Analytical model (straight lines) vs. TCAD simulations (dashed lines) for $Q_i$ and $<\tau_c>$ at both gates vs. $V_g$ for various $V_b$ for a NMOS FET. Traps have more tendencies to be filled (small $\tau_c$) at strong inversion on both gate sides.

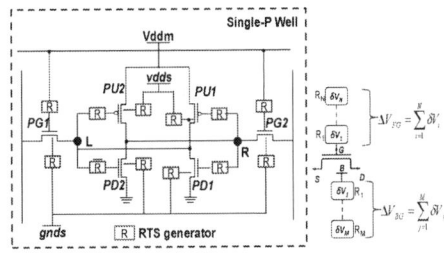

Fig. 5: The RTS-aware bitcell netlist (left). For each trap, a RTS generator R is connected to the device gate. This example has 1 FG and 1 BG trap on each device. A device schematic with multiple traps is also shown (right).

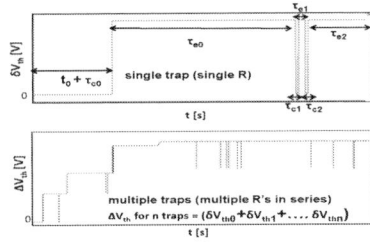

Fig. 6: Vth fluctuations waveform generated in MATLAB for a single trap (top), and for multiple traps (bottom). Fluctuations are added to the RTS-aware netlist through voltage source connected to the device gate.

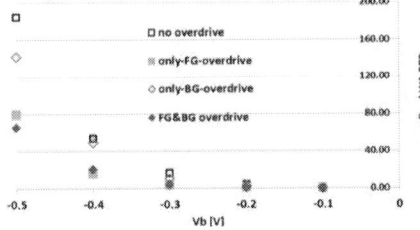

Fig. 7: The RTS aware simulation flow. The bias-and time dependency is extracted through a nominal simulation and used as inputs for the MATLAB code.

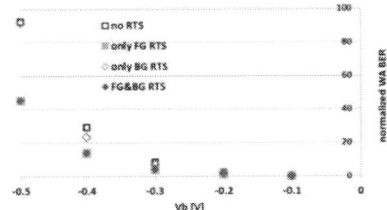

Fig. 8: WA measurement normalized BER vs. $V_b$ at 0.34 Vddm and 80ns world line pulse width. The BER is reduced after overdrive due to RTS in PU1. BG traps' (empty diamonds) impact remains as a second-order.

Fig. 9: WA simulation normalized BER vs. $V_b$ at 0.34 Vddm and 80ns world line pulse width. The BER decreases with overdrive, as on silicon. The RTS noise is dominated by FG traps.

# MIXED-SINGLE WELL 8T SRAM BITCELL FOR WIDE VOLTAGE RANGE IN 28NM FDSOI

Adam Makosiej[1], N. Planes[2], R. Ranica[2], L. Ciampolini[2] and Olivier Thomas[1]

[1]Univ. Grenoble Alpes; CEA, LETI, MINATEC Campus, Grenoble, France, [2]STMicroelectronics, Crolles, France

Email: adam.makosiej@cea.fr

**Abstract:** Enabling high speed SRAM operation at low voltage is typically limited by variability and low device drivability. Most of the reported low-voltage SRAM bitcells show significant area penalty and low performances. This paper proposes a Mixed-Single Well (MSW) 8T SRAM bitcell, which takes advantage of wide back bias voltage range capability of ultra-thin body and box (UTBB) FD-SOI technology The bitcell is evaluated in 28nm using the read-after-write (RAW) dynamic metric which overcomes the limitations of standard readability (RA) and writeability (WA) approaches. It is demonstrated that body biasing tuning enables operation of the proposed bitcell at approx. $5.6\,\sigma$ at 0.4V with over 100MHz.

## INTRODUCTION

Constant increase in mobile market demand for longer battery life time and higher performance forces the development of energy efficient memory design. Moreover, finding a balance between power consumption and high speed operation is challenging and requires a good understanding of SRAM bitcell dynamic behavior.

Several ultra-low voltage SRAM bitcells were reported [1-4] and mostly evaluated through static noise margins (SNM). To satisfy the stability requirements, the proposed solutions use more than 8 transistors leading to a significant area and speed penalty. [5] demonstrates a single P-Well 6T bitcell in UTBB FD-SOI with improved writeability (WA) for high density cache. This paper extends the work done in [5] by proposing a Mixed Single-Well (MSW) 8T SRAM bitcell for wide voltage range high speed cache. The bitcell is evaluated in 28nm UTBB FD-SOI using the read-after-write (RAW) metric [6] and considering the impact of back bias adjustment.

## 28NM UTBB FDSOI MSW 8T SRAM BITCELL

The MSW 8T SRAM bitcell in FDSOI technology (Fig.1) is depicted in Fig.2. The bitcell features a single P-Well (SPW) 6T core (low $V_T$ PMOS, regular $V_T$ NMOS) and single N-Well read port (low $V_T$ NMOS). The SPW 6T core provides a good WA, while the LVT read buffer offers a fast bitline discharge providing a good solution for high speed read. In a design with deep N-Well both P-Well and N-Well can be adjusted in a wide range, allowing both optimizing the 6T core for write speed and stability and switching the read port between standby (N-Well at GND) and high speed (N-Well at $V_{DD}$) modes.

## DYNAMIC BITCELL CHARACTERISATION

Standard approach of extracting readability (RA) and WA is depicted in Fig.4 and Fig.5. While typically they are both considered separately, this may lead to erroneous conclusions, in particular if the clock period is short. This is linked to the 2 phenomena affecting the WA: (i) the bitcell node discharge and (ii) write completion (Fig.3). The former is responsible for successful writing and depends on the commonly known pull-up ratio. The latter not only affects the

WA itself, but may also adversely impact the read in consecutive cycle and cause WA estimation dependency on the $\%V_{DD}$ WA extraction value.

The potential impact of write on the following read is depicted in Fig.6. Clearly while for the weak discharge (Fig.6a) the read is impacted only when write is likely to fail, for slow completion (Fig.6b) the read is strongly affected even when write is successful. In order to overcome this issue, the bitcell functionality can be evaluated using the RAW approach [6], i.e. evaluate the RA in the cycle preceded by a write operation.

## RESULTS

Fig.7 (a) depicts bitcell failure probability ($P_{FAIL}$) in function of P-Well bias for standard RA, WA and RAW with wordline pulse frequency of 125MHz at 0.4V. Fig.7 (b) shows the sensitivity of RAW to each bitcell device variation depending on the $V_B$ value. The simulations were performed with negative bitline (NBL) write assist technique of $-20\%V_{DD}$, which proved required to enable low WA $P_{FAIL}$ at 0.4V. Analysis of Fig.8 results in a few key observations: (i) in weak discharge region, standard write failures dominate the $P_{FAIL}$ (Fig.7b- PUL,PGL dominant, negative $V_{BB}$ resulting in high pull-up ratio); (ii) in weak completion region even if WA $P_{FAIL}$ is low, the following read is likely to fail- high RAW $P_{FAIL}$, dominated strongly by PUR- high $V_{BB}$, hence weak PMOS (Fig.7b) and low pull-up ratio; (iii) in balanced region even though at the end of write cycle the WA criterion is not met, the completion is strong enough to recover the '1' state before the following read operation, hence RAW reaches its optimum and becomes RA limited; (PUR and PGRD dominate RAW indicating simultaneous impact of write and read operations) (iv) there exist an optimum bitcell condition for minimum WA $P_{FAIL}$, (v) WA can either overestimate (weak completion region) or underestimate bitcell stability (balanced region), (vi) outside of weak discharge region WA is strongly dependent on $\%V_{DD}$ extraction value (RAW removes this extraction ambiguity) and (vii) in the optimum case ($V_{BB}\sim$ 0-0.1V) the bitcell operates at approx. $5.6\,\sigma$.

## CONCLUSIONS

The 28nm UTBB FD-SOI MSW 8T bitcell enables wide back range tuning compelling to trade-off bitcell yield and performances. Combination of P-Well bias adjustment, read port operating in high speed mode and application of 20% $V_{DD}$ NBL assist technique enables operation as low as 0.4V with $5.6\,\sigma$ yield over 100MHz making it an attractive solution for wide voltage range high speed cache. RAW dynamic metric, provides accurate $P_{FAIL}$ estimation for high speed operation. Weak completion mechanism is an important part of write process and cannot be overlooked due to its impact on following read.

978-1-4799-7440-5/14 $31.00 © 2014 IEEE

## REFERENCES

[1] J. Mezhibovsky et al., SOCC 2011
[4] B. H. Calhoun et al., JSSC 2007

[2] I. J. Chang et al., JSSC 2009
[5] O. Thomas et al., SOICONF 2012

[3] Q. Li et al., ESSDERC 2012
[6] J. Wang et al., ISLPED 2008

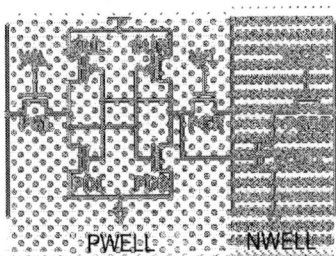

Fig.1 Overview of a 28 nm FDSOI transistor structure

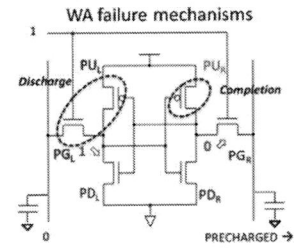

Fig.2 28nm Mixed Single-Well (MSW) 8T SRAM bitcell FDSOI Single P-Well 6T core and N-Well read port

Fig.3 Key devices contributing to write failure mechanisms: bitcell node discharge and write completion

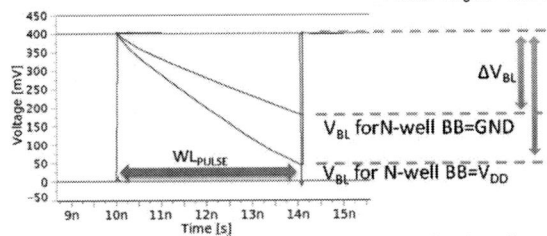

Fig.4 Standard readability assessment with the demonstration of read port forward bias impact on bitline discharge speed; stable if at the end of $WL_{PULSE}$ $\Delta V_{BL}$ drops past a certain target value

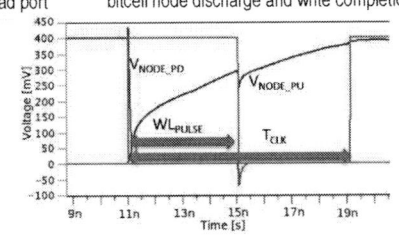

Fig.5 Writeability evaluation- voltage of the node being pulled up ($V_{NODE\_PU}$) reaching a certain %$V_{DD}$ at the end of wordline pulse ($WL_{PULSE}$)

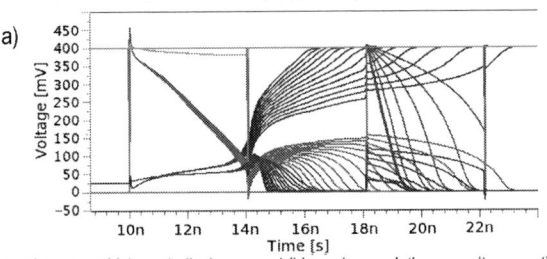

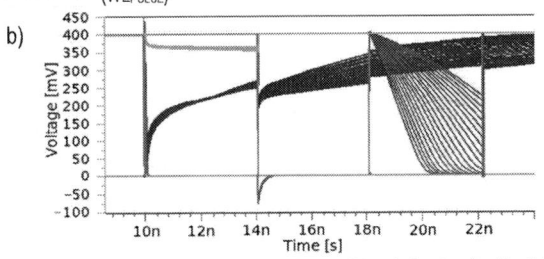

Fig.6 Impact of (a) weak discharge and (b) weak completion on write operation and on read in the following clock cycle. Curves obtained by unbalancing the bitcell through P-Well bias voltage ($V_{BB}$) and performing a small scale sweep on $V_{BB}$ for both cases

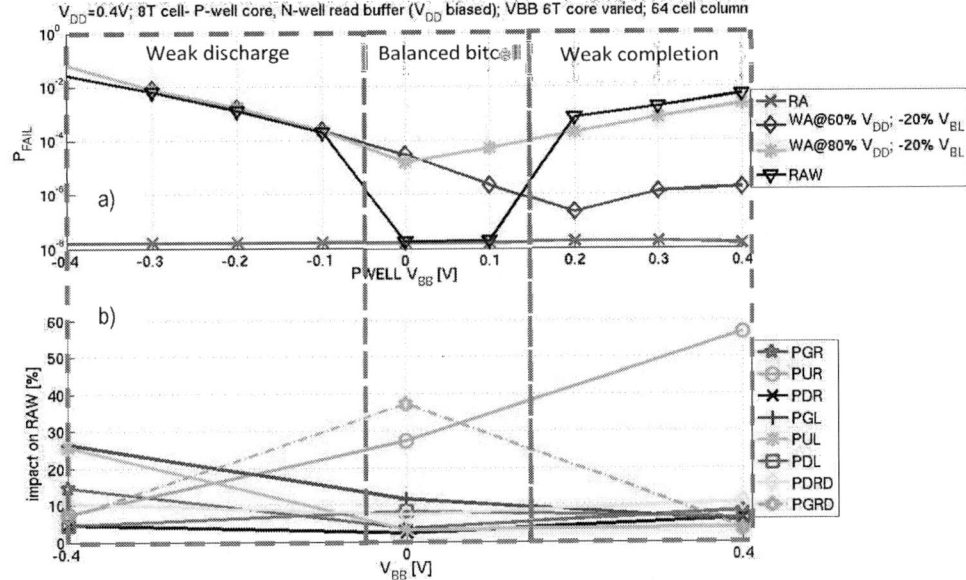

Fig.7 (a) Readability (RA), writeability (WA) extracted at 60% and 80% of $V_{DD}$ and read-after-write (RAW) of the 28nm MSW bitcell at 0.4V for 64 bitcell column with N-Well in forward body bias condition ($V_{DD}$) and with 20% $V_{DD}$ negative bitline assist technique in function of P-Well bias $V_{BB}$; (b) % impact on RAW metric of each device (as in Fig.1) in function of P-Well bias $V_{BB}$; both figures simulated with 125MHz wordline pulse frequency in TT corner using importance sampling; post-layout bitcell parasitics were included in the simulation

978-1-4799-7440-5/14 $31.00 © 2014 IEEE

# A Tunnel-FET SRAM Array for Energy-Efficient Embedded Memory Blocks in Reconfigurable Computing Platforms

Mohammad Faisal Amir, Amit Ranjan Trivedi and Saibal Mukhopadhyay

School of Electrical and Computer Engineering, Georgia Institute of Technology, Atlanta, GA, USA

Email: mamir3@gatech.edu; atrivedi31@gatech.edu; and saibal@ece.gatech.edu, Phone: 404-894-2688

## I. INTRODUCTION

Ultra-low-power reconfigurable computing can benefit various applications with relaxed performance but tight energy constraints like biomedical signal processing, wearable electronics, and remote monitoring. Programmable interconnects (PI) degrade power efficiency of FPGAs that utilize many small and distributed lookup tables (LUTs) [1,2], which can be improved by using larger LUTs to map larger functions in embedded memory blocks (EMBs) [3-5]. The EMBs with large LUTs also find applications in alternative spatiotemporal reconfigurable hardware such as Memory Based Computing with much reduced PI overheads [5,6].

The large, dense, and ultra-low-power EMBs are critical for reconfigurable platforms with tight energy budget. This paper studies the potential of sub-thermal switching slope transistors, namely, Tunneling Field Effect Transistors (TFETs) in designing ultra-low-power SRAM arrays for EMBs in reconfigurable platform. The paper studies the interactions between application properties of EMBs, namely, leakage dominated power and read heavy operation with infrequent write, and characteristics of TFETs. The energy-efficiency and stability of TFET and FinFET based EMBs are compared for applications with relaxed performance constraints.

## II. POWER CHALLENGES IN FINFET EMBs

The memory array plays the key role in determining the energy-efficiency of an EMB. The primary memory acccess in an EMB storing large LUT is 'read' during function evaluation; writing is required only for reconfiguration. While evaluating y=f(x), the input 'x' is the address of the LUT and 'y' is read as the output. Consider the example of a 16kB (256 rows by 512 columns) minimum sized SRAM array in 22nm ASU PTM FinFET[7]. Fig 1 shows the array power dissipation during read assuming 1% activity factor. As the frequency target decreases (through reduced supply voltage), leakage power becomes the dominant component of the array power (Fig 2). Conventional power gating techniques for SRAM, which exploit address locality of consecutive accesses, are difficult to implement for MBC, where consecutive accesses can be random and non-local as they are defined by the input data. Hence low-power EMBs require inherently low-leakage SRAM cells. Although low-voltage can reduce leakage when read performance is not critical, the read stability (Static Noise Margin (SNM)) degrades at reduced voltage (Fig. 3).

## III. TUNNEL FETs

Tunnel FETs demonstrate switching slopes lower than 60mV/decade (the MOSFET limit) and extremely low leakage current (~in the 10s of fA) [8,9]. Although the on current of TFETs can be increased through appropriate material choices and band-gap engineering [10], this paper focuses on Si-Ge nanowire TFETs, which exhibit lower leakage and better switching slope [8], for EMB with relaxed performance but tight energy constraints. The schematic of the SiGe nanowire TFET used, similar to [8], is shown in Fig. 4 with process/geometry specifications in Fig. 5. SPICE compatible lookup table models were generated using Sentaurus. Fig. 6 shows the $I_D$-$V_{GS}$ characteristics.

## IV. TFET BASED EMBEDDED MEMORY BLOCKS

Fig. 7 contains a possible configuration for the TFET SRAM cell [11], TFETA. Fig. 1 shows that even though the TFETA SRAM cell consumes more power at higher frequencies (because of higher supply voltages required to achieve those frequencies), its power consumption does not become limited by the leakage power like FinFETs. TFET power continues to scale at low frequencies, thus providing good energy efficiency. Also, the leakage power in TFETA contributes to a significantly lower fraction of the total power (Fig. 2). However, although SNM of TFETA is higher for equivalent low power due to increased supply voltage (Fig. 8), the SNM of the TFETA cell is quite poor for equivalent VDD (Fig. 3), and will be prone to failure when run at low supply voltage. For equivalent frequency, the SNM of the TFETA cell is poor as well in the low frequency region of interest (Fig. 9).

To improve stability, the TFETA cell is modified to arrive at the structure TFETB in Fig. 10 [12], where the read SNM is simply equal to the much higher hold SNM of the cell. This structure also has faster read due to a smaller number of devices connected to read bitline. In addition, use of uni-directional TFETs in the read path eliminates sneak current problem[13]. From Fig. 11, with $WL_0$ selected, RBL starts to discharge through M1. When bitline swing crosses $V_{th}$, M2 starts to conduct in reverse direction. This causes current flow from $WL_1$ to $WL_0$, dissipating short circuit power. Since TFETs are asymmetric, they do not suffer from this problem, and Fig 12 shows that TFET reverse conduction current is very low (~fA). When compared to the other cells, TFETB provides even lower power, lower leakage and better SNM at equivalent frequency, as discerned from Figures 1, 2, 3, 8 and 9.

The write performance of the cells is compared in Fig. 13; the TFET cell writes are slower due to lower device current. In addition, the single-ended write for TFETB cell requires wordline boosting write-assist[14]. However, since writing is required only for reconfiguration in MBC, the system performance equates to read performance, where TFET cells show numerous advantages, making them an attractive candidate for designing reconfigurable systems.

## V. SUMMARY

This paper studies the potential of Si-Ge TFET for low-power embedded memory blocks in reconfigurable platforms. The key observations from the comparative analysis of FinFET and TFET based EMB are summarized in Fig. 14. At low frequency, switching to the TFET cell from FinFET provides lower read power but degrades read stability, which can be improved through circuit techniques (TFETB). However, as the frequency increases, the TFET advantages begin to decrease, and eventually for high frequency target TFET may become more power hungry than FinFET. The analysis shows the potential of using TFET for designing memory for low-power reconfigurable platform with relaxed performance targets.

978-1-4799-7440-5/14 $31.00 © 2014 IEEE

### ACKNOWLEDGMENT

This work is based on material supported by National Science Foundation (#ECCS-1002090)

### REFERENCES

[1] K. Compton and S. Hauck, *ACM Comput. Survey*, 2002.
[2] V. Betz, et al., *Architecture and CAD for Deep-Submicron FPGAs*. New York: Springer, 1999.
[3] T. Lin et al, *IEEE TVLSI*, vol.20, no.11, Nov. 2012
[4] S. Paul et al., *JETCAS*, 2011, vol.1, no.3, pp.369,380, Sept. 2011
[5] J. Cong et al., *IEEE TCAD* vol19, no11, pp1268,1281, Nov 2000
[6] C. Wang et al., *ISSCC 2014*, pp.460,461, 9-13 Feb. 2014
[7] S. Sinha et al., *DAC 2012*, pp.283,288, 3-7 June 2012.
[8] A.R. Trivedi et al. *DATE 2014*, pp.1-6, 24-28 March 2014
[9] K.K. Bhuwalka et al, *IEEE TED*,vol51,no2,pp279,282,Feb. 2004
[10] D.K. Mohata et al., *IEDM 2011*, pp.33.5.1,33.5.4, Dec. 2011
[11] V. Saripalli et al., *NANOARCH* 2011, pp.45,52, June 2011
[12] S. Paul et al., *ICCAD 2009*, pp.109,112, 2-5 Nov. 2009
[13] K.C. Chun et al., *IEEE JSSC*, vol.47,no.2,pp.547,559, Feb. 2012
[14] V. Chandra et al. *DATE 2010*, pp.345,350, 8-12 March 2010

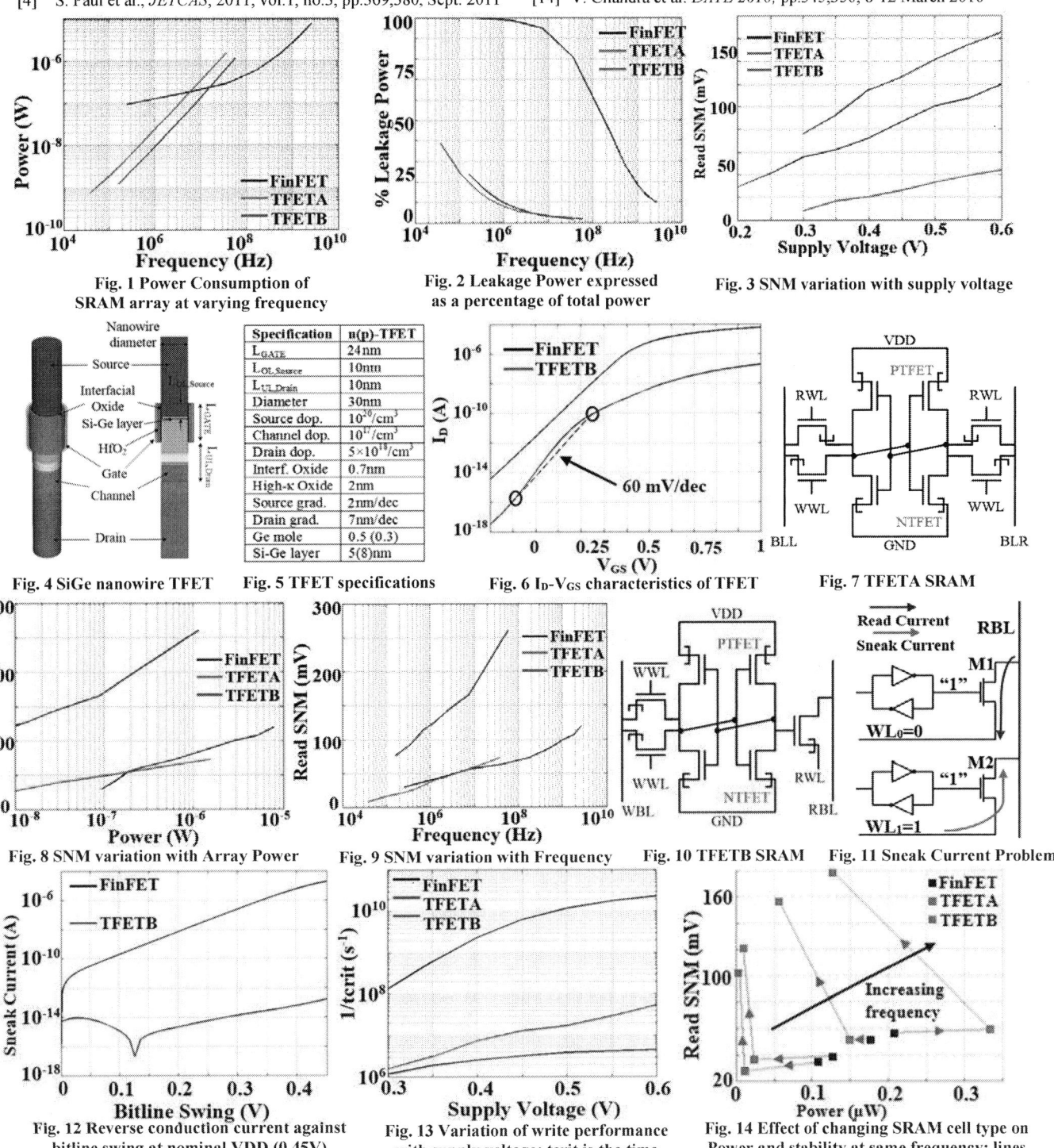

Fig. 1 Power Consumption of SRAM array at varying frequency

Fig. 2 Leakage Power expressed as a percentage of total power

Fig. 3 SNM variation with supply voltage

Fig. 4 SiGe nanowire TFET

Fig. 5 TFET specifications

Fig. 6 $I_D$-$V_{GS}$ characteristics of TFET

Fig. 7 TFETA SRAM

Fig. 8 SNM variation with Array Power

Fig. 9 SNM variation with Frequency

Fig. 10 TFETB SRAM

Fig. 11 Sneak Current Problem

Fig. 12 Reverse conduction current against bitline swing at nominal VDD (0.45V)

Fig. 13 Variation of write performance with supply voltage; tcrit is the time taken during write to flip stored value

Fig. 14 Effect of changing SRAM cell type on Power and stability at same frequency; lines represent iso-frequency plots

# Impacts of Work Function Variation and Line-Edge Roughness on TFET and FinFET Devices and Logic Circuits

Chien-Ju Chen, Yin-Nien Chen, Ming-Long Fan, Vita Pi-Ho Hu, Pin Su and Ching-Te Chuang

Department of Electronics Engineering & Institute of Electronics, National Chiao Tung University, Taiwan

E-Mail: zuzu322.ep97@g2.nctu.edu.tw ; ctchuang@mail.nctu.edu.tw

*Abstract*—In this paper, we analyze the variability of III-V homojunction tunnel FET (TFET) and FinFET devices and logic circuits operating in near-threshold region. The impacts of work function variation (WFV) and fin line-edge roughness (fin LER) on TFET and FinFET device Ion, Ioff, Cg, two-way NAND delay, switching energy and leakage power are investigated and compared using 3D atomistic TCAD mixed-mode Monte-Carlo simulations. The results indicate that WFV and fin LER have different impacts on $I_{ON}$ and $I_{OFF}$. The delay variability of two-way NAND is aggravated by the Miller capacitance of TFET and FinFET devices.

## I. INTRODUCTION

Recently, tunnel field-effect transistor (TFET) has been proposed as a promising device for ultra low-voltage operation due to its steep sub-kT/q sub-threshold slope [1]. With scaled devices, device variations and the resulting impacts on circuit operation become more significant. The major variation sources of TFET and FinFET devices have been shown to be work function variation (WFV) and fin line-edge roughness (fin LER) [2]. In this work, we present a comparative analysis on the variability of TFET and FinFET device characteristics and logic circuits considering WFV and fin LER using 3D atomistic TCAD mixed-mode Monte-Carlo simulations.

## II. DEVICE DESIGN AND SIMULATION METHODOLOGY

In this work, we consider the $In_{0.53}Ga_{0.47}As$ homojunction TFET for its high ON current ($I_{ON}$), while $In_{0.53}Ga_{0.47}As$ N-FinFET and Ge P-FinFET are considered for comparison. Fig. 1 shows the 3D TFET and FinFET device structures constructed for simulation. The relevant parameters are shown in Table I. We use the non-local band-to-band tunneling model which is applicable to arbitrary tunneling barrier with non-uniform electric field for TFET simulations [3], and the parameters used in the model are calibrated with [4]. Fig. 2 shows the $I_{DS}$-$V_{GS}$ characteristics of the $In_{0.53}Ga_{0.47}As$ TFET (N/P) and FinFET (Si-N/P and $In_{0.53}Ga_{0.47}As$-N/Ge-P) with comparable OFF current ($I_{OFF}$) at $V_{DS}$=0.3V. Notice that for III-V TFET, the P-TFET $I_{OFF}$ is lower than other cases due to lower density of states in the conduction band of III-V materials [5]. The threshold voltage ($V_t$) of III-V P-TFET is too high to set equal $I_{OFF}$ with N-TFET with adequate work function, therefore the source doping is adjusted for higher drive current and sizing with wider III-V P-TFET is employed in logic circuits.

To assess WFV, we use the Vonoroi grain pattern [6] for TiN gate material, which has two different grain orientations <200> and <111> with the probability of 60% and 40%, respectively, as shown in Fig. 3(a) by the yellow and orange regions, and the relevant parameters are shown in Table II. To assess fin LER, the rough line edge patterns are generated by Fourier synthesis approach [7] with correlation length ( $\Lambda$ ) =20nm and root-mean-square amplitude ( $\Delta$ ) =1.5nm. The schematic is shown in Fig. 3(b). Finally, we analyze the impacts of WFV and fin LER on logic circuits using 3D atomistic TCAD mixed-mode Monte-Carlo simulations with 100 samples, respectively.

## III. DEVICE VARIABILITY DUE TO WFV AND FIN LER

Fig. 4 shows the impacts of WFV and fin LER on $I_{DS}$-$V_{GS}$ characteristics of TFET and FinFET devices, and Fig. 5 illustrates the probability distributions of $I_{ON}$ (at $V_{DS}$ = 0.3V) and $I_{OFF}$ (at $V_{DS}$ = 0V). For FinFET, WFV causes a $V_t$ shift of $I_{DS}$-$V_{GS}$ curves in subthreshold region with almost equal subthreshold swing (S.S.). On the other hand, Fin LER influences the effective fin width and electrostatic integrity, thus impacting both $V_t$ and S.S.. For TFET, the $I_{OFF}$ distribution with WFV is boarder than that with fin LER since WFV leads to fluctuation in the energy bands and alters the critical tunneling path, and the effect decreases with increasing $V_{GS}$. Therefore the variability of $I_{OFF}$ is larger than $I_{ON}$, and the correlation between $I_{ON}$ and $I_{OFF}$ is weak (Fig. 5(c)). Compared with FinFET, the TFET $\mu$ ($I_{ON}$) is about 2x of that for FinFET, its $\mu / \sigma$ (LER) is slightly better than FinFET, while its $\mu / \sigma$ (WFV) is significantly better ( > 2x) than FinFET. On the other hand, for $I_{OFF}$, both $\mu / \sigma$ (LER) and $\mu / \sigma$ (WFV) of TFET are worse than that for FinFET.

## IV. IMPACTS OF WFV AND FIN LER ON LOGIC CIRCUITS

Fig. 6 shows the probability distributions of two-way NAND delay, switching energy and leakage power for TFET and FinFET considering WFV and fin LER at $V_{DD}$=0.3V. The variability of delay correlates with drive current variations from aforementioned device simulations. Furthermore, the output voltage overshoot caused by Miller capacitance also significantly affects delay and its variation, as illustrated by the transient waveforms for bottom switching shown in Fig.7. The impact of LER on $C_g$ is more significant than WFV, as shown in the probability distribution of $C_g$ in Fig. 8. Notice that, the voltage overshoot is larger for the TFET case due to its larger Miller capacitance, therefore its $\mu$ (delay) is slightly larger than the FinFET $\mu$ (WFV), while for both TFET and FinFET $\mu / \sigma$ (WFV) is significantly better than the $\mu / \sigma$ (LER). Compared with FinFET, the TFET $\mu$ (switching energy) is larger due to its higher drive current. The TFET $\mu$ (leakage power) is smaller due to the lower $I_{OFF}$ of P-TFET while the FinFET $\mu / \sigma$ (LER) is worse due to its larger leakage power.

### ACKNOWLEDGMENT

This work was supported in part by the Ministry of Science and Technology in Taiwan under Contract MOST 101-2221-E-009-150-MY2, and by the Ministry of Education in Taiwan under the ATU Program. The authors thank the National Center for High-Performance Computing in Taiwan for the software and facilities.

### REFERENCES

[1] A. M. Ionescu et al., *Nature*, pp.329-337, 2011.

[2] U. E. Avci et al., *IEDM Tech. Dig.*, pp.777-780, 2012

[3] *Sentaurus TCAD User's manual*, 2011

[4] L. Liu et al., *IEDM TED*, pp.902-908, 2012.

[5] U. E. Avci et al., *IEEE-NANO*, pp.869-872, 2011.

[6] S.-H. Chou et al., *IEEE TED*, pp. 1485-1489, 2013.

[7] A. Asenov et al., *IEEE TED*, pp.1254-1260, 2003.

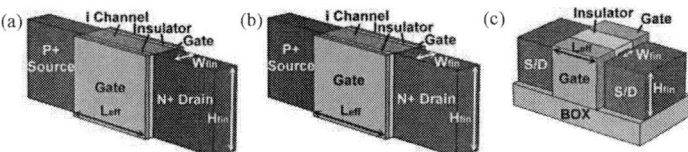

Fig. 1. Structure of (a) In$_{0.53}$Ga$_{0.47}$As homojunction N-TFET, (b) In$_{0.53}$Ga$_{0.47}$As homojunction P-TFET, and (c) FinFET.

TABLE I. PARAMETERS FOR TFET AND FINFET DEVICE

| TFET and FinFET | | | |
|---|---|---|---|
| L$_{eff}$ = 25nm | W$_{fin}$ = 7nm | H$_{fin}$ = 20nm | EOT = 0.65nm |
| | *nTFET* | *pTFET* | *FinFET* |
| Nch (cm$^{-3}$) | undoped | undoped | 1E17 |
| Ns (cm$^{-3}$) | 4.5E19 (p-type) | 1E20 (n-type) | 1E20 |
| Nd (cm$^{-3}$) | 2E17 (n-type) | 2E17 (p-type) | 1E20 |

TABLE II. PARAMETERS FOR WFV SIMULATION

| Gate Material = TiN | | Grain Size = 5nm | |
|---|---|---|---|
| Work function (eV) | *Nominal* | *<200> (60%)* | *<111> (40%)* |
| N-TFET | 4.53 | 4.61 | 4.41 |
| P-TFET | 5.2 | 5.28 | 5.08 |
| InGaAs N-FinFET | 4.88 | 4.96 | 4.76 |
| Ge P-FinFET | 4.27 | 4.35 | 4.15 |

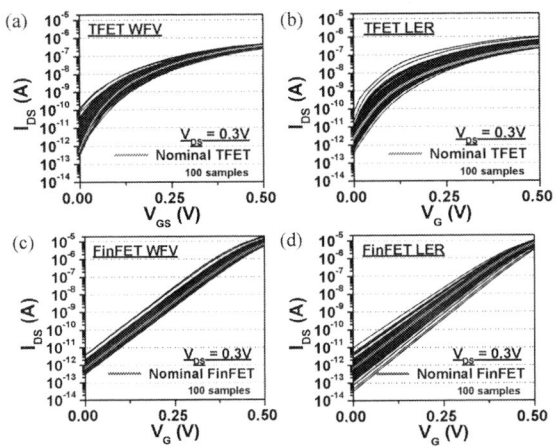

Fig. 4. Simulated I$_{DS}$-V$_{GS}$ characteristics at V$_{DS}$=0.3V for TFET with (a) WFV (b) fin LER; and FinFET with (c) WFV (d) fin LER.

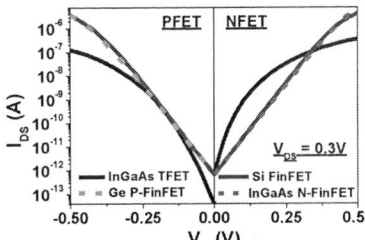

Fig. 2. I$_{DS}$-V$_{GS}$ characteristics at V$_{DS}$=0.3V of In$_{0.53}$Ga$_{0.47}$As homojunction N/P-TFET, Si N/P-FinFET, In$_{0.53}$Ga$_{0.47}$As N-FinFET and Ge P-FinFET.

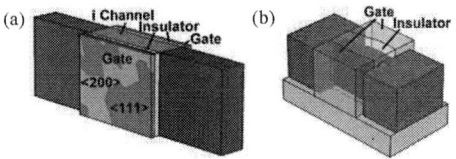

Fig. 3. Example of (a) TFET structure with WFV and (b) FinFET structure with fin LER.

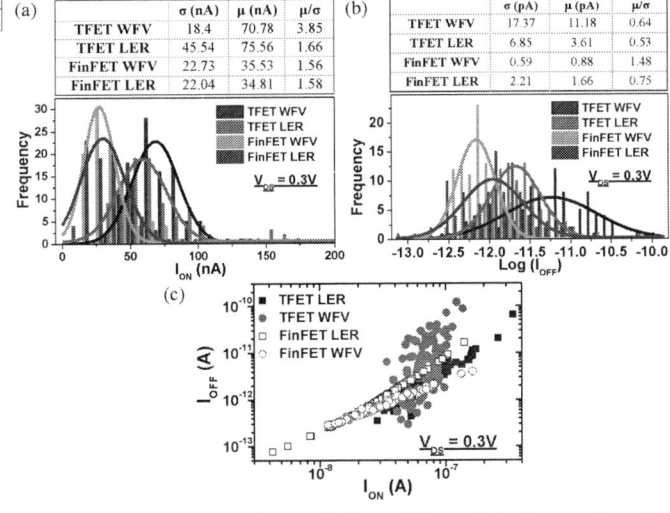

Fig. 5. Probability distribution of (a) I$_{ON}$ (b) I$_{OFF}$ and (c) correlation between I$_{ON}$ and I$_{OFF}$ for TFET and FinFET considering WFV and fin LER.

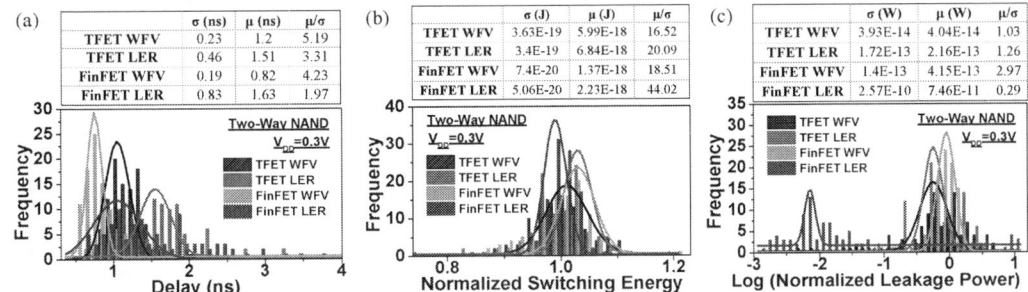

Fig. 6. Probability distribution of (a) delay, (b) switching energy and (c) leakage power for 2W NAND using TFET and FinFET with WFV and fin LER.

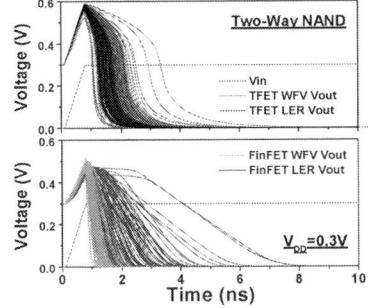

Fig. 7. Transient waveforms for bottom switching of two-way NAND at V$_{DD}$=0.3V for TFET and FinFET with WFV and fin LER.

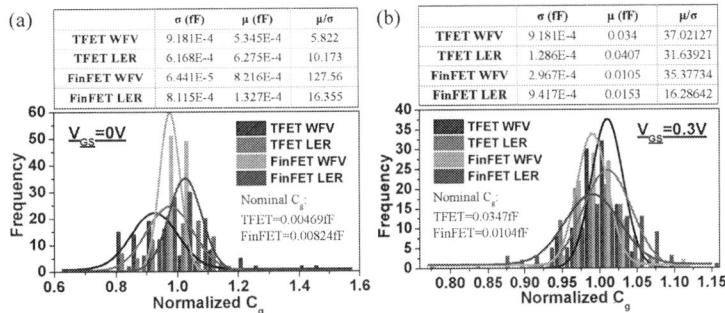

Fig. 8. Probability distribution of C$_g$ at (a) V$_{DS}$= 0V (b) V$_{DS}$= 0.3V for TFET and FinFET considering WFV and fin LER.

978-1-4799-7440-5/14 $31.00 © 2014 IEEE

# OxRAM-based Pulsed Latch for Non-Volatile Flip-Flop in 28nm FDSOI

Alexandre Levisse[1], Natalija Jovanović[1], Elisa Vianello[1], Jean-Michel Portal[2], Olivier Thomas[1]

[1]Univ. Grenoble Alpes, F-38000 Grenoble, France; CEA, LETI, MINATEC Campus, F-38054 Grenoble, France

[2]IM2NP, UMR CNRS 6242, Aix-Marseille Université Marseille, France

Email: alexandre.levisse@cea.fr

**Abstract-** Emerging connected devices operating on battery or harvested energy sources highlight the need for ultra-low standby power design. Including non-volatility in flip-flops (FF) allows nullifying the power consumption in sleep mode, while maintaining the system state. Most of the reported solutions require FF modifications while increasing their complexity. This paper presents a non-volatile flip-flop (NVFF) designed as an OxRAM-based pulsed latch tied to a regular FF for ultra-wide voltage range applications. In 28nm CMOS FDSOI, adding non-volatility cut-off the FF leakage at the cost of 63pJ of data store and restore energy and less than 15% of delay penalty.

## INTRODUCTION

Integrating non-volatile memory (NVM) in logic opens the way to zero consumption in sleep mode by saving the system context while enabling a fast wake-up transition. Among NVM, oxide-based ReRAM technology (OxRAM) appears as a key enabling technology for low-cost applications (high scalability, CMOS back-end of line compatibility and fast switching). One of the main challenges is to enable non-volatility with no modification of regular flip-flop architecture (FF) and ensure a straightforward implementation in integrated circuits. Several non-volatile FF (NVFF) architectures are reported [1]. Most of them integrate NVM directly into the FF leading to master or slave modifications. A solution with unmodified FF using unipolar ReRAM relying on multiple programing voltages ($V_{set}$, $V_{reset}$) and cycles (programing-after-reset) is also reported [2].This paper presents a NVFF designed as a bipolar OxRAM-based non-volatile pulsed latch (NVL) tied to a regular FF cell. Designed to operate within an ultra-wide voltage range, it ensures minimum FF performance penalty and energy for store and restore operations. This architecture is implemented in FDSOI 28nm [3] and HfO2/Ti-based stack bipolar OxRAM [4] technologies.

## ARCHITECTURE

Fig.1 shows the proposed non-volatile pulsed latch. It consists of three main parts: (1) a hybrid OxRAM-CMOS inverter connected to (2) a pulsed restore circuit and (3) a NVM programming circuit. The latch is hooked up to a regular FF as depicted in Fig. 2. The latch is controlled by two signals, RESTORE and STOREb, and works as illustrated in Fig. 3. The *store mode* stands for saving the FF output Q in the NVL. The feedback loop and T4 are turned off while T1 is switched off by switching on T2. The NVM programming circuit set (SN, SP: ON) or reset (RN, RP: ON) the OxRAM cell according to Q. The *restore mode* stands for recovering the saved context from NVL to the FF input, NVD. The restore operation is performed in two steps. First, T1 is turned on by switching on

T3 and switching off T2. T4 is activated by STOREb. NVD node reaches an amplitude voltage relative to the OxRAM resistance state. Then, T3 is turned off while the three-state inverter is enabled (EN) activating the feedback loop. For a low resistive state (LRS), the feedback loop turns off T1 setting NVD node to 0V. For a high resistive state (HRS), T1 is kept on maintaining NVD close to $V_{DD}$.PRE and EN signals are generated by the pulsed restore circuit (Fig.3). The *standby mode* indicates that all power supplies are turned down and can be entered after storing the data. The *operating mode* stands for normal digital circuit operation. T1 is switched off by switching on T2. The feedback loop is turned off minimizing the power consumption of NVL.

## RESULTS

The NVFF was evaluated using the 28nm CMOS FDSOI [3] and bipolar OxRAM [5] models calibrated on silicon. 10.000-sample Monte-Carlo simulations have been run for yield analyses. Considering 28nm thin gate oxide CMOS reliability requirements [6], a maximum supply voltage of 1.5V is used.

For restore operation, Fig.4 shows the minimum HRS and maximum LRS values versus VDD satisfying 3σ yield for various sizing of T1 and T4. [4] demonstrates that higher LRS improves the device endurance. 4μm width for T1 and T4 enables near-threshold restore operation for a minimum LRS of 6KΩ leading to endurance higher than $10^8$ cycles as demonstrated in [4]. Besides, the low memory window required (HRS/LRS = 5) makes the NVL robust against HRS OxRAM variability. Fig.5 shows estimated store energy versus HRS/LRS ratio for different programming voltages and times conditions, starting from the same LRS value. As one can notice, for a same ratio, higher voltage minimizes store energy. Compare to a regular FF, the delay penalty is less than 15% at 0.4V decreasing with higher output drivability and load capacitance (Fig.6), making it competitive with balloon latch solution. A store-restore cycle consumes 63pJ. The time for which regular and balloon latch FF sleep energy is equivalent to the NVFF sleep energy is shown Fig.7.

## CONCLUSION

The proposed NVL design provides a simple standard-cell based non-volatile latch solution fully compatible with thin oxide digital design. The low HRS/LRS ratio required for restore operation ensures good design margins against OxRAM variability and allows the use of high LRS which increases OxRAM endurance. Near-threshold voltage restore operation enables ultra-wide voltage range NVFF design. Low penalty delay makes the NVFF competitive with balloon latch.

978-1-4799-7440-5/14 $31.00 © 2014 IEEE

References: [1] N. Jovanović et al. "OxRAM-based Non Volatile Flip-Flop in 28nm FDSOI", NEWCAS 2014, [2] JM. Portal et al. "Non-Volatile Flip-Flop Based on Unipolar ReRAM for Power-Down Applications" Journal of Low Power Electronics 2012, [3] N. Planes et al., "28nm FDSOI technology platform for high-speed low-voltage digital applications," VLSI Technology Dig., p133-134, 2012. [4] A. Benoist et al. "Advanced CMOS Resistive RAM Solution as Embedded Non-Volatile Memory" accepted to IRPS 2014, [5] M. Bocquet et al. "Robust compact model for bipolar oxide-based resistive switching memories." TED, 2014. [6] X.Federspiel et al. "28nm node bulk vs FDSOI reliability comparison" accepted at IRPS 2012.

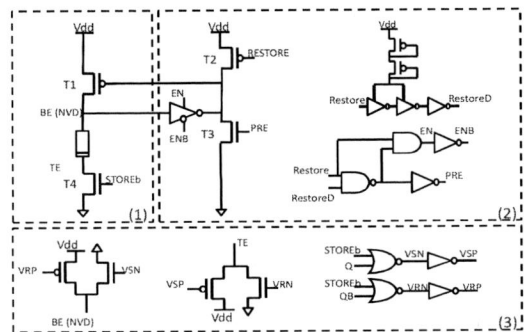

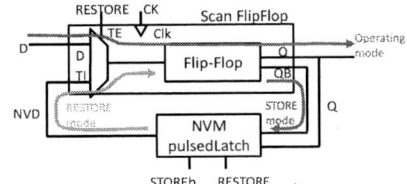

| MODE | STOREB | RESTORE |
|---|---|---|
| Standby | 1 | 0 |
| Active | 1 | 0 |
| Store | 0 | 0 |
| Restore | 1 | 1 |

Fig.1: OxRAM-based Pulsed latch architecture: (1) hybrid OxRAM-CMOS inverter, (2) pulsed restore circuit (3) NVM programming circuit.

Fig.2: Non-volatile Scan Flip-Flop with detailed operating modes.

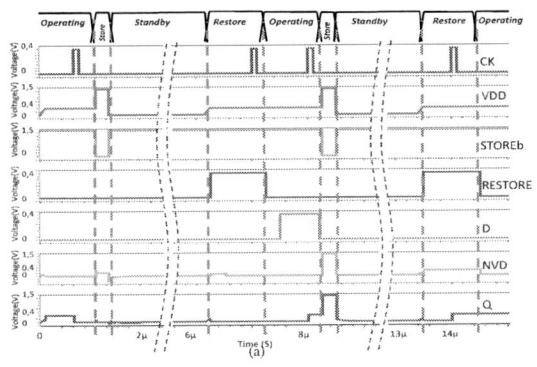

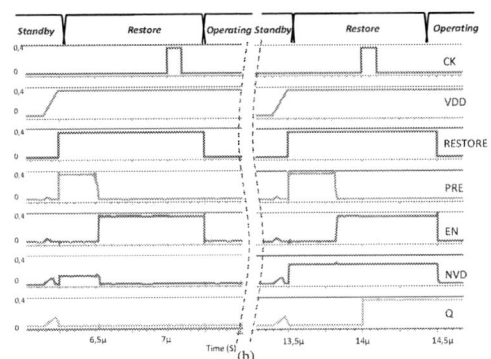

Fig.3: (a) Waveforms of the NVFF during operating, store, standby and restore modes. (b) Detailed restore mode operation for LRS and HRS.

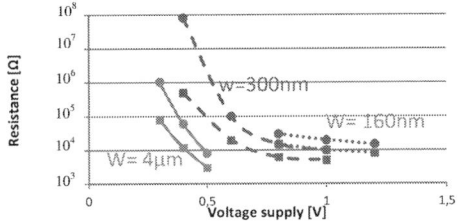

Fig.4: Memory window (minimum HRS, maximum LRS) versus restore supply voltage for various T1 and T4 sizing leading to 3 $\sigma$ yield of 0 (squares) and 1 (circles) restore operation.

Fig.5: Store energy versus HRS/LRS ratio for various store voltages.

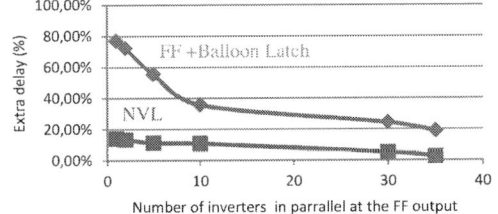

Fig.6: Delay overhead of NVL and balloon latch versus regular FF at 0.4V.

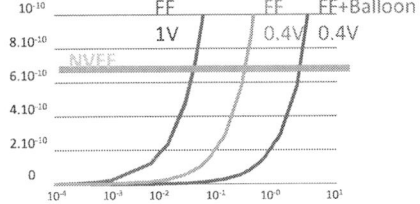

Fig.7: Sleep mode energy for NVFF and FF supplied with different voltages.

978-1-4799-7440-5/14 $31.00 © 2014 IEEE   105

# Electron-Hole Bilayer Deep Subthermal Electronic Switch: Physics, Promise and Challenges

(Invited Paper)

Adrian M. Ionescu, Cem Alper,
Jose L. Padilla, Livio Lattanzio
NANOLAB, Ecole Polytechnique Federale de Lausanne
Lausanne, Switzerland

Pierpaolo Palestri
DIEGM, University of Udine
Udine, Italy

*Abstract*— **This paper overviews the physics and promised performance of electron hole bilayer TFETs (EHBTFET) as deep subthermal electronic switch for ultra-low voltage operation. We provide a first complete roadmap for optimizing its design for combined high performance and low leakage. Based on advanced quantum mechanical (QM) simulation methods, it is shown that the major issue with the EHBTFET is the wavefunction (WF) penetration into the underlap region. Various solutions with different varying complexity are proposed and it is shown that steep slope (SS<<60mV/dec) over a few decades of drain current can be attained using these solutions..**

*Keywords—EHBTFET, subthermal electronic switch, quantum mechanical simulation, low leakage*

## I. INTRODUCTION

While the MOSFET geometrical device scaling and corresponding performance boosting according to the Moore's law have been successfully achieved over many decades, a corresponding scaling of supply voltage $V_{DD}$ for integrated circuits (IC) has been significantly delayed. This supply voltage limitation is caused by thermionic emission of carriers which limits the switching slope (SS) to a theoretical bound of 60mV/dec at room temperature.

To overcome this issue, the search for a 'steep slope switch' has been increasingly gaining momentum. Tunnel Field Effect Transistors (TFETs) are among the most promising candidates as the deep sub-thermal switch (SS<<60mV/dec). TFETs utilize band-to-band tunneling (BTBT) mechanism to effectively filter out the thermionic emission tail [1]. However, the early generations of TFETs [2], [3] has been plagued by very low OFF currents and poor SS due to sensitivity of BTBT to the traps and junction abruptness [4] as well as the gate induced electric field being not aligned with the tunneling direction.

These drawbacks of early generation TFETs have been alleviated by utilizing more optimized structures which offer tunneling in-line with gate induced field effect through the use of an epi-layer in between the source and gate stack [5] or simply by overlapping the gate with source [6]. These proposed structures utilized BTBT between a quantized 2-D inversion layer and a bulk source.

As a natural extrapolation to the second generation of TFETs, the electron hole bilayer TFET (EHBTFET) was proposed [7] and studied [8]–[10]. The 2-D nature of both electron and hole inversion layers are extremely favorable [11] as the step-like density of states (DOS) of 2-D carrier gases allow a very steep switching slope. Historically speaking, BTBT between quantized carrier gases (i.e. bilayers) has been brought up and extensively studied in the case of resonant BJT-type (tunneling parallel to current flow direction) [12] and MOSFET-type (tunneling perpendicular to current flow direction) electron-hole tunnel transistors [13]–[15].

## II. DEVICE STRUCTURE & SIMULATION APPROACH

A schematic depiction of the EHBTFET is given in Fig. 1. The device consists of a semiconductor thin film sandwiched in between two asymmetrically placed gate stacks controlling the quantized levels of the electron and hole gases, separately.

Since the EHBTFET utilizes a very thin semiconductor layer to confine the carriers, a proper treatment of the quantum mechanical effects is necessary. In this work, we utilize two different approaches, one is a fully self-consistent 2-D Schrodinger-Poisson solver based on the effective mass approximation with non-parabolicity corrections using the finite element method (FEM) (referred as 'QM model') [9]. The second approach utilizes semi-classical commercial TCAD tools which are calibrated to the results of a 2-D self-consistent Schrodinger-Poisson simulation (referred as 'QM-calibrated model') [10]. The gate leakage is not taken into account.

According to the simulation results of both approaches, the *most important problem is the wavefunction (WF) penetration into the underlap regions*, which causes a leakage current (schematically shown in Fig.1(c)). The problem can be better seen in Figs. 2, the band profile given for the OFF state of the EHBTFET shows that the barrier between the n+ and underlap regions is too low to prevent WF leakage. This observation is further verified in Fig. 3 by inspecting the subband pair that contributes the majority of the BTBT current in the OFF state, where a clear penetration of the Γ WF is seen.

## III. PREREQUISITES FOR A HIGH PERFORMANCE EHBTFET

Several prerequisites for a high performance are highlighted:

### A. Suppression of Wavefunction Penetration

Structural modifications are needed to be made to initial EHBTFET structure in order to address the aforementioned issue. One such option is to use a hetero-gate (see Fig. 4(a))

978-1-4799-7440-5/14 $31.00 © 2014 IEEE

that uses a different gate metal on the underlap region which, in turn, modifies the potential profile such that the WF penetration is suppressed. As seen in Fig. 5, the hetero-gate solution is able to suppress very effectively the leakage current and the steep slope switching behavior ($SS_{avg} \simeq 30mV/dec$ for about eight decades, $V_{DD} = 0.5V$) is regained [10] with almost no penalty on the ON current.

### B. Low Bandgap Material in the Tunneling Region

Another possible way to achieve a similar effect in terms of WF supression is to use a larger-gap material in the underlap region (or, in other words, use the small bandgap material only in the overlap region). We show a schematic of the proposed hetero-junction structure which in Fig. 4(b). As seen in the Fig. 4(b), the structure utilizes the low bandgap material (Ge) only in the underlap region whereas the remaining semiconductor parts are changed to Silicon.

As seen in Fig. 6, using Silicon in the underlap region very effectively suppresses the direct current contribution, as well as a slight decrease in the indirect current. The reason for this is the direct bandgap in Silicon being ~2.25eV higher than the actual bandgap, hence the direct BTBT barrier is very high and direct BTBT is killed off in the case of transitions from Si conduction band to Ge valence band. However, the opposite case (transitions from Ge $\Gamma$ conduction band to Si valence band), which may occur at certain bias conditions, may not be as effectively suppressed since the only increase in barrier height will be the difference between the valence band edges of Si and Ge $\Delta E_V \simeq 0.53eV$ and therefore needs further investigation.

Material selection is one of the most critical parameter that affects the EHBTFET performance. Using Silicon or SiGe alloys are certainly not the optimal choices for a high-performance EHBTFET due to several reasons.

Using Silicon in a tunneling device immediately suffers from the relatively large and indirect gap, which cripples the BTBT rate, and hence the ON current. Furthermore, in the specific case of EHBTFET, since both electron and hole levels are quantized; the potential required to align the subbands reach to very high values, which is clearly undesirable for a low supply voltage device.

In Germanium, on the other hand, both direct and indirect BTBT can be observed due to the small $\Delta E_C$ ($\simeq 0.14eV$) [16] which allows much higher tunneling currents to be obtained. However, it suffers from the indirect (phonon-assisted) BTBT that occurs at lower voltages than direct BTBT; due to lower bandgap and higher effective masses. Since the BTBT rate for an indirect transition is much lower compared to a direct one, it will effectively act as a 'leakage' mechanism which degrades device figures of merit. Fig. 6 clearly exemplifies this situation.

Finally, the incorporation of III-V materials such as InAs which has a relatively low direct bandgap is critical in obtaining high ON currents. Although the very low effective mass in InAs might indeed be an issue [17], the high non-parabolicity of the conduction bands in InAs will alleviate the impact on the quantized energy levels. Finally, it seems almost indispensable to combine InAs with a large bandgap material

(e.g. in a configuration given in Fig. 4(b)) in order to keep OFF current levels at check.

Another important aspect concerns the channel thickness. It should be duly taken into account that using too narrow channels seems very appealing since it would increase the coupling between the hole and electron layers; however, using a too narrow channel would result in too high bias requirements to align the electron and hole quantized energy levels as noted in [17], an extensive study is given in [18].

### C. Gate Stack Engineering

As the case for all semiconductor devices, the coupling of the gate to the channel is of utmost importance in the EHBTFET, hence the effective oxide thickness (EOT) is needed to be made as small as possible. However, this also results in excessive gate leakage. Therefore, significant improvements in controlling the gate leakage for extremely scaled gate stacks are critical for EHBTFET performance.

Another important observation is that the confinement of carriers in a very thin potential well (approximately half the channel thickness for each carrier type) will result in relatively high quantized subband energies. In practical terms, this will result in an increase in the voltage difference needed between the two gates to align the subbands. Considering that EHBTFET is aimed to be a ultra-low supply voltage device, improvements will be required to employ metals (or other suitable materials) with high and low workfunctions (for p and n gates, respectively) in the gate stacks.

1. J. Appenzeller, M. Radosavljević, J. Knoch, and P. Avouris, *Phys. Rev. Lett.*, vol. 92, no. 4, pp. 2–5, Jan. 2004.
2. C. Sandow, J. Knoch, C. Urban, Q.-T. Zhao, and S. Mantl, *Solid. State. Electron.*, vol. 53, no. 10, pp. 1126–1129, Oct. 2009.
3. M. Fulde, A. Heigl, M. Weis, M. Wimshofer, K. v. Arnim, T. Nirschl, M. Sterkel, G. Knoblinger, W. Hansch, G. Wachutka, and D. Schmitt-Landsiedel, in *2008 2nd IEEE International Nanoelectronics Conference*, 2008, pp. 579–584.
4. M. G. Pala, D. Esseni, F. Conzatti, G. Inp, pp. 135–138, 2012.
5. L. De Michielis, L. Lattanzio, P. Palestri, L. Selmi, and A. M. Ionescu, in *69th Device Research Conference*, 2011, pp. 111–112.
6. P. A. Patel, "Steep Turn On / Off ' Green ' Tunnel Transistors," 2010.
7. L. Lattanzio, L. De Michielis, and A. M. Ionescu, *Solid. State. Electron.*, vol. 74, pp. 85–90, Aug. 2012.
8. C. Alper, L. Lattanzio, L. De Michielis, P. Palestri, L. Selmi, and A. M. Ionescu, *IEEE Trans. Electron Devices*, vol. 60, no. 9, pp. 2754–2760, Sep. 2013.
9. C. Alper, P. Palestri, L. Lattanzio, J. L. Padilla, and A. M. Ionescu, *Accept. to ESSDERC 2014*, no. 1, pp. 2–5, 2014.
10. J. L. Padilla, C. Alper, and A. M. Ionescu, *Submitt. to APL*, no. 2, 2014.
11. S. Agarwal and E. Yablonovitch, in *69th Device Research Conference*, 2011, vol. 51, no. July 1961, pp. 199–200.
12. H. Goronkin, J. Shen, S. Tehrani, and X. Zhu, *US Pat. 5,489,785*, 1996.
13. J. M. Bigelow and J. P. Leburton, *IEEE Trans. Electron Devices*, vol. 41, no. 2, pp. 125–131, 1994.
14. Y. Tada, *US Pat. 5,365,083*, vol. 11, 1994.
15. H. Goronkin, J. Shen, S. Tehrani, and X. Zhu, *US Pat. 5,414,274*, 1995.
16. K.-H. Kao, A. S. Verhulst, W. G. Vandenberghe, B. Soree, G. Groeseneken, and K. De Meyer, *IEEE Trans. Electron Devices*, vol. 59, no. 2, pp. 292–301, Feb. 2012.
17. A. Revelant, A. Villalon, Y. Wu, A. Zaslavsky, C. Le Royer, H. Iwai, S. Cristoloveanu, A. We, and S. Ge, vol. 61, no. 8, pp. 2674–2681, 2014.
18. J. T. Teherani, S. Agarwal, E. Yablonovitch, J. L. Hoyt, and D. Antoniadis, *IEEE Electron Device Lett.*, vol. 34, no. 2, pp. 298–300, Feb. 2013.

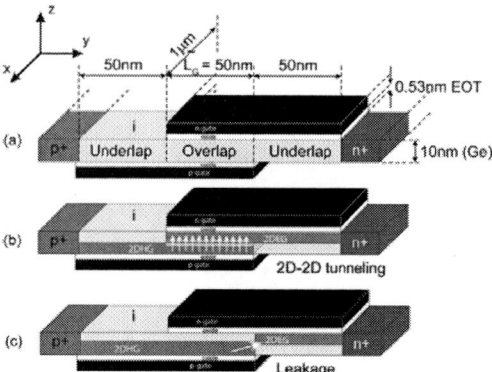

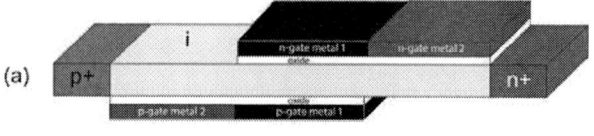

Fig. 4 Proposed solutions for WF penetration (a) Hetero-gate metal. (b) Using a low bandgap material in the channel region.

Fig. 1 EHBTFET device structure: channel thickness $T_{CH} = 10$nm, n+/p+ contact regions extended to 200nm. (a) Indication of overlap and underlap regions (b) 2-D 2-D BTBT between quantized electron and hole gases. (c) Leakage path in the OFF state due to the electron WF penetration [9].

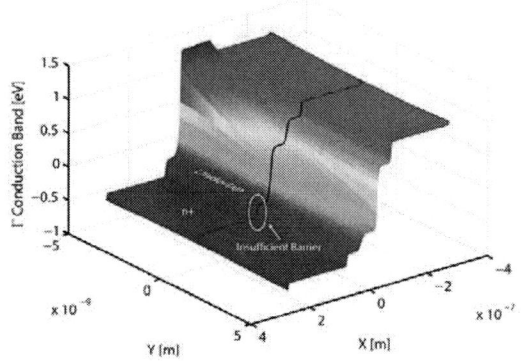

Fig. 2 EHBTFET OFF state electrostatic potential profile calculated by QM method. The barrier height between the p+ (n+) and underlap regions is insufficent and results in excessive WF penetration into the underlap regions causing high leakage current. $\phi_{n-gate1} = \phi_{p-gate1,2} = 4.593$V, $V_{n-gate} = 0$V, $V_{p-gate} = -0.75$V, $V_{DS} = 0.5$V.

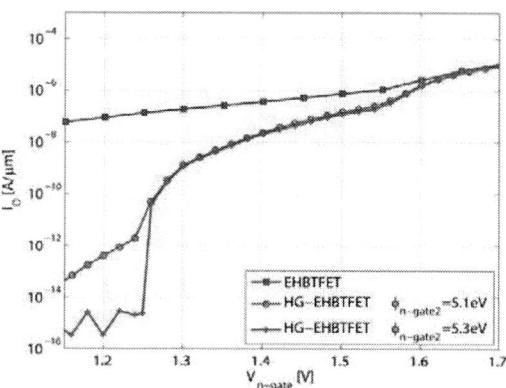

Fig. 5 The transfer characteristics obtained using the QM-calibrated approach for the EHBTFET and the EHBTFET with hetero-gate (HG-EHBTFET) [10]. Very effective suppression of the leakage current is observed for the case with a 2$^{nd}$ metal workfunction of $\phi_{n-gate2} = 5.3$eV in the n-gate. $\phi_{n-gate1} = \phi_{p-gate1,2} = 4.593$V, $V_{p-gate} = -0.75$V, $V_{DS} = 0.5$V.

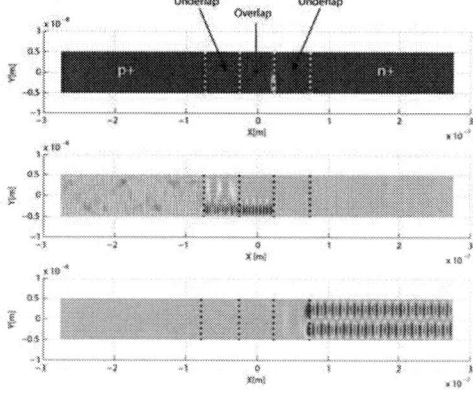

Fig. 3 Top: The BTBT generation rate for the most contributing wavefunction pair in the OFF state. Middle: Heavy hole wavefunction. Bottom: Γ electron wavefunction. The penetration of the Γ WF into the underlap region is clearly visible.

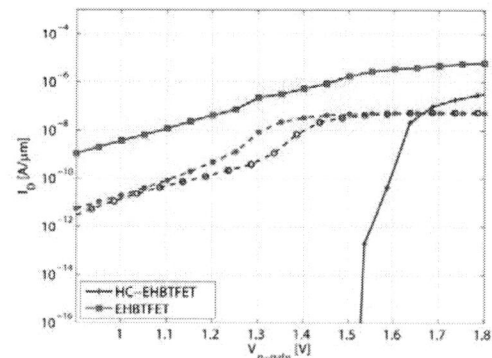

Fig. 6 Transfer characteristics using the QM approach for conventional EHBTFET and EHBTFET with hetero-channel (HC-EHBTFET), with direct (solid) and indirect (dashed) contributions separated. Effective suppression of the direct leakage current is observed, whereas indirect current relatively remains approximately the same, since the indirect current at the investigated voltage range is the ON current generated in the overlap region. For EHBTFET $\phi_{n-gate} = 3.408$eV, $\phi_{p-gate} = 5.642$eV , for HC-EHBTFET $\phi_{n-gate} = 3.57$eV, $\phi_{p-gate} = 5.642$eV. $V_{p-gate} = 0$V, $V_{DS} = 0.5$V.

978-1-4799-7440-5/14 $31.00 © 2014 IEEE

# Beyond TFET: Alternative Mechanisms for CMOS-Compatible Sharp-Switching Devices

S. Cristoloveanu, J. Wan, P. Ferrari, M. Bawedin,
C. Navarro
University of Grenoble-Alpes, CNRS, IMEP-LAHC
Campus Minatec, Grenoble, France

A. Zaslavsky
Brown University, Providence, USA

C. Le Royer, A. Villalon, C. Fenouillet-Béranger
CEA, LETI, Minatec campus, F-38054, Grenoble, France

Y. Solaro, P. Fonteneau
STMicroelectronics, Crolles, France

*Abstract*—Tunneling-based transistors (TFETs) have attracted interest due to their (theoretical) capability of switching more sharply than MOSFETs. However, other mechanisms that take place in SOI devices can provide even more abrupt switching and higher current. We examine the family of emerging TFET-competing devices based on barrier modulation, bipolar amplification and impact ionization. Practical results for devices fabricated in 14-28 nm FDSOI technology will be discussed.

*Keywords—FDSOI; CMOS; TFET, BET-FET; $Z^2$-FET; sharp switching, tunneling*

## I. Introduction

MOSFET scaling requires maintaining both a high ON current $I_{ON}$ for speed and a low OFF current $I_{OFF}$ to limit standby power consumption. The reduction of the supply voltage $V_{DD}$ is impeded by the subthreshold slope which cannot be lowered below S = 2.3(kT/q) ~ 60 mV/decade at room temperature, even in ideal fully-depleted MOSFETs. This explains the growing interest in CMOS-compatible devices that switch more abruptly than MOSFETs [1].

We briefly review the mechanisms that offer abrupt switching between OFF and ON states in SOI transistors: floating body, coupling, band-to-band tunneling and impact ionization. A typical device is the I-MOS [2], studied a decade ago and abandoned because of poor scalability, unacceptably high $V_{DD}$ and lack of reliability.

Tunneling FETs (TFETs) are not limited by S > 60 mV/decade, because the current is carried by $V_G$-controlled tunneling of carriers through the bandgap, rather than injection over a potential barrier. Today, state-of-the-art CMOS-compatible TFETs can provide S < 60 mV/decade over a narrow range of $V_G$ but have difficulty in reaching a competitive $I_{ON}$. We will show recent progress in current drive capability achieved by combining SOI and SiGe technologies (Fig. 1) [3].

In this paper, we focus on alternative sharp switching devices with potential to surpass the performance of TFETs.

## II. BET-FET

We will discuss a proposed innovation: a bipolar-enhanced TFET (BET-FET) structure, in which the tunneling current is amplified by the gain of a heterojunction bipolar transistor [4]. The TFET and the bipolar junction are co-integrated in a single monolithic device (Fig. 2).

The simulated performance in Si/SiGe BET-FET structures is unrivalled by any other type of TFET: $I_{ON}$ higher than 1 mA/um and sub-60 mV/dec subthreshold swing over 7 decades. The BET-FET is, in principle, compatible with other promising material systems (such as staggered GaSb/InAs heterostructures).

## III. Barrier Modulation Devices

A different approach to sharp-switching transistors is based on devices with internal positive feedback combined with potential barrier modulation. Like TFETs, they have a gated-diode configuration, but are operated in forward-bias mode. Electrostatic barriers are formed (via gate biasing or channel doping) to prevent electron/hole injection into the channel until the gate or drain bias reaches a turn-on value. Once injection into the channel begins, the electrons modulate the hole injection barriers and vice versa, leading to abrupt switching (S < 1 mV/decade) to a high $I_{ON}$ corresponding to a forward-biased diode.

The family of feedback barrier-modulation devices includes several SOI designs:

- Field-effect diode (FED) with two adjacent top gates [5],

- Thyristor-like structures with specific dual body doping and control via ground-plane bias [6],

- $Z^2$-FET (zero swing and zero impact ionization FET) which has an underlapped top gate and additional control from the ground plane [7].

We will discuss in detail the device physics, architecture, and technology boosters for the most promising variant ($Z^2$-FET). The $Z^2$-FET features a large hysteresis useful for single-

---

Place2be and MINOS programs are thanked for support. A.Z. also thanks NSF ECCS–1068895.

transistor DRAM and SRAM cells [7], as well as ESD protection [6].

Experimental and simulation results for the 28 nm and 14 nm FD-SOI technology nodes will be documented.

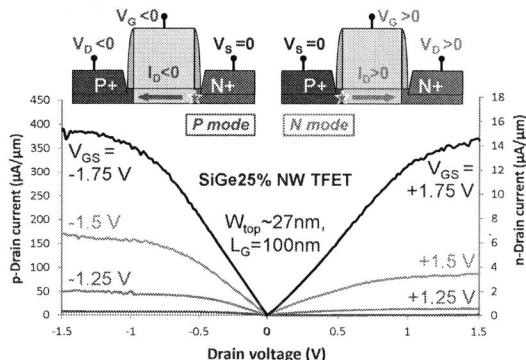

Fig. 1: SiGe/SOI nanowire TFET showing high ON current [3].

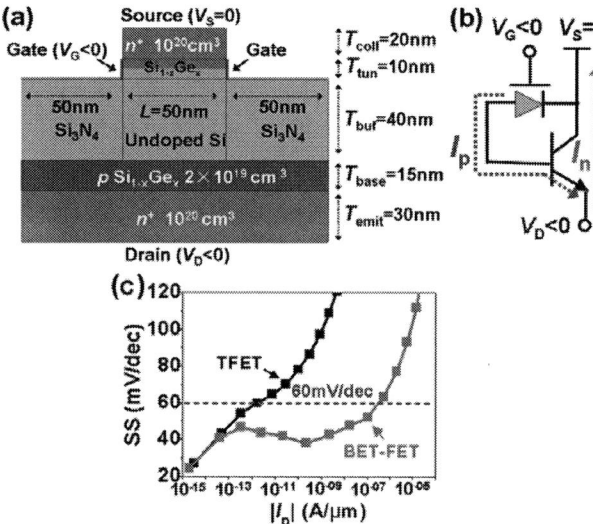

Fig. 2: (a) Configuration of the bipolar-enhanced TFET, (b) equivalent circuit and (c) typical simulated performance [4].

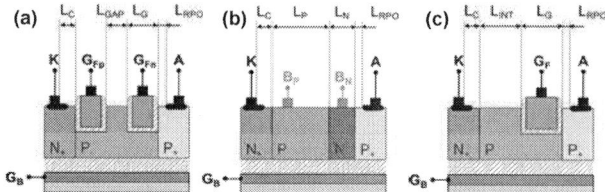

Fig. 3: Configuration of several barrier-modulation devices on FD-SOI: (a) field-effect diode (FED) [5], (b) back-bias-controlled thyristor [6] and (c) $Z^2$-FET [6].

## REFERENCES

[1] A. M. Ionescu and H. Riel, "Tunnel field-effect transistors as energy-efficient electronic switches," *Nature* 479, pp. 329–337, November 2011.

[2] F. Mayer, C. Le Royer, G. Le Carval, C. Tabone, L. Clavelier, S. Deleonibus, "Co-integration of 2 mV/dec Subthreshold Slope Impact Ionization MOS (I-MOS) with CMOS," *Solid-State Device Research Conference, 2006. ESSDERC 2006. Proceeding of the 36th European*, pp. 303-306, Sept. 2006.

[3] A. Villalon, C. Le Royer, P. Nguyen, S. Barraud, F. Glowacki, A. Revelant, L. Selmi, S. Cristoloveanu, L. Tosti, C. Vizioz, J.-M. Hartmann, N. Bernier, B. Prévitali, C. Tabone, F. Allain, S. Martinie, O. Rozeau, M. Vinet, "First Demonstration of Strained SiGe Nanowires TFETs with $I_{ON}$ beyond 700 µA/µm," *VLSI Tech. Symp.*, pp.84-85, June 2014.

[4] J. Wan, A. Zaslavsky, C. Le Royer, S. Cristoloveanu, "Novel Bipolar-Enhanced Tunneling FET With Simulated High On-Current," *Electron Device Letters, IEEE*, vol.34, no.1, pp.24-26, Jan. 2013.

[5] A. A. Salman, S. G. Beebe, M. Emam, M. M. Pelella, D. E. Ioannou, "Field Effect Diode (FED): A novel device for ESD protection in deep sub-micron SOI technologies," *Electron Devices Meeting, 2006. IEDM '06. International*, pp.1-4, Dec. 2006.

[6] Y. Solaro, P. Fonteneau, C.-A. Legrand, D. Marin-Cudraz, J. Passieux, P. Guyader, L.-R. Clement, C. Fenouillet-Beranger, P. Ferrari, S. Cristoloveanu, "Innovative ESD protections for UTBB FD-SOI technology," *Electron Devices Meeting (IEDM), 2013 IEEE International*, pp.7.3.1-7.3.4, Dec. 2013.

[7] J. Wan, S. Cristoloveanu, C. Le Royer, A. Zaslavsky, "A feedback silicon-on-insulator steep switching device with gate-controlled carrier injection", *Solid-State Electronics*, vol. 76, pp. 109-111, Oct. 2012.

# A2RAM: Low-power 1T-DRAM memory cells compatible with planar and 3D SOI substrates

F.Gamiz, N.Rodriguez, C.Marquez, C.Navarro* and S.Cristoloveanu*

Nanoelectronics Laboratory, Department of Electronics, University of Granada 18071 Granada (Spain)
*IMEP, Grenoble INP MINATEC, 3 Parvis Louis Néel, BP 257 38016 Grenoble CEDEX 1(France)

*Abstract-* **A novel concept of multi-body 1T-DRAM cell fully compatible with both planar Silicon-On- Insulator substrates and 3D architectures is presented. Its scalability is ensured thanks to the dedicated body partitioning for hole storage and electron current sensing, suppressing the super-coupling effect and allowing the coexistence of electron and hole layers in very thin silicon films. Numerical simulations of the electrostatics and dynamic operation show attractive performance in terms of state discrimination and retention time. These theoretical results on planar devices have been experimentally validated on structures fabricated at CEA-LETI. Finally, we will demonstrate that this body-partitioning concept is extrapolated to 3D tri-gate structures showing high scalability, low-power consumption, long retention time, nondestructive reading, and wide memory window.**

During the last years the search for alternatives to the standard DRAM and SRAM cells has accelerated due to the drawbacks entailed by the mainstream technologies: area penalty for the SRAM and process complexity for the DRAM. While new emerging technologies like PCM and ReRAM are being suited as future alternatives for storage class memory, the inertia and concept simplicity of DRAM and SRAM make them still the best option for very fast applications compared with emerging, more challenging alternatives [1].

The floating-body (FB) single-transistor (1T) DRAMs are one of these alternatives. The potential use of a Silicon-On-Insulator (SOI) transistor as memory cell with modulated charge inside the body of the device was predicted in the early 90's [2]. However, it was in the past decade, together with the increasing constraints the standard memory cells are facing [3], when the interest around the FB-1T-DRAM concept raised up.

Research groups and semiconductor companies have explored different alternatives to turn a simple SOI transistor into a memory cell. In parallel with the single-gate or double-gate (with independent or interconnected gates) SOI-MOSFET, other proposals, most of them theoretical, have promoted topological modifications in the body of the transistor to enhance the performance. These modifications have primarily been focused on creating dedicated regions to store the charge (holes) while maintaining large concentration of electrons during the reading process of the cell. Typical versions are the A-RAM [4-5] and A2RAM [6] on planar FDSOI substrates.

The concept of the A2RAM memory cell was proposed in 2011 [6] as a new device with potential application as capacitorless DRAM cell. The simulation results predicted large current between states, low-voltage operation and a superior scalability compatible with single-gate operation. Apart from these generic assets, the A2RAM features direct compatibility with SOI and bulk substrates without changes in design. The first preliminary demonstration came out in 2012 [7] on SOI substrate.

The principle of operation of the A2RAM is summarized in Figure 1. A complete description is given in [6]. The device is an SOI transistor (although it may also be a bulk transistor) with a buried N-type layer (N-bridge) which short circuits the source and drain regions. A negative gate bias is applied to accumulate holes at the top interface; the presence of this accumulated charge defines the '1' state bit. Under these circumstances, the gate electric field is screened by the accumulated charge and the carrier concentration in the N-Bridge is hardly affected by the gate bias. The electron concentration in the undepleted N-bridge corresponds to the doping level leading to a possible current flow between source and drain (Figure 1.b). If holes are transitorily removed from the top interface, while negative gate bias is maintained, the gate-induced field is no longer screened affecting the electron concentration of the N-bridge.

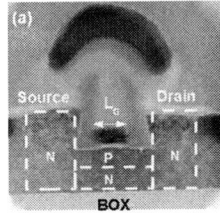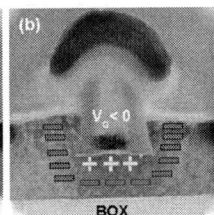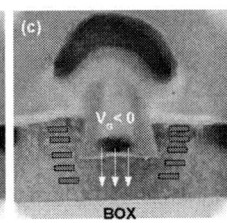

Figure 1: (a) TEM cross section of an A2RAM cell fabricated in a 22nm SOI process with superimposed doping regions in dashed lines. (b) The accumulated holes in the top P-body screen the electric field: an electron current can flow between source and drain. (b) If the P-type body is in deep depletion, the gate electric field is no longer screened, fully depleting the N-bridge from carriers and suppressing the drain current.

If the device dimensions and doping concentration are well calibrated the N-bridge will become fully depleted, therefore preventing any current flow between source and drain ('0' state, Fig.1c). Two main advantages stand out immediately from this concept: i) there is no need for biasing a bottom gate (substrate bias, $V_B = 0V$) and, ii) the reading of the cell state requires very low drain voltage (only needed to test whether the bridge is conductive or not). In Figure 2 we show experimental measurements on the memory cell after applying a voltage pattern to gate and drain terminals. The bias signals (generated with an Agilent B1530 fast measurement unit) are shown in Figure 2.a, whereas the drain current readout is shown in Figure 2.b. It is clear that these devices are able to preserve the memory effect with much reduced operating voltage ($|VG|<1.7V$ and $VD < 1V$). The cumulative retention time plot for the FDSOI A2RAM wafer is shown in Figure 3.a. According to numerical simulations there is room for improvement in the retention time by a more careful source-drain engineering (which was not implemented in these prototypes). Figure 3.b shows the cumulative plots of the current levels in the '0' and '1' states. The distributions do not overlap, leading to a worst case safe-factor of 19.

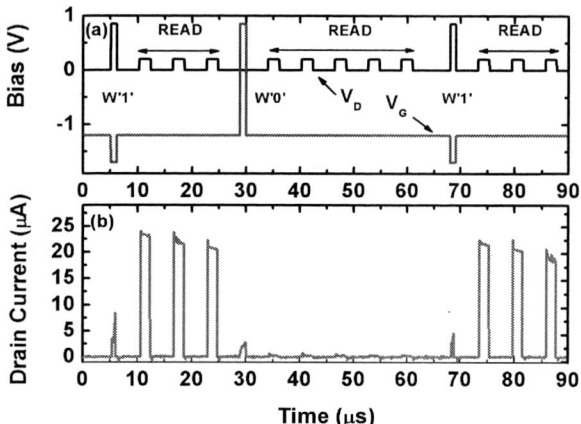

Figure 2: (a) Bias pattern applied to FDSOI A2RAM to demonstrate memory functionality. The pattern consists of a writing '1' event ($V_D$ = 0.85V, $V_G$ = -1.75V), three reads ($V_D$ = 0.2V), a writing '0' event ($V_D$ = 0V, $V_G$ = 0.85V), five reads, a writing '1' event again and three reads. (b) Drain readout current demonstrating that the '1' and '0' states are clearly distinguishable. L = 100nm; W = 10μm; EOT = 3.1nm.

Average retention times of 20ms and 3ms, are obtained at room temperature and 85 °C, respectively, by testing a representative statistical sample of 310 devices.

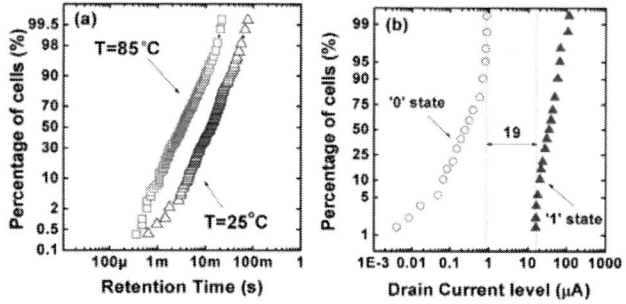

Figure 3: (a) Cumulative retention time plot for a set of 310 devices at 25 °C and 85 °C. The retention time limit is defined when the '0' state recovers 50% of the stable level (state '1') current. (b) Cumulative current level of the '1' and '0' states 200μs after writing. L = 80nm; W = 10μm; EOT = 3.1nm; $V_{G(hold)}$ = -1.2V, $V_{D(read)}$ = 0.2V.

We have also shown that the concept of multi-body partitioning 1T-DRAM cells, demonstrated so far in 2D devices, can be also transferred to 3D structures (FinFET, tri-gate and nanowire transistors), thus enabling memory cells with low voltage operation, energy efficiency, high performance, and fabrication compatibility as embedded memory in next technological nodes. A three dimensional picture of the proposed memory cell, on SOI substrate, is shown in Figure 4. Typical total Fin width can vary from 15 to 25 nm.

We have shown by simulation that the new proposed concept of tridimensional 1T-DRAM cell features a N/P body partitioning that enables the physical separation of hole

storage and electron current. Holes concentration controls the partial or full depletion of the N-core.

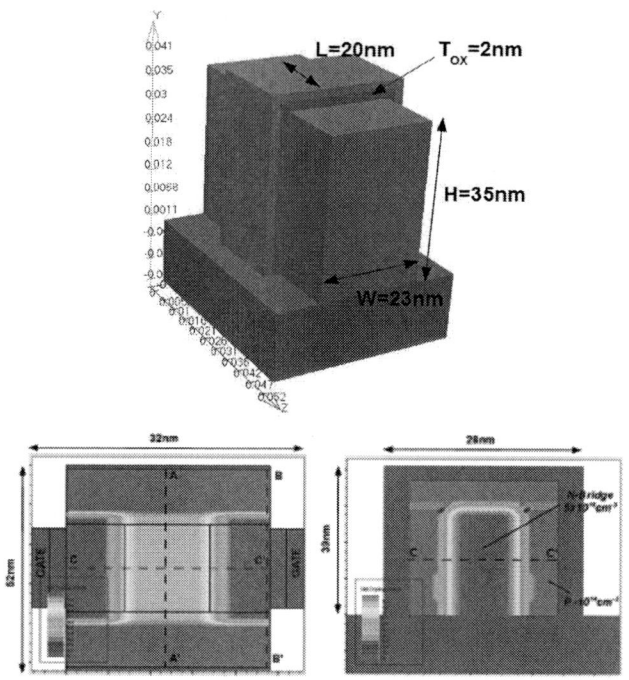

Figure 4. (top) Schematic representation of the nanowire A2RAM memory cell. (bottom-left) Doping cross-section (perpendicular to the BOX, parallel to source an drain) of the nanowire A2RAM. The high doped N-Bridge ($5 \times 10^{18}$cm-3) is surrounded by a low doped P-type layer ($10^{14}$cm-3) (bottom-right). Doping cross-section (parallel to the BOX) of the nanowire A2RAM.

The cell is compatible with ultimate scaling and shows attractive performance (long retention, wide memory window, simple programming, nondestructive reading, and very low-power operation) for embedded systems.

*Acknowledgment*

The authors would like to thank Dr. Maud Vinet and Dr. Francois Andrieu from CEA-LETI for device fabrication and useful discussions.

REFERENCES

[1] K. Kim, U.-I. Chung, Future Trends in Microelectronics. 1. Technology Innovation. Reshaping the Microelectronics Industry, 1st Edition, John Willey and Sons, New Jersey, 2013.
[2] F. Assad-Eraghi, J. Chen, R. Solomon, T. Chan, P. Ko, C. Hu, Time dependence of Fully-Depleted SOI MOSFETs subthreshold current, in: Proceedings IEEE International SOI Conference, Colorado, 1991, pp. 32–33.
[3] K. Kim, Perspectives on giga-bit scaled DRAM technology generation, Microelectronics Reliability 40 (11) (2000) 191–206.
[4] N. Rodriguez, F. Gamiz, S. Cristoloveanu, A-RAM memory cell: Concept and operation, IEEE Electron Dev. Letters 31 (9) (2010) 972–974.
[5] N. Rodriguez, F. Andrieu, C. Navarro, O. Faynot, F. Gamiz, S. Cristoloveanu, Properties of 22nm node extremely-thin-SOI mosfets, in: Proceedings of the 2011 IEEE International SOI Conf., Tempe, Arizona, 2011.
[6] N. Rodriguez, S. Cristoloveanu, F. Gamiz, Novel capacitorless 1T-DRAM cell for 22-nm node compatible with bulk and SOI susbtrates, IEEE Transactions on Electron Devices 58 (5) (2011) 2371–2377.
[7] N. Rodriguez, C. Navarro, F. Gamiz, F. Andrieu, O. Faynot, S.Cristoloveanu, F. Gamiz, Experimental demonstration of capacitorless A2RAM cells on Silicon-On-Insulator, IEEE Electron Device Letters 33 (12) (2012) 1717–1719.

# Bias-Flip Technique for Frequency Tuning of Piezo-Electric Energy Harvesting Devices: Experimental Verification

### Sheng Zhao[1], Yogesh Ramadass[2], Jeffrey H. Lang[3], Jianguo Ma[1] and Dennis Buss[2,3]*

[1] Tianjin University, School of Electronic Information Engineering, 92 Weijin Rd, Nankai Dist., Tianjin, P. R. China;
   E-Mails: shengzhao@tju.edu.cn (S.Z.); majg@tju.edu.cn (J.M.)
[2] Texas Instruments, Inc., 12500 TI Blvd., Dallas, TX 75243, USA;
   E-Mails: Yogesh.Ramadass@ti.com; Buss@ti.com
[3] Massachusetts Institute of Technology, EECS Dept, 77 Mass Ave, Cambridge, MA 02139, USA;
   E-Mail: lang@mit.edu

*Abstract* – Harvesting the maximum energy from mechanical vibration is typically achieved using a High-Q mechanical resonator. A drawback of this approach is that the mechanical resonance frequency $\Omega_t$ must be matched to the source vibration frequency. In [1,2], it was shown through simulation that the Bias-Flip (BF) technique can be used to electronically tune the mechanical spring constant of a Piezo-Electric (PZ) Energy Harvesting Device (EHD) to achieve the match, and thereby achieve substantial power output over a broad range of source vibration frequencies. This paper reports two new results: 1) SPICE simulation confirms that ideal, lossless BF gives output power that is independent of frequency over a wide frequency range, and 2) experiments show that, when the source vibration frequency is far away from $\Omega_t$, BF gives improvement in output power of ~6X.

In [1,2], a simplified model was used to characterize the PZ EHD. In this paper, a Compact Model (CM) is developed for a cantilever PZ EHD shown schematically in Fig. 1. The Mide V25W EHD is an example of such a cantilever structure. This device forms the basis for analysis in this paper.

Shown in Fig. 2 is the output voltage measured as a function of frequency for different resistive loads. For these measurements, the two leafs of the V25W EHD are connected in parallel, as illustrated in Fig 1. These measurements are used to derive the CM parameter values, shown in Table 1. Parameters in the first and second columns are physical CM parameters. Parameters in the third and fourth columns are the equivalent circuit CM parameters, which are illustrated in Fig 6.

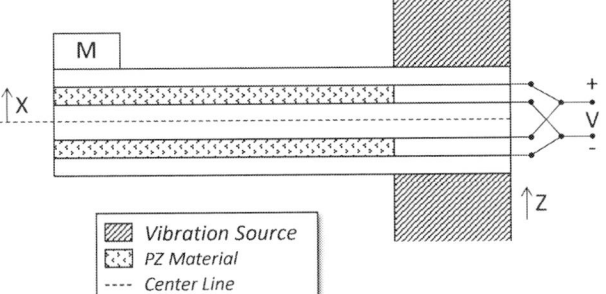

Fig 1: Cantilever EHD structure that is analyzed in this paper

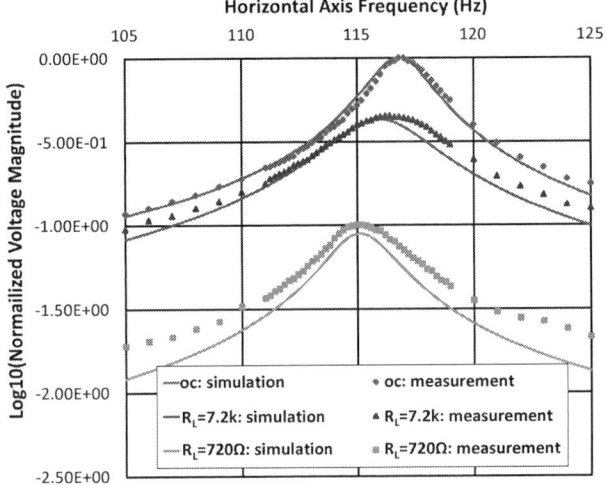

Fig. 2: Magnitude of the sinusoidal voltage at the output of the V25W EHD as a function of resistive load. The input acceleration is 0.2g. The simulation curves are generated from the CM parameters given in Table 1.

### TABLE 1:
### COMPACT MODEL PARAMETERS MEASURED FOR THE MIDE V25W EHD

| $\Omega_t$ | 723.2 sec$^{-1}$ | $L_m = m$ | $4.42\times10^{-3}$ H |
|---|---|---|---|
| $\rho_{\mathit{eff}}$ | 0.03 | $C_m = (L_m \Omega_t^2)^{-1}$ | $4.33\times10^{-4}$ F |
| $C_t$ | $1.91\times10^{-7}$ F | $C_t$ | $1.91\times10^{-7}$ F |
| $Q_t$ | 44.9 | $R_m = m\Omega_t/Q_t$ | $0.071$ $\Omega$ |
| $m$ | $4.42\times10^{-3}$ kg | $A = (\rho_{\mathit{eff}}\Omega_t^2 C_t m)^{1/2}$ | $3.64\times10^{-3}$ |

In [1,2], an equation was derived for the reactive impedance that optimizes the output power at each frequency. The resulting reactive impedance is a complicated function of frequency, but it gives output power that is comparable to the maximum power that can be obtained at resonance. Moreover, this analysis shows that the optimum reactive impedance gives a phase of V, shown in Fig. 3. Note that V is the output voltage of the EHD, defined in Fig 1

978-1-4799-7440-5/14 $31.00 © 2014 IEEE

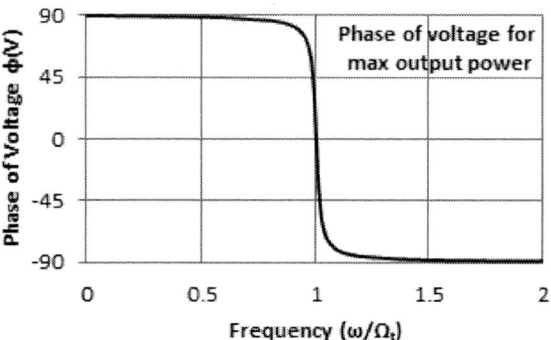

Fig. 3: When the reactance at the output of the EHD is optimized to give max output power, the phase of the voltage shows this behavior independent of the resistive load.

Fig 3 suggests that the BF technique can be used to approximate the max power by flipping the bias polarity according to the following algorithm.

- In Region 1: $\omega < \Omega_t$  $\phi(V) \approx 90^o$  Flip the bias when magnitude of acceleration is max

- In Region 2: $\omega \approx \Omega_t$  $\phi(V) \approx 0^o$  Flip the bias when acceleration is zero

- In Region 3: $\omega > \Omega_t$  $\phi(V) \approx -90^o$  Flip the bias when magnitude of acceleration is max

Note that the BF is non-directional in the sense that, when the BF pulse fires, the voltage flips from positive to negative or negative to positive depending on the initial state. Simulation shows, however, that, in the steady state, V acquires the correct phase with respect to acceleration, if the above algorithm is adopted.

In the linear case when the optimized reactive impedance is used, the resistive load that maximizes output power is given by

$$G_L(\omega) = G_{in}(\omega) = \frac{A^2 R_m}{\left|Z_m(\omega)\right|^2} \qquad \text{Eq(1)}$$

$$Z_m(\omega) = R_m + j\omega L_m + \frac{1}{j\omega C_m} \qquad \text{Eq(2)}$$

In this case, the maximum output power is independent of frequency. Eq(3) gives the max output power when the source acceleration is 1g

$$P_L^{max}(1g) = \frac{1}{8}\frac{mg^2 Q_t}{\Omega_t} = \frac{1}{8}\frac{m^2 g^2}{\eta} \qquad \text{Eq(3)}$$

where η is the mechanical damping coefficient

$$\eta = \frac{m\Omega_t}{Q_t} \qquad \text{Eq(4)}$$

For the CM parameters in Table 1, $P_L^{max}(1g) = 3.3mW$. Eq(3) predicts that max output power can be obtained across a wide range of frequency. However, the voltage required to achieve this power increases away from mechanical resonance, and this puts a practical limit on the frequency range over which the EHD can be tuned.

SPICE simulation has been performed in which the optimized reactive impedance is approximated by an ideal, lossless BF circuit. Fig 4 shows that, the ideal, lossless BF circuit gives output power that is independent of frequency and is within ~10% of the theoretical maximum output power given by Eq 3.

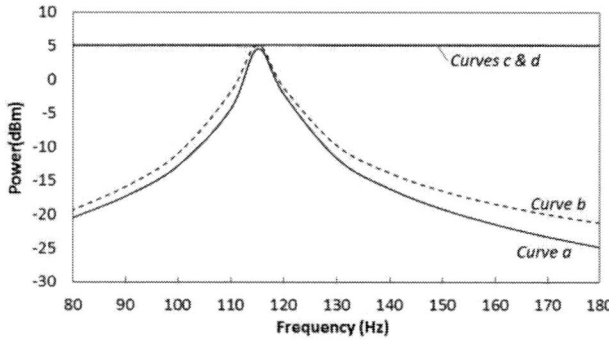

Fig 4: Output power as a function of frequency
a) Neither optimized reactive impedance nor BF is used, and the output circuit consists of a frequency-independent resistive load that is optimized at $\omega = \Omega_t$. The output power decreases sharply away from resonance.
b) The reactive impedance is optimized, but the resistive load is constant with frequency as in curve a. Output power is 3.3mW at resonance, and away from resonance, it is somewhat increased, compared to curve a, as a result of phase alignment.
c) Reactive impedance and resistive load are both optimized. For this case, the output power is 3.3 mW, independent of frequency.
d) The ideal BF is used in place of the optimized reactive impedance, and the resistive load is optimized as in curve c, the output power is ~10% below 3.3 mW. This difference is not discernible on the dbm scale.

The operation of the BF circuit is illustrated in Fig 5 for the case when the EHD is excited at the mechanical resonance frequency. In this case, the equivalent circuit model for the EHD consists of a current source $I_p$ that has the same phase as the acceleration, a capacitor $C_t$, and an internal resistance given by

$$R_{in} = (\rho_{eff} Q_t \Omega_t C_t)^{-1} \qquad \text{Eq(5)}$$

978-1-4799-7440-5/14 $31.00 © 2014 IEEE

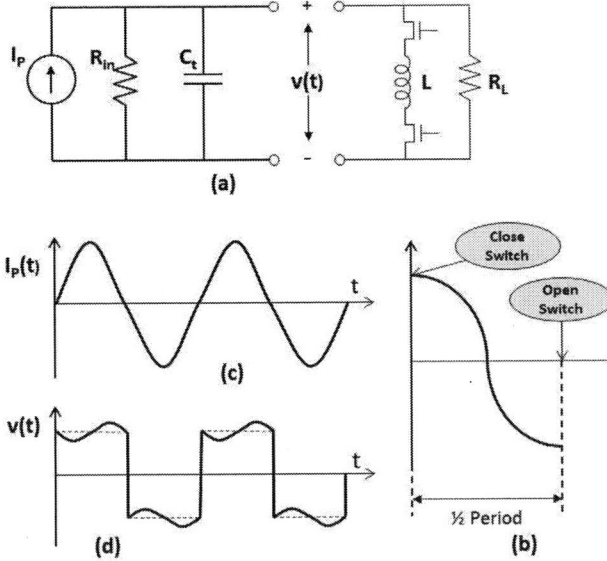

(a)

(c)

(d)

(b)

Close Switch

Open Switch

½ Period

Fig 5: Operation of Bias Flip circuit

(a) A small inductor L is connected to the output through ideal switches

(b) When the switches are closed, the LC tank circuit begins to oscillate. When the switches are opened, half a period later, the sign of the voltage has been adiabatically "flipped"

(c) To achieve maximum power to the load, the bias is flipped when $I_P$ changes sign

(d) The resulting voltage waveform is "in-phase" with the current

Up to this point, we have considered AC power delivered to a resistive load. However, in most applications, we are interested in optimizing DC power delivered to a battery or storage capacitor as illustrated in Fig 6 and Fig 7.

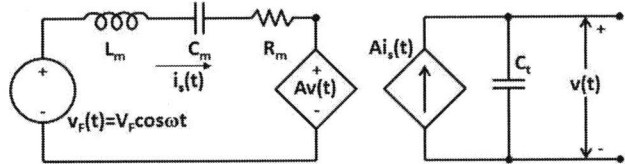

Fig 6: Equivalent Circuit model for the PZ EHD, which is valid for all frequencies. The parameters for this model are given in Table 1.

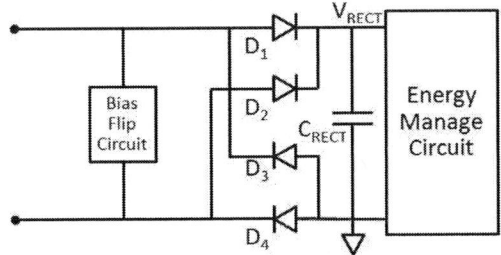

Fig 7: Output circuit for DC rectification and storage

Fig. 8 shows DC power delivered to the storage element with and without the use of BF for three different source vibration frequencies. For the 80Hz and 140Hz measurements, the bias is flipped when the magnitude of the acceleration is maximum. For the 115.1Hz measurements the bias is flipped when the acceleration is zero. The 115.1Hz measurements confirm the results of [3]. However, the 80Hz and 140Hz measurements show an improvement in output power by ~6x.

This is the first time that the BF technique has been experimentally shown to improve output power away from resonance. Improvement in the BF circuitry is in progress. Fig 4 indicates that it should be possible to improve output power by up to 100x power away from resonance.

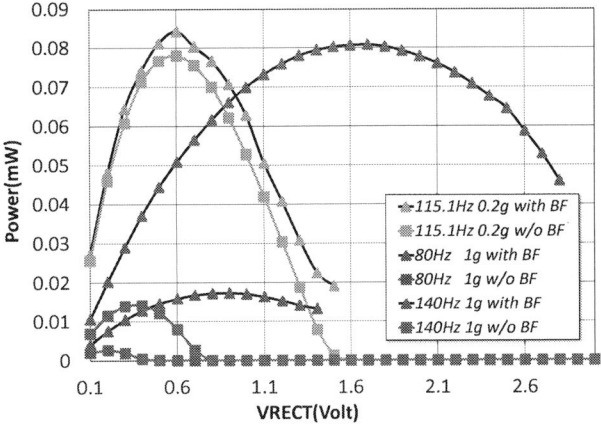

Fig 8: Power delivered to a DC storage element with and without the use of BF. Measurements are made at three frequencies; 80Hz, 115.1Hz and 140Hz. At 80Hz and 140Hz, the output power is improved by 6.5x and 5.7x. At the mechanical resonance frequency (115.1Hz) the improvement is 8%. For the 80Hz and 140Hz measurements, the source acceleration is 1g. For the 115.1Hz experiment, the source acceleration is 0.2g.

## REFERENCES

[1] Zhao, J.; Ramadass, Y.; Buss, D.; Ma, J. Microelectronic Techniques for Frequency Tuning of Piezo-Electric Energy Harvesting Devices. In Proceedings of the 2012 IEEE Subthreshold Microelectronics Conference (SubVt), Waltham, MA, USA, 9–10 October 2012.

[2] Zhao, J; Ramadass, Y; Lang, J; Ma, J; Buss, D. Bias-Flip Technique for Frequency Tuning of Piezo-Electric Energy Harvesting Devices. Journal of Low Power Electronics Applications, May 2013.

[3] Ramadass, Y.; Chandrakasan, A. An Efficient Piezoelectric Energy Harvesting Interface Circuit Using a Bias-flip Rectifier and Shared Inductor. In Proceedings of the IEEE International Solid-State Circuits Conf , San Francisco, CA, USA, Feb 2009.

# Adaptive Subthreshold Switched Capacitor Voltage Boost for Thermoelectric Generation

Rene Brito[1], Mauricio Barba[1], Praveen Palakurthi[1],
David Nemir[2], Eric MacDonald[1]

[1]Electrical and Computer Engineering,
University of Texas at El Paso
El Paso, Texas, USA
[2]TXL Group, Inc.
El Paso, Texas USA

*Abstract* – **Thermoelectric (TE) generation often provides voltages that are not directly usable by traditional electronics as levels are too low from the TE cell. By implementing charge pumps optimized for these ultra-low levels, a proposed circuit technique is described that can boost the TE output to levels that can used for energy harvesting applications. To improve efficiency, the architecture included Maximum Power Point Tracking and ultra-low power digital control, which can be implemented with subthreshold logic.**

*Index Terms – Thermoelectric, Energy Harvesting, Subthreshold Logic*

## I. INTRODUCTION

In many thermoelectric generation applications, the generated voltage is unusably low (< 100 mV) and a DC-DC boost converter is required. This report describes a proposed boost converter that uses switched capacitors to convert a subthreshold voltage levels to higher, more appropriate voltages. By eliminating the relatively bulky and expensive magnetics of existing techniques and by employing low power control, the converter size and cost can be dramatically reduced. The primary deployment is for thermoelectric modules that generate an open circuit voltage of 50 mV to 500 mV and have an internal resistance of between 1 to 10 ohms. By adaptively changing the circuit topology with subthreshold controlled digital logic and the rate at which that topology is switched, a match of source impedance to converter impedance can be made, effectively implementing maximum power point tracking (MPPT) and improving the overall system efficiency as shown in Fig. 1.

Spice modeling demonstrates the feasibility of the approach. Two separate integrated circuits have been designed and fabricated (half micron ON semiconductor) in which a digital core provides all sensing

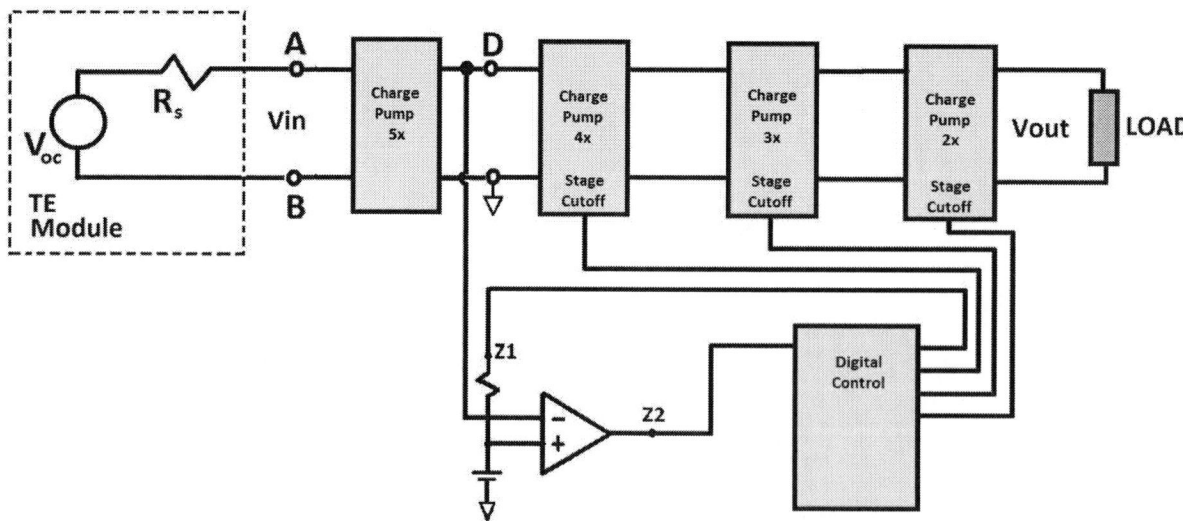

**Fig. 1.** Thermoelectric (TE) System Architecture Block Diagram

978-1-4799-7440-5/14 $31.00 © 2014 IEEE

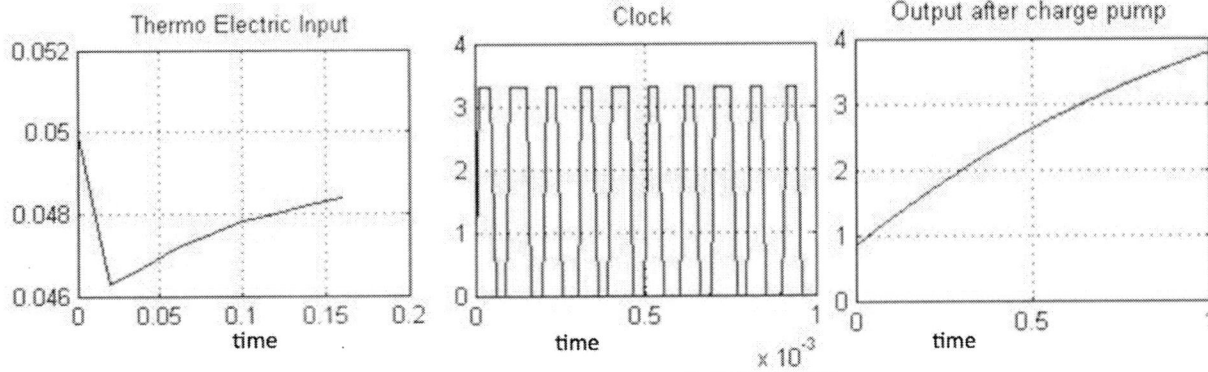

**Fig. 2.** Spice Simulation Results

and control and switching clocks and a second chip which includes the four separate switched capacitor charge pumps. The first chip was designed with the possibility of being operated in the subthreshold for some low performance application points. External capacitors are used to provide charge storage for the charge pumps without requiring any magnetic components, which are large and expensive.

## II. RESULTS

Fig. 2 illustrates the basic waveforms of the two chips working in conjunction to boost a 50 mV output of a two-ohm series resistance thermoelectric cell. By sensing the input, the control determines which of the charge pumps need to be operating to provide to a usable voltage between 3.0 and 3.6 volts. If the input increases, charge pumps are inactivated and circumvented as required to maintain the correct output voltage.

Fig. 3 shows the CAD and actual die photos of the two chips – both of which were 1.5 mm per side and where the digital chip had 28 pins and the analog chip which connected to external capacitors included 40 pins. Both chips were packaged in open-lid ceramic DIP packages.

## REFERENCES

[1] A. Chavan, E. MacDonald, B. Graniello, "Optimized Charge Pumps for Subthreshold Operations", Proceedings of 13th NASA Symposium on VLSI Design, University of Idaho, 2007.

[2] B. Graniello, B. Rodriguez, A. Chavan, E. MacDonald, "Optimized Circuit Styles for Subthreshold Logic", *Electro 2005 Conference*, Oct. 2005.

[3] A. Chavan, E. MacDonald, "Ultra-Low Voltage Level Shifters to Interface Sub and Super Threshold Reconfigurable Logic Cells," *IEEE Aerospace Conference*, March 2008.

[4] A. Chavan, P. Palakurthi, E. MacDonald, J. Neff, E. Bozeman, "Heavy Ion Characterization of a Radiation Hardened Flip-Flop Optimized for Subthreshold Operation," *Journal of Low Power Electronics and Applications*, March 2012.

[5] E. Carlson, S. Kai, and B. Otis. "A 20 mV input boost converter with efficient digital control for thermoelectric energy harvesting." *Solid-State Circuits, IEEE Journal of* 45.4 (2010): 741-750.

[6] J. Damaschke, "Design of a Low-Input-Voltage Converter for Thermoelectric Generator", *IEEE Transactions on Industry Applications* 33.5 (1997): 1203-1207.

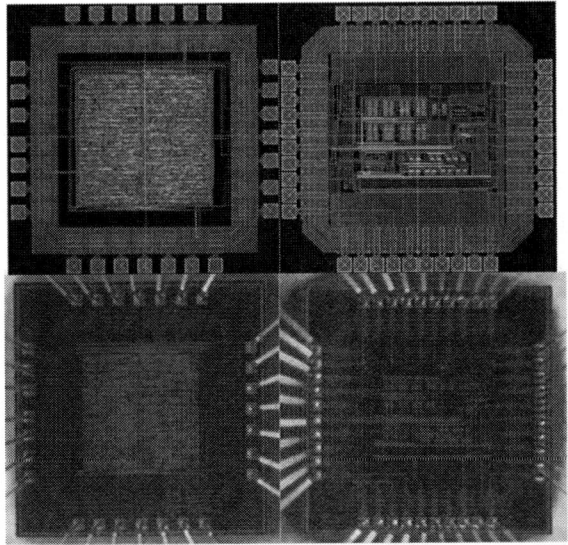

**Fig. 3.** Digital Chip (left) and Charge Pump Chip (right) – CAD top and Pictures below

# SOI Substrate Solutions for Recent Advanced Device Applications

N. Noto, O. Ishikawa, H. Aga, T. Ishizuka, I. Yokokawa and M. Nakano
Technology and Development Division, Shin-Etsu Handotai Co. Ltd.,
6-2 Otemachi 2, Chiyoda, Tokyo, Japan
E-mail n.noto@seh.jp

## 1. Introduction

SOI substrate technology for recent advanced device applications is reviewed from a standpoint of a dedicated wafer supplier. A structure of Si thin film on sapphire (SOS) has been used for device development and manufacturing from 1970's. Currently a SOI substrate of Si/SiO2/Si structure is widely used for industrial applications.

As for the SOI substrate with Si/SiO2/Si structure, a thickness combination of each layer is selected according to their applications. A thickness of a device layer (surface Si layer) is varied within very wide range from several 10nm to over 10um according to their applications. It can be classified either thin-SOI or thick-SOI according to a device layer thickness, which is less than or larger than 1um in general. A thickness of SiO2 (buried oxide: BOX) layer, which existence characterizes SOI structure, has also very wide range from several 10nm to several um according to their device requirements.

Thick-SOI substrate is used for the application of BCD devices, high voltage devices, sensors and various types of MEMS. Thin-SOI is mainly used for an application of high-end CMOS as PDSOI devices. It has been used for applications of high-end servers, MPUs for PC and consumer games. These devices have been produced in mass production by using 300mm thin-SOI from 2002.

There has been developed many SOI substrate fabrication technologies such as BG (bonding, grinding and polishing) method [1, 2], SIMOX [3], SmartCut® [4], Eltran® [5] and rT-CCP® [6]. Each technology has unique advantages and there have been great useful discussions in academic conferences and industrial meetings.

A BG (bonding, grinding and polishing) method is mainly used for fabricating thick-SOI. Furthermore, epitaxial growth on thin-SOI is also used for fabricating thick-SOI substrate, in case that tight thickness control and/or resistivity control are required for device layer.

A device layer thickness under 100nm is used for the PDSOI device applications, in general. A layer transfer technology of SmartCut® is mainly used as a fabricating method of thin-SOI substrate and it was reported in 1995 in this conference. This layer transfer technology made it possible to supply the device-demanded quality of 300mm thin SOI substrate with high volume production level.

From the recent progress of mobile technologies and cloud computing, some of new SOI device applications such as FDSOI devices, RF devices and Si photonics have generated an outstanding interest. SOI substrate solutions for such newly applications are reviewed.

## 2. FDSOI

Recently there is a very attractive progress in FDSOI device development for SOC application, which realizes high speed, lower power consumption and low VDD operation.

There has been a big difficulties for scaling in CMOS logic around 20nm design rule (DR), 3D type of bulk-FINFET technology with FD operation has announced to start mass production in 2011 and a further development has pushed forward to 14/16nm, and beyond, design rules. FDSOI is an evolutionary innovation because it has an advantage of being a planar transistor with FD operation. A mass production stage of 28nm DR devices has already developed and 14nm and 10nm devices are under development.

In case of SOI substrate structure for this FDSOI application, the initial thickness of device layer is less than 15nm for 28nm DR. An important device factor Vth is directly affected by a channel thickness of FDSOI device, therefore a very tight control of device layer thickness is required for SOI substrate.

A thickness variation of +/- 0.5nm in entire 300mm SOI substrate is already able to supply at mass production level. Fig.1 shows cross section TEM image and thickness uniformity of 300mm SOI substrate for FDSOI application. In order to use of hybrid system of bulk-devices with SOI-devices and to use an effective body biasing technology, there is a requirement of thin BOX structure with less than 30nm. A 300mm thin SOI substrate which can meet tight thickness control of device layer and BOX layer has been already developed.

## 3. RF-SOI

RF devices using an SOI substrate have been developed as an attractive technology for explosive expansion of smartphone applications. Besides SOS (silicon on sapphire) and/or GaAs are being used for the fabrication of RF devices, SOI substrate has advantages of large diameter, low cost and utilizing accumulated Si CMOS technology.

A RF front-end part of smartphone has been developed aggressively using SOI CMOS technology which enables an integration of not only RF switches but also power amplifiers and antenna tuners. It will be an attractive technology in considering a spreading of IOT/M2M technology.

Thin-SOI substrate structure is used for RF applications. It has a 1-2um of thick-BOX and high resistivity handle wafer, realizing high performance of RF characteristics.

A Si wafer technology, with over a resistivity of 1000 ohm-cm with low oxygen content, allows the production of a SOI substrate for a RF device application at production level. For the better RF harmonics performance, to use high resistivity substrate is the key technique. The element to impediment a stable high resistivity in substrate is oxygen donor. To minimize this effect it can be one solution to use low oxygen concentration substrate.

In order to improve the RF characteristics (crosstalk, nonlinearity) by suppressing the parasitic surface condition (PSC) effect, a new SOI structure with so-called "trap-rich layer", which is installed beneath the BOX layer, has been developed. This layer can be achieved with several methods such as Si amorphous layer, implanted damage layer and poly Si crystalline layer. A SOI substrate with trap-rich structure can realize a similar RF performance to the devices manufactured from a SOS substrate and it is easily applicable to a large-diameter (200mm/300mm).

Since the SOI substrate characteristics (resistivity of handle wafer, w/wo trap-rich layer) directly affects the device performance, a design of SOI substrate structure is required excellent performance and stable quality in production level.

## 4. Si Photonics

A development of Si Photonics for fusing the optical technology and Si CMOC technology has been studied actively. A progress of related technologies (an achievement of CMOS scaling, a dramatic improvement of device performance and a realizing of several 100 nm thin SOI substrate) and demand for handling big data in cloud computing is the background of this development.

Thin SOI substrate allows a fabrication of fine waveguide by using a refractive index very different SiO2 and Si. Board-to-board communication is mainly being developed currently. Furthermore the technology of chip-to-chip and inter-chip communication has also been studied for future high speed and lower power applications.

SOI substrate structure for Si Photonics is thin-SOI with around 200nm thick of device layer and around 2um thick of BOX layer. As thickness variation and micro roughness of device layer affects the optical loss of waveguide, a further improvement has been discussed.

Three of recent topics were reviewed. The fabrication of the device using an SOI substrate has attracted attentions as a technology to meet the demands of the mobile markets in recent years.

## References

[1] T.Abe et al., The Fourth Internal Symposium on Silicon on Insulator technology and Devices, PV90-6, p61, 1990
[2] K.Mitani et al., The Fourth Internal Symposium on Semiconductor Wafer Bonding: Science, Technology, and Applications, PV97-36, p1, 1997
[3] K.Izumi et al., Electronics Letters, vol.14 p593, 1978
[4] M.Bruel, Electronics Letters, vol31 No.14 p1201, 1995
[5] T.Yonehara et al., Applied Physics letter Vol.64, p2108, 1994
[6] S.Farrens et al., The Ninth Internal Symposium on Silicon on Insulator technology and Devices, PV99-3, p122, 1999

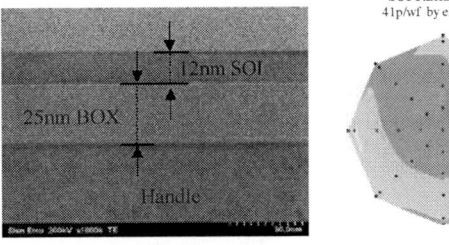

Fig.1 TEM photograph and SOI thickness disribution for FDSOI application (300mm SOI Substrate).

# The Role of Radiation Effects in SOI Technology Development

Les Palkuti[1], Michael Alles[2], Harold Hughes[3]

[1]Defense Threat Reduction Agency, Ft Belvior, VA
[2]Vanderbilt University, Nashville, TN
[3]Naval Research Laboratory, Washington, DC
Les.palkuti@DTRA.mil 703 767-7455

*Abstract— The technical development of radiation resistant CMOS/SOI technologies is reviewed. Inherent hardness of SOI to dose rate upset and latchup has leveraged major developments of SOI technologies. TID hardness for up to the 150nm node was addressed by process hardening. Inherent hardness of 45nm and 32nm technologies reduced the need for TID hardening. As technology is scaled to 28nm and 14nm nodes TID hardening is again required. SEU hardening is addressed by circuit design for a wide range of technologies with significant SEU improvement for 28nm and 14nm is observed.*

*Keywords—Radiation resistant; CMOS/SOI; TID effects; SEE effects*

## I. INTRODUCTION

Silicon-on-insulator (SOI) electronics has a history of about 50 years,, including primary embodiments as dielectric-isolation(DI), silicon-on-sapphire (SOS) [1], and thick and thin silicon on silicon-dioxide (SOI) [2]. SOI has been, and continues to be, used for bipolar, LDMOS, CMOS, optical, and other device configurations. SOI wafers have evolved from small diameter (100 mm) to 300 mm diameter, with 450 mm in development. Commercial SOI technology nodes have moved from >250nm to 14nm with 10nm in development - over two orders of magnitude reduction. This paper will present the target requirements for rad-hard applications, connect the attributes of SOI to these requirements, and discuss the role of these requirements in driving the development of digital CMOS SOI technology.

## II. BACKGROUND

The target requirements for rad-hard applications are outlined in Table I. These requirements are given for total dose (TID), single-event upset (SEU), latchup (SEL), transient upset (DR Upset). As SOI technologies are being developed for rad-hard applications it is typical to specify levels that are required for generic missile and space applications.

TABLE I. RADIATION CRITERIA

| PARAMETER | GOAL | REQUIREMENT |
|---|---|---|
| TID (krad(Si)) | 1,000 | 300 |
| SEU (errors/bit-day) | $< 10^{-11}$ | $< 10^{-10}$ |
| SEL | NONE | NONE |
| DR Upset (radSi)/s | | $> 10^8$ |

In addition, these development projects also specify goals to meet the needs of systems that operate in extreme environments.

SOI is well suited to meet rad-hard requirements due to the substrate isolation. One of the driving factors throughout the history of SOI has been the benefit of reduced volumes for collection of charge from transient radiation, including terrestrial alpha particles and neutrons, as well as protons and heavy ions in space, and gamma pulses in weapons environments. The amplification of charge induced by the parasitic bipolar effects in SOI CMOS is a somewhat offsetting factor, and has played a role in the development of approaches to mitigate floating-body effects (body ties, etc.). The isolation of devices from one another eliminates the parasitic path for latchup (including that due to heavy ions in space). The inherent dose-rate hardness of the SOI technology leveraged for circuit operation was a major consideration for rad-hard applications. These attributes have been responsible for a substantial portion of investment in rad-hard CMOS/SOI over a time span of decades, particularly by the U.S. government.

## III. TID EFFECTS

A primary challenge for rad-hard CMOS/SOI is to mitigate the effects of TID namely the SOI-device edge leakage, front and backchannel coupling effects (from charge trapping in the buried oxide) and floating body effects. For partially depleted (PD) SOI devices at the 150nm node substantial effort has gone to development of process optimization particularly modification of the front and backchannel insulator properties, adjustment of the doping profiles along the device body (well)/device edge and development of special body contacts. These process modifications were particularly successful for the CMOS/SOI 150nm technology node where Honeywell developed a series of SOI products including memories, processors, ASICs and communication (SERDES) devices that met the TID stretch goals and far exceeded the capability of commercial devices (see Fig.1) and were also QML qualified for space applications [3]. At this time, developments are underway to enhance the 150nm technology to a partially-scaled 90nm technology using 1D-layout methods to meet low-power defense applications.

Unfortunately, the TID hardening of 130nm, 90nm and 65nm commercial SOI nodes using minimally invasive process modifications (extensively used for bulk technologies) was not successful and thus no TID-hard CMOS/SOI technologies at these nodes are available. As shown in Fig. 1 inherent resistance to TID effects was observed for PD-SOI

978-1-4799-7440-5/14 $31.00 © 2014 IEEE

technologies at the 45nm to 32nm nodes. Therefore extensive developments of rad-hard ASIC by design and layout methods are being pursued using commercial processes without the need for TID hardening. With the intendant use of fully-depleted (FD) planar and FinFET technologies at and below the 28 nm nodes, the issue of TID has to be revisited since the use of un-(or lightly)-doped active regions can allow efficient coupling of trapped charge to the device threshold voltage. Scaling to thinner buried oxide regions, changes in device architecture, use of minimally-invasive process changes and substrate engineering, are approaches to mitigate these effects.

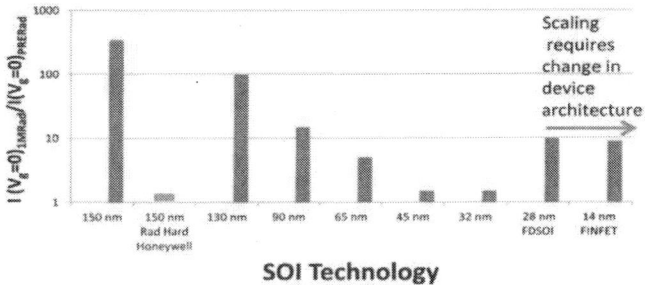

Fig. 1. TID hardness of commercial (red) and rad-hard (green) CMOS/SOI devices.

## SEE EFFECTS

Single events in SOI devices are reduced because the SOI film thickness limits the charge collection. For this reason, SOI devices are consistently less sensitive than bulk devices. For devices at the 150nm node, design and layout optimization (body ties) and introduction of extra capacitors at sensitive nodes resulted in the successful development of circuits with SEU resilience that exceeds the error rate goal. In addition, error rates are also reduced by error correction, memory scrubbing and redundant circuits. Using the 150nm SOI technology, SRAMS were fabricated that have single-bit and multi-cell error rates that are essentially immune to SEU. [3]

The SEU cross sections for commercial SOI SRAMs for the technologies from 90nm and below are shown in Fig. 2 [4]. The limiting cross section shows an order of magnitude reduction between the 90nm and 65nm nodes because of operating voltage reduction and changes in the ratio of collected charge to critical charge. From the 65nm to the 32nm nodes the SEU cross sections are relatively constant indicating similarities in technology and the change from a $SiO_2$ gate stack to a high-k/metal gate stack had only minor effect on the error rates. Another order of magnitude drop in the error cross section is seen as scaling from 32nm PD technology to the 28nm FD technology and also for the 14nm FinFETs [5]. Using TCAD modeling that compares the charge collection to the critical charge clarifies the cross sections shown in Fig. 2.

Note that these cross sections were measured on SRAM's that are used as process monitors and thus the cross section includes both single and multiple-bit upsets. In SRAMs designed for circuit implementation the cell bits are interleafed and physically separated so that multicell upsets are rare. Double bit error detection and single bit error correction along

with scrubbing are typically used to reach the required error rate. Addition of capacitance into the memory cell can also significantly improve the threshold LET and thus reduce the required error correction.

The single event cross section for standard latches are typically similar (within a factor of 2) to those shown in Fig. 2. The limiting cross sections for 45nm and 32nm standard latches are approximately $10^{-9}$ cm$^2$/latch. Several radiation-hard latches were developed using device stacking, redundant nodes (DICE) and layout modifications that incur power, delay or area penalties. It is important to characterize these SEE hard latches over a wide range of incident ion angles. A reduction of approximately $10^3$ in the total error rate can be achieved with these rad-hard latches [6]. It is also important to harden to single-event transients by incorporating delay elements in the hard latches, correctly sizing latches and adding capacitive filters.

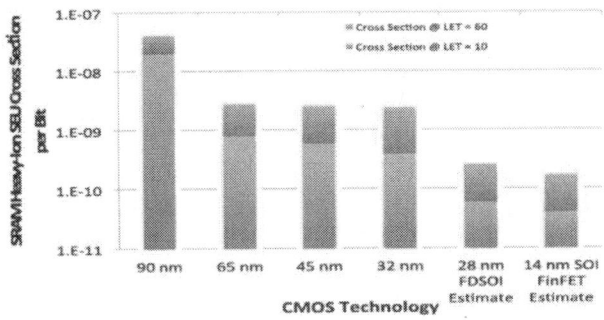

Fig. 2 Examples of SRAM cross sections data (at both LET values of 10 and 60 MeVcm/mg for advanced SOI technologies.

## SUMMARY

Rad-hard electronics has played a major role in the advancement of SOI technology, including development of SOI materials, devices and designs. The inherent resistance to latchup, and small volumes for transient photocurrent generation motivated the support of development of SOI for many years prior to the expansion into commercial applications for reduced power and soft-error resistance.

### ACKNOWLEDGMENT

The authors wish to acknowledge the support of the DTRA radiation technology program

### REFERENCES

[1] L. J. Palkuti, et. al, IEEE Trans. Nucl. Sci., 23 (6), 1715 (1976).

[2] J. R. Schwank, et. al, IEEE Trans. Nucl. Sci., 50(3), 522 (2003).

[3] http://aerospace.honeywell.com/products/communication-nav-and-surveillance/navigation/space-navigation-and-microelectronics/microelectronics/radiation-hardened-electronics-and-technology

[4] K. Rodbell, NSREC Short Course 2013

[5] P Roche et.al IEDM Tech Digest, 2013

[6] J. S. Kappila, et. al, IEEE Rel Phy Sym, 2014

# A very low power CMOS 28FDSOI programmable fractional frequency divider for Wifi-WiGig

Mathieu Vallet, Olivier Richard[1], Yann Deval[2], Didier Belot[1]

STMicroelectronics, 850 Rue Jean Monnet, 38920 Crolles, France[1]
IMS Bordeaux, 351 Cours de la Libération, 33405 Talence, France[2]
mathieu.vallet@st.com

*Abstract*—**A 2.4 GHz very low-power programmable fractional frequency divider is presented in this work. A new kind of pulse swallowing architecture is exposed, offering news possibilities in terms of frequency range, frequency step and power consumption. This circuit was designed and implemented in 28nm FDSOI CMOS from STMicroelectronics. The divider consumes 300 µA over 1 V.**

*Keywords— fractional frequency divider; low power; 28nm FDSOI CMOS; Wifi-WiGig convergence*

## I. INTRODUCTION

The fractional frequency divider presented in this paper is integrated into a WiGig-Wifi Transceiver. The divider must be able to meet the requirements of the Wifi and WiGig standards, forcing the operating frequency range to be bigger than the one required for fractional frequency divider dedicated to Wifi standard applications. Finally the division rank is comprised between 89 and 110, allowing the phase synchronization of the output signal with a 26MHz clock reference frequency.

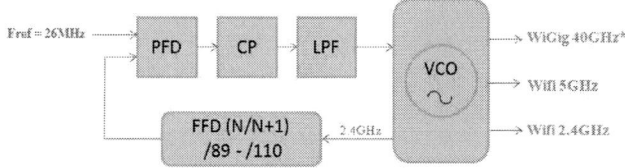

*Fig. 1. Fractional frequency divider context*

## II. CIRCUIT DESCRIPTION AND DESIGN

### A. 28nm FDSOI CMOS technology

The 28nm FDSOI CMOS is a Silicon on Insulator Fully Depleted technology, relying on ultra-thin layer of silicon over a buried oxide. This technology allows a global control of the MOS threshold voltage, via the bulk, offering new possibilities in terms of transistor control. Low power capability is attained taking advantage of the threshold voltage variation, enabling the design of ultra-low power circuits. As shown in this paper, the high frequency ability offered by this 28nm FDSOI CMOS technology gives the opportunity to do a fractional frequency divider without pre-division stage, directly driving a 2.4GHz counter. Thus, it is possible to consider new architecture perspectives taking advantage of the benefits given by this technology. The circuit has been designed in 10 metals levels in order to reach higher RF performances.

### B. Overall functioning

A pulse swallowing architecture has been selected to realize the fractional division. This kind of architecture is very close to the integer-N frequency divider, including a divider modulus shift from N to N+1. A division rank (N) and a rehearsal of division by N or by N+1 are selected by the user depending on the desired division ratio, knowing that the final division is corresponding to the average division ratio on 26 cycles define by :

$$Fractional\ division = \frac{N \times M + (N+1) \times (26-M)}{26} \quad (1)$$

With N being the desired division rank, and M the number of time that the divider divides by N (rather than N+1).

All the division ratio have been determined previously and have been inserted inside a summary table (Fig. 2).

| band | freq. center (MHz) | 89 | 90 | 92 | 93 | 94 | 95 | 96 | 97 | 98 | 99 | 100 | 101 | 102 | 103 | 104 | 105 | 106 | 107 | 108 | 109 | 110 | division |
|---|---|---|---|---|---|---|---|---|---|---|---|---|---|---|---|---|---|---|---|---|---|---|---|
| Wifi 2.4 GHz | 2412 | 0 | 0 | 6 | 20 | 0 | 0 | 0 | 0 | 0 | 0 | 0 | 0 | 0 | 0 | 0 | 0 | 0 | 0 | 0 | 0 | 0 | 92.769 |
| | 2417 | 0 | 0 | 1 | 25 | 0 | 0 | 0 | 0 | 0 | 0 | 0 | 0 | 0 | 0 | 0 | 0 | 0 | 0 | 0 | 0 | 0 | 92.962 |
| | 2422 | 0 | 0 | 0 | 22 | 4 | 0 | 0 | 0 | 0 | 0 | 0 | 0 | 0 | 0 | 0 | 0 | 0 | 0 | 0 | 0 | 0 | 93.154 |
| | 2427 | 0 | 0 | 0 | 17 | 9 | 0 | 0 | 0 | 0 | 0 | 0 | 0 | 0 | 0 | 0 | 0 | 0 | 0 | 0 | 0 | 0 | 93.346 |
| | 2432 | 0 | 0 | 0 | 12 | 14 | 0 | 0 | 0 | 0 | 0 | 0 | 0 | 0 | 0 | 0 | 0 | 0 | 0 | 0 | 0 | 0 | 93.538 |
| | 2437 | 0 | 0 | 0 | 7 | 19 | 0 | 0 | 0 | 0 | 0 | 0 | 0 | 0 | 0 | 0 | 0 | 0 | 0 | 0 | 0 | 0 | 93.731 |
| | 2442 | 0 | 0 | 0 | 2 | 24 | 0 | 0 | 0 | 0 | 0 | 0 | 0 | 0 | 0 | 0 | 0 | 0 | 0 | 0 | 0 | 0 | 93.923 |
| | 2447 | 0 | 0 | 0 | 0 | 23 | 3 | 0 | 0 | 0 | 0 | 0 | 0 | 0 | 0 | 0 | 0 | 0 | 0 | 0 | 0 | 0 | 94.115 |
| | 2452 | 0 | 0 | 0 | 0 | 18 | 8 | 0 | 0 | 0 | 0 | 0 | 0 | 0 | 0 | 0 | 0 | 0 | 0 | 0 | 0 | 0 | 94.308 |
| | 2457 | 0 | 0 | 0 | 0 | 13 | 13 | 0 | 0 | 0 | 0 | 0 | 0 | 0 | 0 | 0 | 0 | 0 | 0 | 0 | 0 | 0 | 94.500 |
| | 2462 | 0 | 0 | 0 | 0 | 8 | 18 | 0 | 0 | 0 | 0 | 0 | 0 | 0 | 0 | 0 | 0 | 0 | 0 | 0 | 0 | 0 | 94.692 |
| | 2467 | 0 | 0 | 0 | 0 | 3 | 23 | 0 | 0 | 0 | 0 | 0 | 0 | 0 | 0 | 0 | 0 | 0 | 0 | 0 | 0 | 0 | 94.885 |
| | 2472 | 0 | 0 | 0 | 0 | 0 | 24 | 2 | 0 | 0 | 0 | 0 | 0 | 0 | 0 | 0 | 0 | 0 | 0 | 0 | 0 | 0 | 95.077 |
| Wifi 5 GHz | (5180/2) => 2590 | 0 | 0 | 0 | 0 | 0 | 0 | 0 | 0 | 0 | 10 | 16 | 0 | 0 | 0 | 0 | 0 | 0 | 0 | 0 | 0 | 0 | 99.615 |
| | (5200/2) => 2600 | 0 | 0 | 0 | 0 | 0 | 0 | 0 | 0 | 0 | 0 | 26 | 0 | 0 | 0 | 0 | 0 | 0 | 0 | 0 | 0 | 0 | 100.000 |
| | (5220/2) => 2610 | 0 | 0 | 0 | 0 | 0 | 0 | 0 | 0 | 0 | 0 | 16 | 10 | 0 | 0 | 0 | 0 | 0 | 0 | 0 | 0 | 0 | 100.385 |
| | (5240/2) => 2620 | 0 | 0 | 0 | 0 | 0 | 0 | 0 | 0 | 0 | 0 | 6 | 20 | 0 | 0 | 0 | 0 | 0 | 0 | 0 | 0 | 0 | 100.769 |
| | (5260/2) => 2630 | 0 | 0 | 0 | 0 | 0 | 0 | 0 | 0 | 0 | 0 | 0 | 22 | 4 | 0 | 0 | 0 | 0 | 0 | 0 | 0 | 0 | 101.154 |
| | (5280/2) => 2640 | 0 | 0 | 0 | 0 | 0 | 0 | 0 | 0 | 0 | 0 | 0 | 12 | 14 | 0 | 0 | 0 | 0 | 0 | 0 | 0 | 0 | 101.538 |
| | (5300/2) => 2650 | 0 | 0 | 0 | 0 | 0 | 0 | 0 | 0 | 0 | 0 | 0 | 2 | 24 | 0 | 0 | 0 | 0 | 0 | 0 | 0 | 0 | 101.923 |
| | (5320/2) => 2660 | 0 | 0 | 0 | 0 | 0 | 0 | 0 | 0 | 0 | 0 | 0 | 0 | 18 | 8 | 0 | 0 | 0 | 0 | 0 | 0 | 0 | 102.308 |
| | (5500/2) => 2750 | 0 | 0 | 0 | 0 | 0 | 0 | 0 | 0 | 0 | 0 | 0 | 0 | 0 | 0 | 0 | 6 | 20 | 0 | 0 | 0 | 0 | 105.769 |
| | (5520/2) => 2760 | 0 | 0 | 0 | 0 | 0 | 0 | 0 | 0 | 0 | 0 | 0 | 0 | 0 | 0 | 0 | 0 | 22 | 4 | 0 | 0 | 0 | 106.154 |
| | (5540/2) => 2770 | 0 | 0 | 0 | 0 | 0 | 0 | 0 | 0 | 0 | 0 | 0 | 0 | 0 | 0 | 0 | 0 | 12 | 14 | 0 | 0 | 0 | 106.538 |
| | (5560/2) => 2780 | 0 | 0 | 0 | 0 | 0 | 0 | 0 | 0 | 0 | 0 | 0 | 0 | 0 | 0 | 0 | 0 | 2 | 24 | 0 | 0 | 0 | 106.923 |
| | (5580/2) => 2790 | 0 | 0 | 0 | 0 | 0 | 0 | 0 | 0 | 0 | 0 | 0 | 0 | 0 | 0 | 0 | 0 | 0 | 18 | 8 | 0 | 0 | 107.308 |
| | (5600/2) => 2800 | 0 | 0 | 0 | 0 | 0 | 0 | 0 | 0 | 0 | 0 | 0 | 0 | 0 | 0 | 0 | 0 | 0 | 8 | 18 | 0 | 0 | 107.692 |
| | (5620/2) => 2810 | 0 | 0 | 0 | 0 | 0 | 0 | 0 | 0 | 0 | 0 | 0 | 0 | 0 | 0 | 0 | 0 | 0 | 0 | 24 | 2 | 0 | 108.077 |
| | (5640/2) => 2820 | 0 | 0 | 0 | 0 | 0 | 0 | 0 | 0 | 0 | 0 | 0 | 0 | 0 | 0 | 0 | 0 | 0 | 0 | 14 | 12 | 0 | 108.462 |
| | (5660/2) => 2830 | 0 | 0 | 0 | 0 | 0 | 0 | 0 | 0 | 0 | 0 | 0 | 0 | 0 | 0 | 0 | 0 | 0 | 0 | 4 | 22 | 0 | 108.846 |
| | (5680/2) => 2840 | 0 | 0 | 0 | 0 | 0 | 0 | 0 | 0 | 0 | 0 | 0 | 0 | 0 | 0 | 0 | 0 | 0 | 0 | 0 | 20 | 6 | 109.231 |
| | (5700/2) => 2850 | 0 | 0 | 0 | 0 | 0 | 0 | 0 | 0 | 0 | 0 | 0 | 0 | 0 | 0 | 0 | 0 | 0 | 0 | 0 | 10 | 16 | 109.615 |
| WiGig | (37200/16) => 2325 | 15 | 11 | 0 | 0 | 0 | 0 | 0 | 0 | 0 | 0 | 0 | 0 | 0 | 0 | 0 | 0 | 0 | 0 | 0 | 0 | 0 | 89.423 |
| | (39360/16) => 2460 | 0 | 0 | 0 | 0 | 10 | 16 | 0 | 0 | 0 | 0 | 0 | 0 | 0 | 0 | 0 | 0 | 0 | 0 | 0 | 0 | 0 | 94.615 |
| | (41520/16) => 2595 | 0 | 0 | 0 | 0 | 0 | 0 | 0 | 0 | 0 | 5 | 21 | 0 | 0 | 0 | 0 | 0 | 0 | 0 | 0 | 0 | 0 | 99.808 |
| | (43680/16) => 2730 | 0 | 0 | 0 | 0 | 0 | 0 | 0 | 0 | 0 | 0 | 0 | 0 | 0 | 0 | 0 | 26 | 0 | 0 | 0 | 0 | 0 | 105.000 |

*Fig. 2. Summary of the division rank*

## C. Architecture

The proposed fractional frequency divider could be decomposed in 6 parts. Among them, there are two counters, two comparators, one signal output generator and one adder.

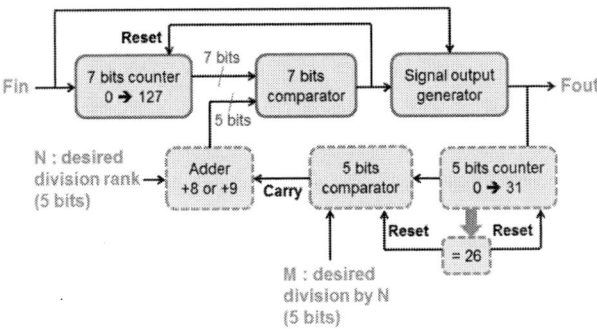

*Fig. 3.* Frequency divider architecture, with elements working at 2.4Ghz (surrounded by bold) and elements working at 26MHz (surrounded by dash)

The first block is a 7 bits counter, counting the number of input signal state changes. The choice to count the state rather than the rising edge of the input signal has been decided in order to offer the possibility to divide by odd number, without the use of pre-scalar divider cell, allowing the generation of an output signal with duty cycle of 50%. A pre-circuit has been added in the front of the first latch allowing a correct reset of the counter regardless the LSB state during the reset action.

The 7 bits comparator is checking the matching between the desired division rank selected by the user (5 bits) and the 5 least significant bits of the counter output. Whenever the desired division rank is reached by the counter, the 7 bits comparator generates a pulse at the output, resulting in a change of the output state and a rising edge on the clock of the second counter (the 5 bits counter).

The signal output generator is composed of two D latches. The first receives the pulses of the 7 bits comparator output as input, thereby generating an output signal of 26MHz. While the second D latch is intended to reduce the divider noise by synchronizing the divider output with the frequency divider.

The 5 bits counter is used to count the number of the time a division by N or N+1 occurred. An automatic reset of this counter appears after 26 cycles, indicating the end of a period where the division average is equal to the desired fractional division. The second counter is directly linked to the 5 bits comparator, via 5 outputs corresponding to the number of division appeared in the cycle.

The role of the 5 bits comparator is to emit a "0" output signal if the number of divisions by N is inferior to the number selected by the user, or to send a "1" output if the counter is over counting (until 26 counting cycles are performed).

The adder +8 or +9 enables an artificial modification of the desired division rank in the aim to carry out a division by N+1 rather than N, when its carry input (the 5 bits comparator output) is taking the value of "1".

One of the main advantages of this new architecture is the low operating frequency of the modulus controller, enabling global low power consumption and highlighting the existence of a single critical path (the 7 bits counter reset path).

## D. Layout

A careful approach of the layout design was required in order to ensure a minimum contribution of the undesired parasites. Dummies had to be implemented in parts of the layout to respect the density rules of the 28nm FDSOI CMOS technology.

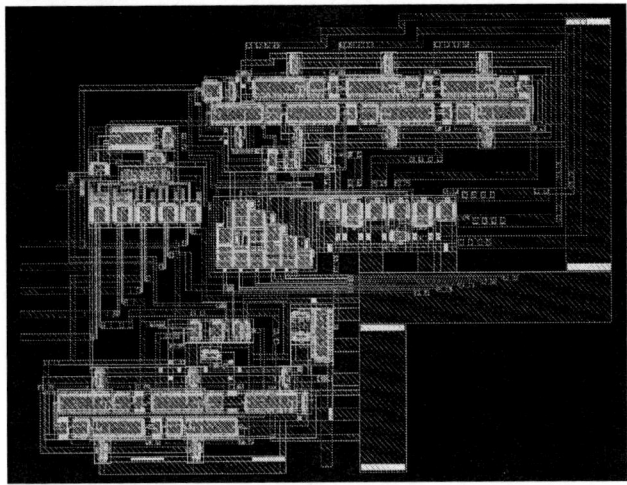

*Fig. 4.* Fractional frequency divider layout

## III. SIMULATIONS

The simulations pointed out the counters 7 bits as the most critical block. The pre-circuit, located at the 7 bits counter entry, ensures proper operation of the divider, as the division rank are even or odd. After analyzing the LSB state value of the counter, which causes the reset, the pre-counter circuit allows an eventual inversion of the LSB state to undo the counter delay. Thus in the case of odd division rank, the counter will gain one state delay, assuring that, the next counting cycle is correct.

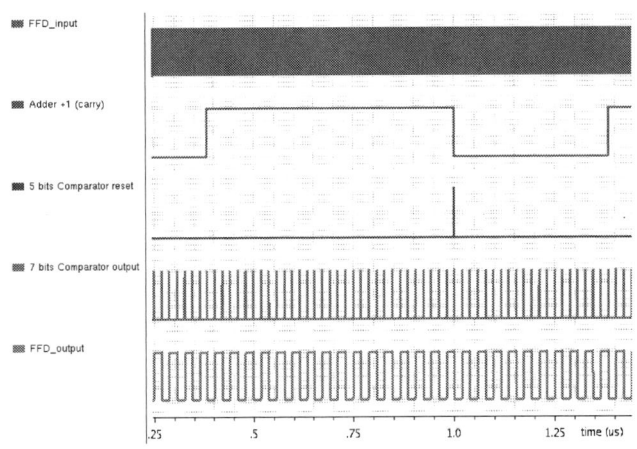

*Fig. 5. Simulation example (Fin=2850MHz)*

## IV. CONCLUSION

A low-power programmable fractional frequency divider design is reported in this paper. Taking the advantage of the 28nm FDSOI CMOS technology, a counters based architecture with a real 50% duty cycle has been presented, offering an interesting alternative to commonly used pre-scalar divider. The frequency divider meets expectations of the Wifi or WiGig standards with a very thin division step (1/26). The low power consumption of the divider (300µW) makes it usable in case of ultra-low power PLL design and is, to the author's knowledge, the state of the art.

TABLE I.     SUMMARY OF THE FRACTIONAL FREQUENCY DIVIDER PERFORMANCES AND COMPARISON WITH PREVIOUS WORKS

| Reference | [1] | [2] | [3] | This work |
|---|---|---|---|---|
| CMOS Technology (nm) | 350 | 180 | 65 | 28FDSOI |
| Input Freq. range (GHz) | 2.24 2.70 | 0.8 3.8 | 2 2.6 | 2.325 2.850 |
| Division ratio | 240 248 | 30.5 510.5 | 8 31.4 | 89 110 |
| Division step | 0.25 | 0.5 | 0.2 | 1/26 |
| Power consumption (mW) | 6 | 9 | 0.9 | 0.3 |

## ACKNOWLEDGMENT

The author would like to thank Milovan Blagojevic, Renald Boulestin, Sébastien Dedieu, Marc Houdebine and Mathilde Sié for their expertise and their knowledge on fractional frequency divider, and Fabien Senilhes (Mentor Graphic) for his software support.

## REFERENCES

[1] Boon, C.C.; Do, M.A.; Yeo, K. -S; Ma, J-G, "Fully integrated CMOS fractional-N frequency divider for wide-band mobile applications with spurs reduction," *Circuits and Systems I: Regular Papers, IEEE Transactions on* , vol.52, no.6, pp.1042,1048, June 2005.

[2] Jing Jin; Xiaoming Liu; Tingting Mo; Jianjun Zhou, "Quantization Noise Suppression in Fractional-$N$ PLLs Utilizing Glitch-Free Phase Switching Multi-Modulus Frequency Divider," *Circuits and Systems I: Regular Papers, IEEE Transactions on* , vol.59, no.5, pp.926,937, May 2012.

[3] Thirunarayanan, R.; Ruffieux, D.; Enz, C., "An injection-locking based programmable fractional frequency divider with 0.2 division step for quantization noise reduction," *ESSCIRC (ESSCIRC), 2013 Proceedings of the* , vol., no., pp.233,236, 16-20 Sept. 2013.

# Recent Advances and Future Trends in SOI for RF Applications

Aniruddha Joshi, Tzung-yin Lee, Yuh-yue Chen and David Whitefield

Skyworks Solutions Inc.

Irvine, CA

aniruddha.joshi@ieee.org

*Abstract*—In recent years, RFCMOS on Silicon-on-Insulator has rapidly evolved as a mainstream technology for switches used in wireless applications. Requirements of lower insertion loss, better isolation and better linearity have driven RFCMOS-SOI roadmap. $R_{on}*C_{off}$, a key figure-of-merit for switch application, has scaled from > 300fs to < 200fs. In this paper, we review commonly adopted techniques to further improve $R_{on}*C_{off}$, along with their limitations

## I. INTRODUCTION

RFCMOS on SOI substrates has become the technology of choice in recent years for wireless applications such as switches, tuners and power amplifiers. The popularity of RFCMOS-SOI is a result of a 'push-pull' effect. The 'push' comes from component providers who prefer SOI over GaAs due to its integration potential (*e.g.* control functions, digital interface, multiple RF functions), benefits of Silicon ecosystem (*e.g.* design kits, ESD) as well as lower cost. The 'pull' comes from wafer manufacturers who can extract higher value from older manufacturing facilities by enabling SOI-based offerings. This symbiotic interaction has driven the RFCMOS-SOI market towards million-wafers-per-year run rate.

Today's typical RFCMOS-SOI process, offered by multiple Silicon foundries, has the following attributes:

- $V_{dd}$= 2.5V. As explained in our previous review [1], this is an optimal region based on low $R_{on}*C_{off}$ requirement and necessity to maintain relatively low number of FETs in a stack to withstand high power.

- Low $R_{on}*C_{off}$ to ensure low insertion loss and better isolation. $R_{on}*C_{off}$ < 200fs is routinely available.

- Optimized SOI substrate to address stringent linearity requirements

- High quality passives (polysilicon resistors, MIM capacitors)

In this paper, we will review commonly adopted techniques to further improve $R_{on}*C_{off}$, along with their limitations.

## II. RECENT TECHNOLOGY IMPROVEMENTS

### A. Channel Length

'On-resistance' ($R_{on}$) of a FET plays a key role in insertion loss of switches. $R_{on}$ is a combination of channel resistance, resistance of source-drain extensions as well as that of interconnects. Among these, channel resistance is the largest contributor. Consequently, aggressive reduction in channel length is a routinely adopted method to reduce $R_{on}$ [2]. As shown in Fig. 1, channel length reduction along with necessary adjustments to doping profile can improve $R_{on}*C_{off}$ towards 150fs. This approach, however, faces multiple limitations. Beyond the obvious limitations such as drain-induced barrier lowering, punch-through and hot electron injection, a more stringent limitation arises due to the nature of the application. To be able to handle high voltage levels (>30V), it is necessary to stack multiple FETs to form a switch arm [1]. Due to gate and substrate losses, the voltage across FETs within an arm can vary significantly; by as much as 35% for an arm with 10 FETs. The amount of imbalance increases rapidly with the stack height (stack height= number of FETs in a switch arm) [3]. Device design needs to take into consideration this imbalance. Secondly, as shown in Fig. 2, high levels of harmonics are generated when the power applied across an arm exceeds a certain threshold; the threshold being lower for shorter channel length. This sharp increase in nonlinearity is a combination of transistor self-biasing and activation of lateral parasitic bipolar [4]. Both these effects are more prominent for shorter channel length. In other words, even if a FET is optimized to minimize short-channel effects under the highest $V_{ds}$ within an arm, high harmonics level can render it useless for RF application. These RF considerations may limit the shortest usable channel length for a given technology.

### B. Silicon Thickness

'Off-capacitance' ($C_{off}$) is normally associated with

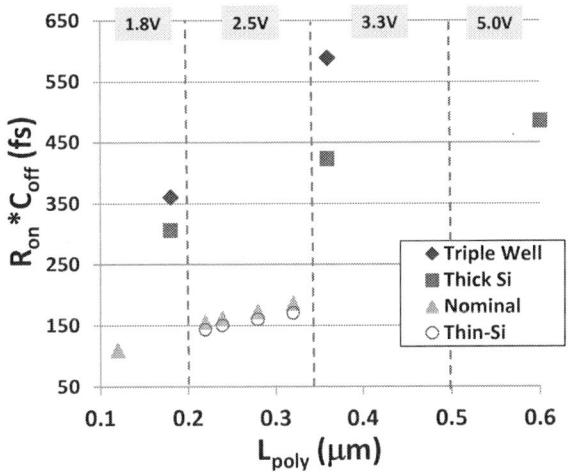

Fig. 1. Benchmark of $R_{on}*C_{off}$

978-1-4799-7440-5/14 $31.00 © 2014 IEEE

isolation between ports of a switch; lower capacitance resulting in better isolation. Typically, switches are designed in a series-shunt configuration wherein the shunt arms provide a low resistance path to minimize RF power leaking into the OFF ports. For switches with a high throw-count, $C_{off}$ from multiple shunt branches of OFF arms leaks a large amount of RF power, resulting in high insertion loss, especially at high frequency [1]. $C_{off}$ is a combination of a number of linear (interconnects) and nonlinear (gate, gate-to-source/drain, junctions, buried oxide) capacitances. Silicon thickness plays a key role in controlling $C_{off}$. A typical RFCMOS-SOI offering today is partially depleted, with ~0.15μm thick Silicon layer. PDSOI reduces/eliminates multiple components of junction capacitances [5], resulting in lower $C_{off}$ than triple-well bulk CMOS and SOI with thicker (~1μm) Silicon. A logical next step would be to further reduce the Silicon thickness, moving towards a fully-depleted solution. An improved $R_{on}*C_{off}$ resulting from thinner Silicon is evident in Fig. 1. In addition to $R_{on}*C_{off}$ reduction, this technique improves linearity because it predominantly reduces nonlinear capacitance of FETs. Unfortunately, there are two key considerations that may limit aggressive scaling of Silicon thickness. Firstly, thinner Silicon increases $R_{on}$ [6]; ~5% increase is seen in going from ~0.15μm to ~0.1μm as illustrated in Fig. 3. This increase partially offsets the benefits of $C_{off}$ reduction. Secondly, since the Silicon volume for a device shrinks proportionately with thinner Silicon, ESD device size needs to grow to maintain the same ESD rating.

*C. Interconnects*

As the device level improvements to reduce $R_{on}*C_{off}$ are reaching a point of diminishing returns, the contribution from interconnects has become even more important. A typical switch can have >30% contribution from interconnect capacitance, and this relative contribution increases as the intrinsic FET continues to improve. Whereas Cu and Low-K interconnects are common in sub-130nm nodes, these are not routinely available in older nodes (e.g. 180nm) where most of today's RFCMOS-SOI technologies are offered. Due to its lower resistivity, Cu (*vs.* Al) enables use of thinner conductors for a given resistance value. Thinner metal layers, in turn,

result in lower metal-to-metal capacitance, which translates into lower $R_{on}*C_{off}$. Likewise, use of low-K inter-metal dielectrics can help reduce $R_{on}*C_{off}$. Unfortunately, use of Cu and/or Low-K potentially increases the wafer cost that can be prohibitive for this highly cost-sensitive application. As a compromise, 'hybrid' solutions that use Cu only for critical layers are gaining popularity.

*D. Substrate*

Presence of a passivation layer at the interface between buried oxide and handle wafer is critical to minimize nonlinearity contributed by SOI substrates [7,8]. Such passivation layer can be introduced during substrate or wafer manufacturing. Composition of the passivation layer has a significant impact on the effective resistivity of the handle wafer [9]. As the linearity of FETs improves, the relative contribution of substrate nonlinearity on the overall switch nonlinearity becomes even more important. We look for innovations from substrate and foundry suppliers to meet demanding product level challenges.

### III. FUTURE DIRECTIONS

As outlined above, traditional means to reduce $R_{on}*C_{off}$ face a variety of limitations; technological as well as commercial. Moving to 1.8V or lower voltages can improve $R_{on}$ substantially. However, use of these FETs will force an increase in stack height which leads to higher imbalance within a switch arm. This, in turn, can degrade power handling capability. Techniques to alleviate voltage imbalance will open up an opportunity for using lower voltage FETs, thereby allowing an aggressive $R_{on}*C_{off}$ scaling beyond 100fs. While many of the routinely adopted techniques in advanced CMOS (Cu, Low-K, mobility enhancement *etc.*) may further improve $R_{on}*C_{off}$, applicability of these to RFCMOS-SOI may be limited due to extreme cost-sensitivity of the application.

### REFERENCES

[1] A. B. Joshi, S. Lee, Y. Y. Chen, and T. Y. Lee, "Optimized CMOS-SOI Process for High Performance RF Switches," IEEE SOI Conference, Oct 2012.

[2] T. McKay, M. Carroll, D. Kerr and J. Costa, "Advances in silicon-on-insulator cellular antenna switch technology," Silicon Monolithic

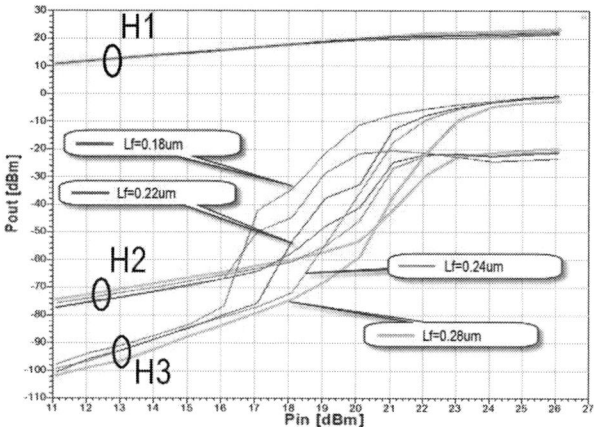

Fig. 2. Impact of channel length reduction on 2$^{nd}$ (H2) and 3$^{rd}$ (H3) Harmonics

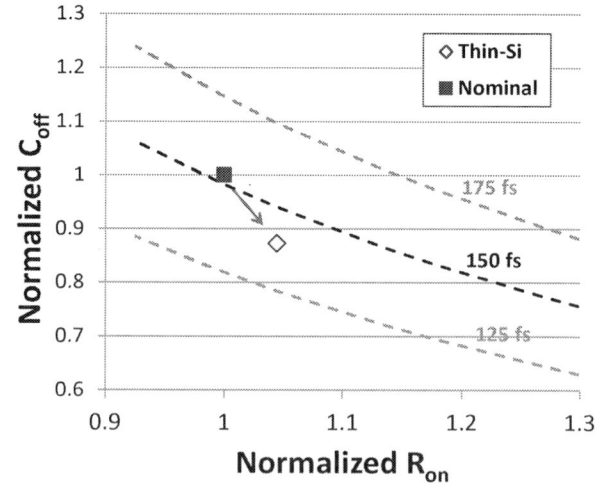

Fig. 3. Effect of Silicon thickness reduction on $R_{on}$ and $C_{off}$

Integrated Circuits in RF Systems (SiRF), 2009.

[3] T. Lee and S. Lee, "Modeling of SOI FET for RF switch applications," IEEE RFIC Symp, 2010, pp. 479–482.

[4] J. Chen, F. Assaderaghi, P. K. Ko, and C. Hu, "The Enhancement of Gate-Induced-Drain-Leakage (GIDL) Current in Short-Channel SO1 MOSFET and its Application in Measuring Lateral Bipolar Current Gain β," IEEE Electron Dev. Lett., vol. 13, pp. 572–574, Nov 1992.

[5] G. Shahidi, "SOI Technology for the GHz Era," IEEE International Symp. VLSI Technology, Systems and Applications, 2001, pp. 11–14.

[6] O. Bon, O. Gonnard, L. Boissonnet, F. Dieudonne, S. Haendlerl, C. Raynaud, and F. Morancho, "RF Power NLDMOS Technology Transfer

Strategy from the 130nm to the 65nm node on thin SOI," IEEE SOI Conference, Oct 2007.

[7] C. Neve and J.-P. Raskin, "RF Harmonic Distortion of CPW Lines on HR-Si and Trap-Rich HR-Si Substrates," IEEE Trans. Electron Devices, vol. 59, pp. 924–932, April 2012.

[8] D. Lederer and J.-P. Raskin, "RF Performance of a Commercial SOI Technology Transferred Onto a Passivated HR Silicon Substrate," IEEE Trans. Electron Devices, vol. 55, pp. 1664–1671, July 2008.

[9] D. Lederer, R. Lobet and J.-P. Raskin, "Enhanced High resistivity SOI wafers for RF applications," IEEE SOI Conference, Oct 2004.

# SEU Hardening: Incorporating an Extreme Low Power Bitcell Design (SHIELD)

Ariel Pescovsky*, Oron Chertkow*, Lior Atias* and Alexander Fish[†]

*VLSI Systems Center, Ben-Gurion University of the Negev, Be'er Sheva
[†]Faculty of Engineering, Bar-Ilan University, Ramat Gan
Email: arielpes@post.bgu.ac.il ,chertkow@post.bgu.ac.il, lioratia@ee.bgu.ac.il, alexander.fish@biu.ac.il

*Abstract*—The pursuit of continuous scaling of electronic devices in the semiconductor industry has led to two unintended but significant outcomes: a rapid increase in susceptibility to radiation induced errors and an overall rise in power consumption. Operating under low voltage to reduce power only aggravates radiation related reliability issues. In this paper, a novel "SEU Hardening Incorporating Extreme Low Power Bitcell Design" (SHIELD) is proposed to attend these two major concerns simultaneously. The SHIELD bitcell tolerates upsets with charge deposits over 1 pC when operated at a scaled 700 mV supply voltage utilizing a 65 nm process. Simulations confirm its advantages in terms of leakage power, with more than twofold lower leakage currents than previous solutions.

## I. INTRODUCTION

THE scaling of transistor dimensions in recent years has led to an increase in power consumption. Memory arrays constitute up to 70% of the die area, making memory power dissipation one of the key concerns of the semiconductor industry. Low-power (LP) operation is of particular importance in VLSI chips for space applications, where available energy resources are limited. Due to direct association between power consumption and voltage, the most attractive solution is to scale the supply voltage. This reduces parasitic leakage currents, which define static power consumption [1]. However, low-voltage circuits are much more susceptible to radiation effects than circuits powered at nominal supply voltages.

Radiation induced errors in electronic circuits are caused by an energy transfer when a radiation particle is knocked to the substrate, resulting in the excitement of electron-hole pairs. If the impact occurs at a transistor's reversed biased drain junction, these carriers will drift into the junction, resulting in a transient current pulse [2], [3]. The prime indicator of such a single event upset (SEU) is called the 'critical charge' ($Q_{crit}$) and represents the maximal charge deposition that a node can withstand without changing its logical state. $Q_{crit}$ is defined as the multiplication of the node's voltage with its capacitance. For this reason, lowering the supply voltage increases susceptibility to SEU.

Typically, SEU is handled at the system level by Error Correction Codes (ECC) and Triple Modular Redundancy (TMR) [4], [5] although these methods have many drawbacks in terms of additional delays and decreased efficiency related to the increased probability of multi-bit errors in small size scales [6]. At the device level, special process techniques such as triple-well and Silicon-On-Isolator (SOI) are used. However, these are very expensive to manufacture and do not guarantee immunity to SEU [7]. Circuit level solutions such as DICE [8] and the Quatro 10 T [9], can efficiently increase SEU tolerance.

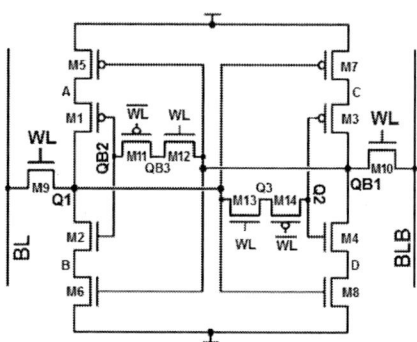

Fig. 1: The proposed bitcell design - SHIELD

Both low voltage bitcells or radiation hardened bitcells have been designed, but an efficient solution covering all issues has yet to be developed. Here we present a novel bitcell dubbed SHIELD, which is specifically designed for low-voltage operation, and is based on the principles of gating the familiar cross-coupled inverters while introducing a novel 'cut-off' network. This creates redundant storage nodes and eliminates the internal feedback loop during radiation particle impact.

## II. THE NOVEL SHIELD BITCELL

### A. Proposed bitcell structure

The proposed bitcell is presented in Fig. 1. To mitigate SEU susceptibility, SHIELD uses gated inverters (M5-M1-M2-M6 and M7-M3-M4-M8). A gated inverter is an inverter with an additional input gate. If both inputs are in the same logical state, the output will be equivalent to that of a regular inverter. When the inputs differ from each other, the output floats with the logical state of the previous output. In the SHIELD bitcell, a novel radiation tolerant 'cut-off' network (M11-M12 and M13-M14) is located between the two gates of each gated inverter. The SHIELD bitcell has two of these upgraded gated inverters, which are cross-coupled. This topology manifests as two sets of separate dual data nodes.

In the normal operation mode (i.e. no radiation impact), the $Q_1$, $Q_2$ pair has the same logical state ('0' or '1') while the $QB_1$, $QB_2$ is in the opposite state. To illustrate, let us assume the bitcell holds logical '1', that is, $Q_1$ ='1', turning M8 ON and M7 OFF, $Q_2$ ='1', turning M4 ON and M3 OFF, $QB_1$ ='0', turning M5 ON and M6 OFF and finally $QB_2$ ='0', turning M1 ON and M2 OFF. This topology results in a low resistance path from the supply rails to the main data storing nodes $Q_1$ and $QB_1$, which allows them to replenish voltage levels, as seen in Fig. 2a. Two additional access NMOS

978-1-4799-7440-5/14 $31.00 © 2014 IEEE
128

transistors are added for read and write purposes as in the standard 6T SRAM.

The Write operation resembles the standard 6T SRAM. It pre-charges the differential bit-lines (BL and BLB) signals to the desired data followed by the assertion of the word-line (WL) signal and its complement (WLB). This opens both of the access and cut-off network transistors, as seen in Fig. 2b. After a short transient time, the voltage set at the '0' holding main data storage node ($Q_1/QB_1$) is a result of the opposition between the pull up network (PUN) and the access transistor. For the write operation to be successful, the PUN must be weaker than the access transistor. At standard 6T SRAM, this condition is met by implementing special sizing to satisfy the 'write constraint'. Our design achieves the same outcome at minimal sizing since its PUN consists of two serially connected transistors that double its resistance path. Since the cut-off network transistors are open, the same desired logical state will permeate to the secondary data storage node ($Q_2/QB_2$) as well. After the data are written, WL and WLB are turned off, thus returning to the hold state.

For the Read operation, BL and BLB are pre-charged to $V_{DD}$ prior to WL assertion. A sense amplifier, connected to BL and BLB senses the voltage drop at the '0' holding node to determine which data the bitcell holds. The read operation might result in a bit-flip if the PDN is weaker than the access transistor at the '0' holding node. For standard 6T, this condition is denied by implementing special sizing to satisfy the 'read constraint'. The SHIELD bitcell obviates the need for such sizing as it is inherently protected against read bit-flips. Unlike the write operation, for the read operation, only the WL is asserted and not WLB. This results in transistors M9, M10, M12, M13 turning ON, yet M11 and M14 stay turned OFF as seen in Fig. 2c. Thus, in the case where the main storage node ($Q_1/QB_1$) is flipped as a result of a read operation, the secondary storage node ($Q_2/QB_2$) will not be affected. This state resembles aftermath of a particle impact at the main storage node, and therefore the bitcell will recover its original state using the same built-in SEU recovery mechanisms described below.

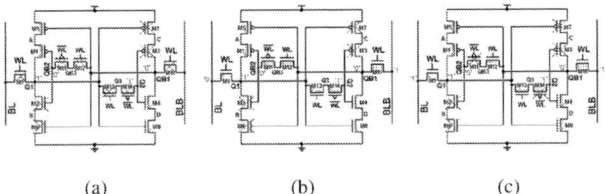

Fig. 2: a) SHIELD bitcell in hold state b) SHIELD in write operation c) SHIELD in read operation

*B. SEU hardening*

The cut-off network is comprised of PMOS and NMOS devices connected in series. The impact of a particle can induce negative voltage at the source of an NMOS and turn it on even when the gate is biased GND, since $V_{GS} > 0$. The same logic applies to PMOS, in that a particle impact can induce voltage levels higher than $V_{DD}$ at its source node and turn it on even when the gate is biased $V_{DD}$, since $V_{GS} < 0$.

The chained formation ensures tight sealing of the cut-off network by guaranteeing that when one of the transistors fails the other will stay closed and an SEU transient will not permeate to the other end of the cut-off network. The PMOS transistor was chosen to encapsulate the secondary data storage node ($Q_2,QB_2$) because it is less susceptible to radiation impact induced transients [10].

An analysis of three different potentially hazardous cases where the bitcell might exhibit a bit-flip is presented. Due to the bitcell's symmetry, particle impacts at $Q_1$ and $Q_2$ alone will be discussed. Impacts at $QB_1$ and $QB_2$ are identical but operate on the opposite nodes.

*1) Particle impact at node $Q_1$ from '0' to '1':* When an energized particle hits the reversed biased PMOS drain at the '0' holding $Q_1$ node, a current transient changes its logical state from '0' to '1'. The 'cut-off' network prevents the transient from propagating over to $Q_2$. This results in M7 turning OFF and M8 turning ON. However, at the same time, M3 is turned ON and M4 is turned OFF due to the '0' stored in $Q_2$. Thus, $QB_1$ does not change its logical state so M2 and M6 remain turned ON, enabling the struck node $Q_1$ to replenish its original state.

*2) Particle impact at node $Q_1$ from '1' to '0':* In the event of a particle hitting the reversed biased NMOS drain at the '1' holding $Q_1$ node, a current transient changes its state from '1' to '0'. The 'cut-off' network prevents the transient from propagating over to $Q_2$. This results in M7 turning ON and M8 turning OFF. However, at the same time, M3 is turned OFF and M4 is turned ON due to the '1' stored in $Q_2$. Thus $QB_1$ does not change its logical state so M1 and M5 remain turned ON, enabling the struck node $Q_1$ to replenish its original state.

*3) Particle impact at node $Q_2$ from '0' to '1':* Recall that $Q_2$ is affected only when the drain is reversed biased, i.e., the logical '0' at $Q_2$ as it is enclosed by a PMOS transistor. The current transient changes its state from '0' to '1'. This results in M3 turning OFF and M4 turning ON. However, at the same time, M7 is turned ON and M8 is turned OFF due to the '0' stored in $Q_1$. Thus $QB_1$ does not change its logical state. While $Q_2$ does not have a direct path to the supply rails, it will eventually restore its original state by leakage currents through the 'cut-off' network transistors M13 and M14. This process takes longer than the restoration of the main storage node $Q_1$ after being struck, but it prevents an immediate flipping effect of the bitcell's logical state. Only the combination of a strike at $Q_1$ when $Q_2$ has not sufficiently regained its logical state will result in a bit-flip. Nonetheless, the odds of the same bitcell being struck twice within this small restoration time window such that the two critical nodes will change their state simultaneously are fairly small.

To demonstrate the cell operation described in this section, representative write, read, and upset suppression events are shown in Fig. 4 with the minimal 700 mV supply voltage.

## III. SIMULATIONS AND RESULTS

To simulate an SEU impact on the bitcell's sensitive nodes, a current pulse was injected into these nodes. The mathematical function traditionally used to portray the transfiguration of

978-1-4799-7440-5/14 $31.00 © 2014 IEEE

the current source over time is the double exponential current model [11]:

$$I(t) = \frac{Q}{\tau_f - \tau_r} \left( e^{-\frac{t}{\tau_f}} - e^{-\frac{t}{\tau_r}} \right) \qquad (1)$$

where $Q$ is the amount of charge deposited as a result of the ion strike, $\tau_r$ is the collection time constant and $\tau_f$ is the ion track establishment constant. For the simulations conducted, the $\tau_r$ value was set to be 45 ps, and $\tau_f$ value to be 145 ps as in [11]. $Q_{\text{crit}}$ was calculated as the time integral over the minimum parasitic current source that cause a bit-flip.

The bitcell was simulated and tested against the standard 6T SRAM bitcell and competitive circuit level radiation hardened bitcells. All the bitcells were implemented using a standard 65 nm technology process. Particle strike suppression according to the double-exponential model was tested across statistical Monte Carlo (MC) simulations for every type of disrupt event.

Table I summarizes SEU robustness by measuring the Qcrit values of several bitcells at 700 mV supply voltage. The results show that the proposed SHIELD bitcell is fully immune to SEU induced in a natural space environment, which equals a charge deposition of 1pC [10].

| SEU Simulation | SHIELD | DICE | Quatro-10 T | 6 T |
|---|---|---|---|---|
| $Q_1$:'1' → '0' | >1 pC | >1 pC | >1 pC | 2.2 fC |
| $Q_1$:'0' → '1' | >1 pC | >1 pC | 3.7 fC | 5.6 fC |
| $Q_2$:'1' → '0' | NP | >1 pC | 2.5 fC | - |
| $Q_2$:'0' → '1' | >1 pC | >1 pC | >1 pC | - |
| $Q_{\text{crit}}$ | >1 pC | >1 pC | 2.5 fC | 2.2 fC |

NP - Not Possible (Junction is not in reverse bias)

TABLE I: Comparison of $Q_{\text{crit}}$ values

Traditionally, the prime source of power dissipation has been dynamic consumption. However, as the industry moved into sub-micron technologies, scaling of the transistor sharply increases the sub-threshold leakage currents, resulting in static power consumption as the dominant cause of power dissipation [12]. Fig. 3 compares leakage currents across the simulated bitcells at different supply voltages. All curves represent mean values interpolated from 1000 MC samples. For the simulation, bitlines were pre-charged for worst case scenario (voltage level opposite to those in the adjacent data node). The findings show that the SHIELD has the lowest power consumption at all voltages. These results stem from the use of gated inverters in the design, which provide the bitcell with internal leakage suppression capabilities through

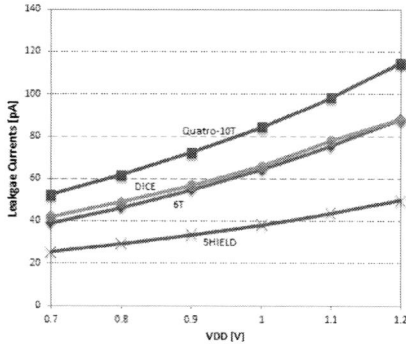

Fig. 3: Comparison of leakage currents versus supply voltage

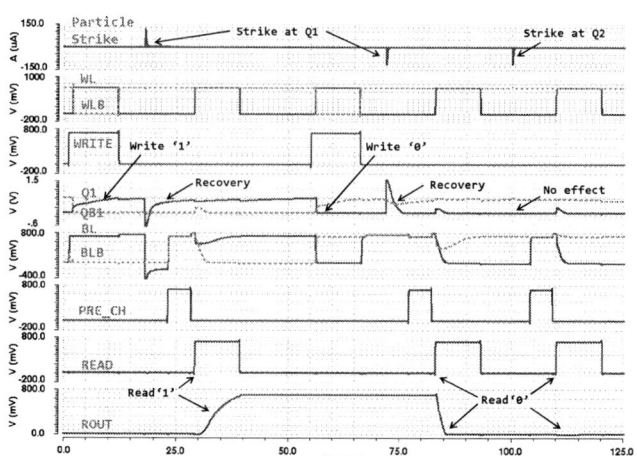

Fig. 4: Waveform of write-strike-read operations.

additional serial resistance in the bitcell's leakage path to the supply rails.

## IV. CONCLUSIONS

A novel SEU immune bitcell featuring extreme low power capabilities dubbed SHIELD was proposed. Simulations show complete immunity to SEU induced in a natural space environment. SHIELD provides significantly lower power characteristics than previously proposed SEU tolerant bitcells and the standard 6T bitcell. The SHIELD demonstrates functionality at a minimal supply voltage of 700 mV. This ability to operate regularly at low supply voltages and withstand any radiation particle impact makes it compatible as a low power bitcell in radiation abundant environments where energy sources are limited, such as space. SHIELD operates regularly at unified minimal sizing, thus reducing its overall area and power consumption. The layout design, presented in Fig. 5, incurs an area penalty of only 2.07× in comparison to the standard 6T bitcell. Future plans are to fabricate a test chip containing the proposed bitcell memory array in standard CMOS technology and to test it under standard space radiation conditions.

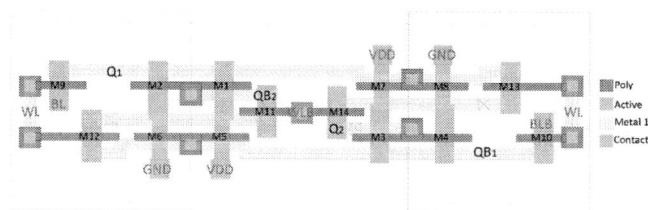

Fig. 5: Layout Design of the SHIELD bitcell

## REFERENCES

[1] K. Roy, Mukhopadhyay et al., Proceedings of the IEEE, 2003.
[2] G. Messenger, Nuclear Science, 1982.
[3] T. Karnik et al., Dependable and Secure Computing, 2004.
[4] Bajura et al., Nuclear Science, IEEE Transactions on, vol. 54, 2007.
[5] Sterpone et al., Nuclear Science, IEEE Transactions on, vol. 52, 2005.
[6] C.-L. Chen et al., IBM Journal of Research and Development, 1984.
[7] J. Schwank et al., Nuclear Science, IEEE Transactions on, 2003.
[8] T. Calin et al., Nuclear Science, IEEE Transactions on, 1996.
[9] S. Jahinuzzaman et al., Nuclear Science, IEEE Transactions on, 2009.
[10] Hass et al., Circuits and Systems, 42nd Midwest Symposium on, 1999.
[11] Garg et al., Very Large Scale Integration (VLSI) Systems, 2009.
[12] N. S. Kim et al., Computer, 2003.

978-1-4799-7440-5/14 $31.00 © 2014 IEEE

# Impact of Ultra-Low Voltages on Single-Event Transients and Pulse Quenching

J. R. Ahlbin, *Member, IEEE* and P. Gadfort, *Member, IEEE*

*Abstract*— Single-event transients (SET) and pulse quenching are analyzed at sub-Vt and super-Vt voltage levels. Two different inverter designs are simulated for their single-event response. These simulations show that SET pulse widths become longer with decreasing voltage for the inverter designed to work at sub-Vt voltage levels. Additionally, pulse quenching becomes significant between 0.5 V and 0.7 V for the standard 2-T inverter.

*Keywords- sub-threshold, low-power, soft errors, single events, single-event transient, single-event upset, charge sharing*

## I. INTRODUCTION

A key reliability concern for CMOS processes operating at nominal voltages is soft errors [1]. One of the major factors that affect the soft error vulnerability of a circuit are single-event transients (SETs), which can be caused by an energetic particle strike on an Integrated Circuit (IC). As the pulse width of SETs increases and assuming the clock rate is constant, the soft error failures in time (FIT) rate also increases because the probability to latch the SET as a soft error has a direct correlation with pulse width [2].

With the necessity to design circuits that use minimal voltage to conserve energy for space, medical, and Internet-of-Things (IoT) applications, circuits are running at sub-threshold (sub-Vt) and super-threshold (super-Vt) voltage levels. Previous work shows that single-event transient pulse widths increase with decreasing voltage [3]. An issue that leads to a higher soft-error rate for low-power circuits compared to nominal-power circuits. With circuits operating at low voltages and an increase in sensitivity to energetic particles [4], circuits designed in sub-130 nm bulk CMOS technologies may be even more susceptible to pulse quenching [5].

## II. BACKGROUND

### A. Description of Pulse Quenching

Pulse quenching is the result of multiple nodes in a circuit collecting charge [6-7] and interacting with the intended electrical signals in the circuit. To understand how pulse

Manuscript received July 28, 2014. This work was supported in part by the DTRA Radiation-Hardened Microelectronics Program.

J. R. Ahlbin, and P. Gadfort are with Information Sciences Institute, University of Southern California, Arlington, VA, USA. (email: jahlbin@isi.edu).

quenching occurs in a circuit, it is necessary to be aware of the changing electrical states of the transistors in the circuit as a signal propagates.

Fig. 1 shows a schematic of a three-stage inverter chain. The output nodes of each inverter (Inv1, Inv2, and Inv3) in the figure are designated by Out1, Out2, and Out3, respectively. Assume a LOW input to the chain. Initially, the pMOS transistors associated with Inv1 and Inv3 are ON, the pMOS transistor of Inv2 is OFF, and Out2 is LOW. If an ion strikes the OFF pMOS transistor of Inv2, the logic LOW state at Out2 is driven HIGH by the charge collection and an SET pulse HIGH is generated at Out2.

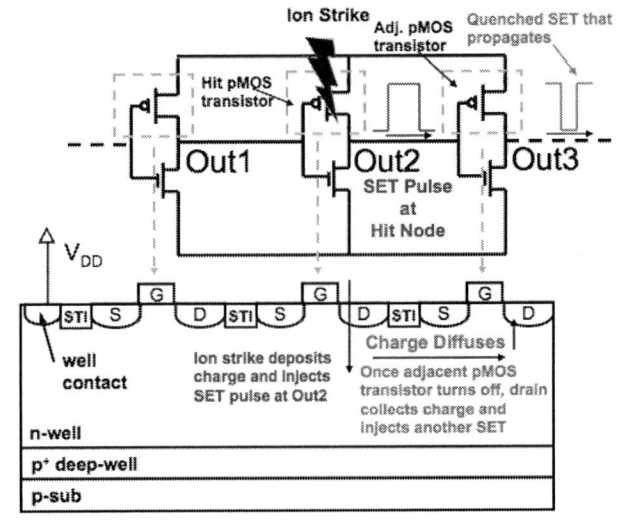

Fig. 1. Two-dimensional slice of three pMOS transistors depicting the electrical signal and the charge sharing signal caused by an ion strike.

This SET pulse will propagate to the gates of Inv3, resulting in a HIGH-to-LOW transition at Out3 and a change of state of the pMOS transistor of Inv3 to OFF. When the pMOS transistor of Inv3 turns OFF, it is then susceptible to charge collection. Charge from the single-event strike on Inv2 can diffuse to Inv3 (through the mechanisms of charge sharing) and be collected by the pMOS transistor of Inv3, driving the state of Out3 HIGH. This LOW-to-HIGH transition at Out3 "resets" the node voltage to the pre-event state, effectively truncating the SET pulse. Thus, there are two factors affecting the dynamic node voltage at Out3. The first is the electrical signal propagation of the SET pulse from Out2; the second is the delayed collection of charge at Out3 due to charge sharing with the struck node of Out2. The propagating SET electrical signal arrives first at Out3, resulting in a HIGH-to-LOW transition; the charge sharing signal arrives second,

978-1-4799-7440-5/14 $31.00 © 2014 IEEE

prematurely forcing Out3 back to a HIGH state. The pulse at Out3 is truncated, or quenched by the delayed charge collection.

### B. Design of a Schmitt Trigger Inverter

The 2-T inverter is a traditional inverter design with pMOS transistor sized twice the size of the nMOS transistor. The Schmitt trigger inverter, shown in Fig. 2a with a standard 2-T inverter design shown in Fig. 2b for comparison, is a unique inverter design that works well at ultra-low voltages. This is because the Schmitt trigger maintains a suitable on/off current ratio at lower voltages than a regular 2-T inverter is capable of at equivalent voltages [8]. The feedback transistors, $M_{fp}$ and $M_{fn}$, provide the mechanism to reduce the leakage currents by applying approximately 0 V across either $M_2$ or $M_3$, depending on the state of the inverter.

At super-Vt operation, this operation manifests itself as a regular Schmitt trigger with hysteresis. Therefore, at regular voltage operation, a Schmitt trigger is designed to have a higher noise margin, compared to a regular inverter, effectively meaning the Schmitt trigger will filter out SETs. If we consider a regular inverter designed to switch at 0.6 V and a Schmitt trigger designed to switch at 0.9 V, as long as the peak voltage of the SET does not exceed 0.9 V, the Schmitt trigger is not upset by this event, while the regular inverter would. This means that the Schmitt trigger offers an additional 0.3 V margin protecting it against SETs that are less than that. However, if the Schmitt trigger is upset, the resultant pulse width is significantly longer because $M_3$ and $M_{fn}$ will both be on, reducing the effective pull-down current for the Schmitt trigger.

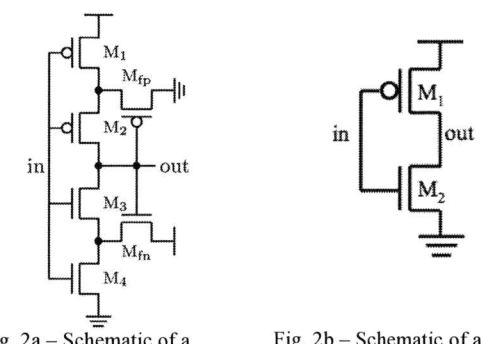

Fig. 2a – Schematic of a Schmitt trigger inverter.

Fig. 2b – Schematic of a standard 2-T inverter.

## III. SINGLE-EVENT ANALYSIS

### A. SPICE Analysis of Standard and Schmitt Trigger Inverters

The two inverters are designed in a 65 nm bulk process and both were simulated with Cadence HSPICE. The Schmitt trigger inverter is simulated with the single-event strike modeled as a current source connected to the $M_2/M_3$ drain. The transistors are sized with $M_1$ & $M_2$ at 300/60 nm, $M_3$ & $M_4$ at 150/60 nm, and $M_{fp}$ & $M_{fn}$ at 450/60 nm. While the standard inverter is simulated with the current source connected to the

"out" node of the inverter with $M_1$ sized 1.0/0.06 μm and $M_2$ sized 0.5/0.06 μm. This source injects a current of 100 μA for 6.17 ps, effectively simulating 6.17 fC of deposited charge at a Vdd from 0.0 V to 1.2 V. Each inverter is part of a chain of 10 inverters with the SET pulse widths measured after the 10th inverter.

Fig. 3 shows the SET pulse widths measured at each Vdd for both types of inverters. No SETs are observed for the Schmitt trigger above 0.8 V and no SETs are able to propagate through the 10 standard inverter chain below 0.7 V. As the Vdd decreases, the SET pulse width increases because the amount of charge deposited is staying constant while the amount of drive current for recovery decreases with Vdd. These simulation results show that even inverters designed to function at sub-Vt Vdd levels do not have the drive current to recover as quickly as inverters operating at nominal voltage.

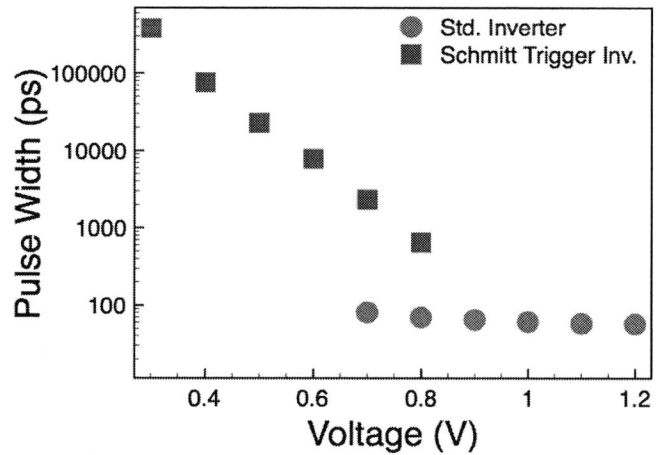

Fig. 3 – SET pulse widths for a Schmitt trigger inverter and a standard 2-T inverter operating at different voltages simulated in HSPICE.

### B. 3D Mixed-Mode TCAD Analysis of Standard Inverter

To understand the effects of ultra-low voltage on pulse quenching, large-scale simulations were carried out using Synopsys SDevice 3D mixed-mode (i.e. combination of TCAD finite-element transistor models and SPICE-like admittance-matrix compact models) tools on a nominal voltage inverter chain. Mixed-mode simulations allow specific transistors in a circuit to be modeled fully in 3D TCAD while the rest of the circuit can be implemented using compact models. Such an approach is very efficient and effective in understanding various mechanisms involved in single-event effects.

A TCAD model was developed that matched the layout and design of a minimum sized standard inverter in the 65 nm process with spacing of 750 nm from gate to gate. These transistors were calibrated to match dc and ac electrical characteristics (e.g., $I_d$-$V_d$ and $I_d$-$V_g$ curves) based on the 65 nm process design kit. The remainder of the simulation setup included calibrated compact models for a chain of inverters also matching the sizing of the target inverters for the SPICE-based portion of the simulation.

978-1-4799-7440-5/14 $31.00 © 2014 IEEE

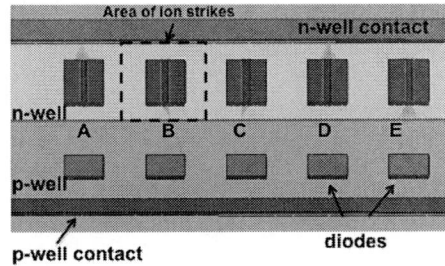

Fig. 4. TCAD model of 65 nm inverters with pFETs modeled in TCAD and nFETs as compact models.

All simulations were carried out with a 9-stage inverter chain with the 3D TCAD model representing the pMOS transistors in the 2nd through 6th stages. All simulations used the following physical models: Fermi-Dirac statistics, SRH and Auger recombination, and the Carrier Carrier Scattering mobility model. The incident heavy-ions were modeled using a Gaussian radial profile with a characteristic 1/e radius of 50 nm, and a Gaussian temporal profile with a characteristic decay time of 2 ps. In each simulation, the ion strike occurred at 1 ns.

A 58 MeV-cm²/mg ion strike that deposited approximately 5.8 pC of charge was simulated at nominal incidence striking the drain center of pMOS transistor B as shown in Fig. 4. Notice the n-diffusion p-body diodes. The actual nMOS transistors are not modeled in TCAD, and are modeled as compact models because of computing limitations. The inverter string was statically biased such that the pMOS transistors A, C, and E were on and the remaining pMOS transistors were off.

While the Vdd is adjusted in 100 mV steps decreasing from 1.0 V to 0.0 V, the strike location and ion strike energy are kept the same for each simulation. The results of these simulations are displayed in Fig. 5. As Vdd of the inverter chain decreases from 1.0 V to 0.8 V, no significant pulse quenching is observed. However at 0.7 V to 0.5 V, pulse quenching is observed. These results suggest that at 0.7 V to 0.5 V the pulse width of the SET and the timing of the inverter chain meet the pulse quenching condition. Charge is collected by the adjacent node (pMOS transistor C) while it is in the "OFF" state, and it is induced by the collected charge to turn "ON" and shorten the SET. Below 0.5 V, SETs are induced on the struck transistor, but they are unable to propagate through the chain because the transistors are operating near or below the threshold voltage.

## IV. CONCLUSION

In this work, two types of inverters are designed and simulated to analyze their response to single-event transients. A Schmitt trigger inverter designed to operate at sub-threshold voltages has the same characteristic response as nominal Vdd inverters to SETs that occur at decreasing Vdd – the SET pulse width is inversely related to Vdd. Additionally, TCAD simulations show that pulse quenching does occur for inverters that operate at ultra-low voltages even though pulse quenching is not observed at nominal Vdd.

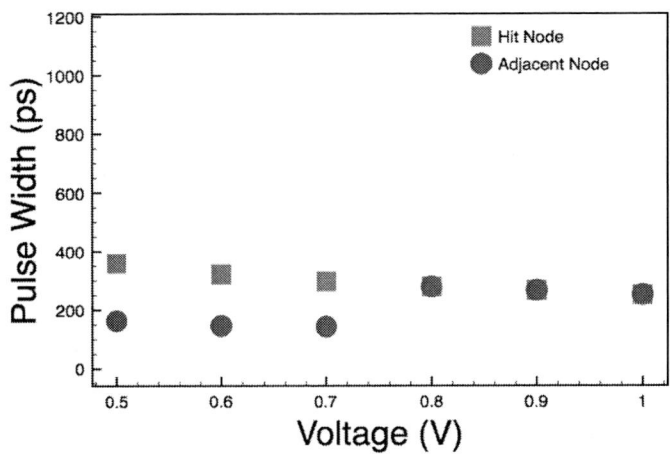

Fig. 5 – SET pulse widths observed in TCAD for an inverter chain operating at different voltages for a LET 58 MeV-cm²/mg.

## ACKNOWLEDGMENTS

The authors would like to thank the Defense Threat Reduction Agency for their support of this effort. The authors would also like to acknowledge Michael Fritze and Michael Bajura of USC's Information Sciences Institute for their suggestions about this work. The authors would also like to thank Dennis "Scooter" Ball of Vanderbilt University's Institute for Space and Defense Electronics for his help and guidance on the TCAD model.

## REFERENCES

[1] International Technology Roadmap for Semiconductors, 2007 Edition[Online]. Availablehttp://www.itrs.net/Links/2007ITRS

[2] S. Buchner, K. Kang, D. Krening, G. Lannan, and R. Schneiderwind, "Dependence of the SEU Window of Vulnerability of a Logic Circuit on Magnitude of Deposited Charge," *IEEE Trans. Nucl. Sci.*, vol. 40, no. 6, pp. 1853-1857 Dec. 1993.

[3] J.M. Benedetto, et al., "Digital Single Event Transient Trends with Technology Node Scaling," *IEEE Trans. Nucl. Sci.*, vol. 53, no. 6, pp. 3462-3465, Dec. 2006.

[4] M. P. King, et al., "Electron-Induced Single-Event Upsets in Static Random Access Memory," *IEEE Trans. Nucl. Sci.*, vol. 60, no. 6, pp. 4122-4129. Dec. 2013.

[5] J. R. Ahlbin, et al., "Single event transient pulse quenching in Advanced CMOS logic," *IEEE Trans. Nucl. Sci.*, vol. 56, no. 6, pp. 3050-3056, Dec. 2009.

[6] B. D. Olson, D. R. Ball, K. M. Warren, L. W. Massengill, N. F. Haddad, S. E. Doyle, and D. McMorrow, "Simultaneous Single Event Charge Sharing and Parasitic Bipolar Conduction in a Highly-Scaled SRAM Design," *IEEE Trans. Nucl. Sci.*, vol. 52, no. 6, pp. 2132-2136, Dec. 2005.

[7] O.A. Amusan, A. F. Witulski, L. W. Massengill, B. L. Bhuva, P. R. Fleming, M. L. Alles, A. L. Sternberg, J. D. Black, and R. D. Schrimpf, "Charge Collection and Charge Sharing in a 130 nm CMOS Technology," *IEEE Trans. Nucl. Sci.*, vol. 53, no. 6, pp. 3253-3258, Dec. 2006.

[8] N. Lotze and Y. Manoli, "A 62mV 0.13 μm CMOS standard-cell-based design technique using schmitt-trigger logic," *IEEE J. Solid-State Circuits*, vol. 47, no. 1, pp. 47-60, Jan. 2012.

# Compensation of Total Ionizing Dose Effects in ULV SoCs Through Adaptive Voltage Scaling

Julien De Vos, Valeria Kilchytska, Denis Flandre, David Bol,

ICTEAM Institute, Université catholique de Louvain (UCL), Louvain-la-Neuve, Belgium

{julien.devos,valeriya.kilchytska,denis.flandre,david.bol}@uclouvain.be

*Abstract*—Total ionizing dose (TID) jeopardizes the operation of ULV circuits by shifting the threshold voltage of the devices. Measurements on a 65nm SoC show that it modifies the output of an on-chip 4-T voltage reference by 3.5% and the gate delay at ULV by 17%. This harms the timing closure of ULV digital systems based on a conventional power management architecture generating constant clock frequency and supply voltage. We show by experimental measurements that the use of an on-chip adaptive voltage scaling system efficiently cancels these effects of TID for robust timing closure at ULV.

## I. INTRODUCTION

Ultra-low voltage (ULV) operation of system-on-chip (SoCs) in advanced CMOS technologies allows record energy efficiency [1]. However, in harsh environments ULV operation is jeopardized by radiations including total ionizing dose (TID) effects. Indeed TID harms the isolation between transistors and induces threshold voltage ($V_T$) shifts. When $V_T$ is lowered, TID induces an increase in leakage currents and of the stand-by power. Furthermore, in ULV digital systems based a conventional power management architecture (Fig. 1 (a)), a slight shift in $V_T$ can also harm the timing closure of synchronous designs as it modifies the delay of logic gates while the clock frequency (usually generated by a PLL) remains constant. Finally TID affects on-chip voltage reference used by the voltage regulators supplying the ULV circuits such as a CPU.

In Section II, we study the impact of TID on timing closure of digital systems based on conventional power management architecture. Therefore we perform radiation measurements of the SunPixer SoC [2]–[4] in 65nm CMOS. The measurements assess the deviation with TID of a 4-T on-chip voltage reference. We also evaluate the direct impact of TID on the gate delay in the critical path of a ULV CPU in the SoC. Finally, Section III presents an adaptive voltage scaling (AVS) system shown in Fig. 1 (b) that is able to generate the supply voltage required for the CPU to meet a given target clock frequency, without requiring a voltage reference. This system compensates fluctuations in gate delay to robust proper timing closure even at high TID.

## II. EFFECTS OF TID ON CONVENTIONAL ARCHITECTURE

The SoC including a critical-path replica (CPR) ring oscillator, the voltage reference, the linear regulator and the AVS system is shown in Fig. 2. Three dies are measured before irradiation, after a 250krad TID (meas. session 1) and a 1.25Mrad TID (meas. session 2). Two other dies serve as reference to track the stability of environmental conditions (temperature) between measurements. The irradiation was performed during

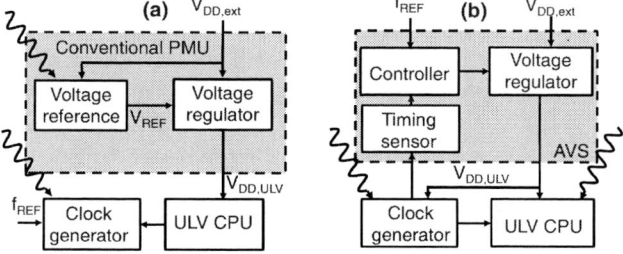

Fig. 1. (a) In conventional SoCs, TID affects both the gate delay of the ULV CPU and the voltage reference output. (b) An AVS system cancels both the impact of TID on the clock generator and on the ULV CPU.

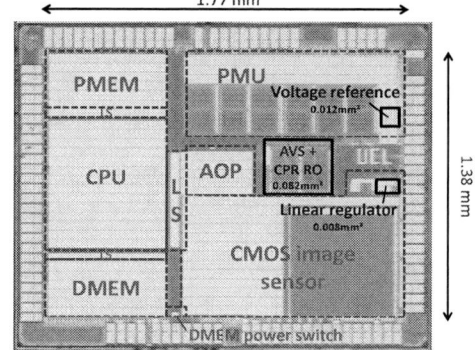

Fig. 2. Die microphotograph of the SunPixer SoC including the measured CPR ring oscillator, the voltage reference and the AVS system.

19 days with gamma rays from a Cobalt 60 source. It was paused for two hours to perform the meas. session 1.

The voltage reference features a new 4-T architecture delivering an output that is proportional to the $V_T$ difference between two core NMOS transistors: $V_{REF} \propto V_{T,LP} - V_{T,GP}$ where $V_{T,LP}$ (resp. $V_{T,GP}$) is the threshold voltage of Low-Power (resp. General-Purpose) transistors with thick (resp. thin) gate oxide. Measurement before irradiation for a 1V $V_{DD,ext}$ show that $V_{REF}$ is 111mV and that the power consumption is 72nW. Fig. 3 (a) shows the $V_{REF}$ deviation after irradiation. With 250krad TID, $V_{REF}$ increases by 2.2% and by 3.5% with 1.25Mrad TID. The power consumption of the circuit is reduced by 10% after the 1.25Mrad TID.

The critical path delay of the CPU is evaluated through the measurement of a critical path replica (CPR) ring oscillator. Before irradiation the CPR delay was 23.5ns when supplied at 370mV, which corresponds to the design point of the CPU. Fig. 3 (b) shows that the CPR delay decreases by 2.6% and by 6.4% after 250krad TID and 1.25Mrad TID respectively.

978-1-4799-7440-5/14 $31.00 © 2014 IEEE

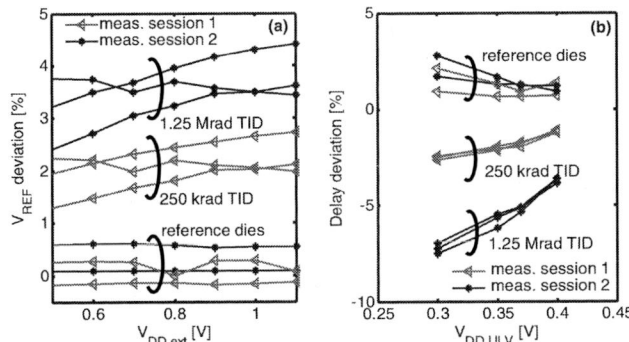 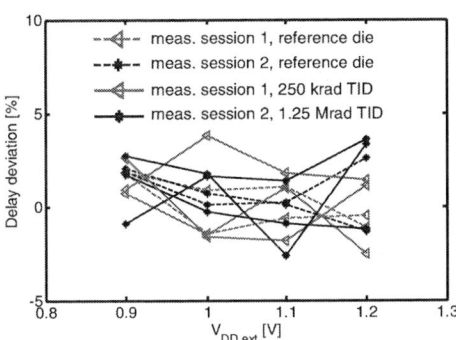

Fig. 3. (a) $V_{REF}$ increases from 111mV by 2.2% (3.5%) after 250krad (1.25Mrad) TID. (b) The CPR ring oscillator delay decreases from 23.5ns by 2.6% (6.4%) after 250krad (1.25Mrad) TID.

|  | $V_{REF}$ | Delay @ 0.37V | Delay @ $V_{DD,ULV}$ with $V_{REF}$ deviation |
|---|---|---|---|
| Mean value before irradiation | 111mV | 23.5ns | 23.5ns |
| Mean deviation of reference dies | 0.3% | 1.2% | 1.2% |
| Mean deviation of irradiated dies @ 1.25 Mrad | 3.8% | -5.2% | -17.0% |
| Impact of 1.25 Mrad TID | 3.5% | -6.4% | -17.1% |

Fig. 4. The delay of the CPR is strongly dependent on its supply voltage deviation and thus decreases by 17% with the 1.25Mrad TID.

This delay varies significantly with the CPU supply voltage $V_{DD,ULV}$ because it is close to $V_T$. Therefore, when supplied by the voltage regulator in the conventional power management architecture, the delay also highly dependent on $V_{REF}$ deviations. Fig. 4 shows a CPR delay reduction by 17.1% at 1.25Mrad TID when supplied at 383mV $V_{DD,ULV}$, accounting for the corresponding $V_{REF}$ deviation.

To qualitatively explain these effects, two phenomena allows us to make hypothesis on the impact of TID on the MOS transistors. First, the power consumption of the voltage reference decreases. As this circuit only features NMOS transistors, we assume that the NMOS $V_T$ increases. Due to the presence of electron traps in nitrided gate oxides [5], NMOS transistors may suffer from a radiation-induced $V_T$ increase due to the dominance of negative charge buildup. Because the gate oxides are very thin (< 1.8nm), radiation-induced charge buildup is low and observed effects on $V_{REF}$ are rather small. The increase in $V_{REF} \propto V_{T,LP} - V_{T,GP}$ could be explained by the stronger negative charge buildup in the thick-oxide LP transistors than in the thin-oxide GP ones. Second, the CPR ring oscillator delay decreases. With negative charge buildup, the $V_T$ of PMOS (resp. NMOS) transistors is decreased (resp. increased) which speeds up (resp. slows down) the rising (resp. falling) gate output transitions. Simulations of logic gates at ULV show a $V_T$ imbalance between NMOS and PMOS leading to a significantly longer delay for rising output transitions driven by PMOS stacks than falling transitions driven by NMOS stacks. Rising transitions thus contribute to the total CPR delay above 50%, which can explain why negative charge buildup speeds up the total CPR delay.

## III. COMPENSATION THROUGH AVS

The purpose of the adaptive voltage scaling (AVS) system is to adapt the supply voltage of a digital circuit to environmental

Fig. 5. When supplied by the AVS system, the CPR delay deviation is within ± 3%, corresponding to the resolution of the AVS control loop, for both reference and irradiated dies.

conditions so that timing failure cannot occur at the target operating frequency [1], [6]. Fig. 1 (b) shows the architecture of an AVS system. Information about the delay of the CPU critical path is gathered by the timing sensor. The controller then generates control signals for the regulator to deliver to the CPU and its clock generator a safe adapted supply voltage. In the proposed AVS system, the timing sensor and the clock generator are implemented by a single CPR ring oscillator. By doing so, the clock generator circuit is affected by TID in the same way as the ULV CPU. Fig. 5 shows that the deviation of the CPR delay of reference dies supplied by the AVS system is within ± 3%, which corresponds to the resolution of the AVS control loop. The CPR delay of irradiated dies remain within ± 3% even after 1.25Mrad TID, thereby demonstrating that the AVS system efficiently compensates TID effects.

## IV. CONCLUSIONS

This paper shows that TID alters the operation of a 4-T voltage reference and the gate delay of ULV CPU. Combined effects of TID on both $V_{REF}$ and the CPU critical path, leads to 17% deviation of the CPU critical path delay after the radiation. It can thus harm the timing closure at ULV. An AVS system is able to compensate the TID effects by tracking it, and by tuning accordingly the CPU supply voltage thereby allowing robust ULV operation against TID.

## ACKNOWLEDGMENT

This work was partially supported by the National Foundation for Scientific Research (FNRS) and by the Walloon region of Belgium under SunPixer SoC project (Fonds de maturation - PoC) and S@T project (SkyWin).

## REFERENCES

[1] D. Bol *et al*, "SleepWalker: A 25-MHz 0.4-V Sub-$mm^2$ 7-$\mu$W/MHz Microcontroller in 65-nm LP/GP CMOS for Low-Carbon Wireless Sensor Nodes", in *IEEE J. Solid-State Circuits*, vol. 48 (1), pp. 20-32, 2013.

[2] F. Botman *et al*, "Bellevue: a 50MHz Variable-Width SIMD 32bit Microcontroller at 0.37V for Processing-Intensive Wireless Sensor Nodes", in *IEEE ISCAS*, pp. 1207-1210, 2014.

[3] David Bol *et al*, "A 65-nm 0.5-V 17-pJ/frame.pixel DPS CMOS Image Sensor for Ultra-Low-Power SoCs achieving 40-dB Dynamic Range", in *IEEE VLSI*, pp. 180-181 2014.

[4] G. de Streel, *et al*, "A 65nm 1V to 0.5V Linear Regulator with Ultra Low Quiescent Current for Mixed-Signal ULV SoCs", in *IEEE FTFC*, 2014.

[5] C. Claeys, E. Simoen, "Radiation effects in advanced semiconductor Materials and Devices", Springer Series in Materials Science, 2002.

[6] J. De Vos, D. Flandre and D. Bol, "Pushing adaptive voltage scaling fully on chip", in *ASP J. Low-Power Electronics*, vol. 8, pp. 95-105, 2012.

978-1-4799-7440-5/14 $31.00 © 2014 IEEE

# Near-threshold voltage operation of a nonvolatile SRAM cell based on pseudo-spin-FinFET architecture

Y. Shuto, S. Yamamoto, and S. Sugahara

Imaging Science and Engineering Laboratory, Tokyo Institute of Technology, Yokohama, Japan.
Tel: +81 (45) 924-5456, Fax: +81 (45) 924-5456, E-mail: shuto@isl.titech.ac.jp

**Introduction:** Near- and sub-threshold voltage operations of CMOS logic systems have attracted considerable attention owing to their ability of dramatic reduction of dynamic and static power dissipation [1]. The proportion of static power to dynamic power increases for the low-voltage operations [1], and thus various reduction techniques for static power are still important even for low-voltage CMOS logic systems. In particular, nonvolatile power-gating (NVPG) [2-9] is expected to be a promising architecture for low-voltage CMOS logic systems, since the NVPG architecture enables spatially and temporally fine-grained power-gating with high energy efficiency. Recently, we proposed the NVPG architecture employing nonvolatile bistable circuits such as nonvolatile SRAM (NV-SRAM) and nonvolatile flip-flop (NV-FF) circuits that can be simply configured by connecting pseudo-spin-MOSFETs (PS-MOSFETs) [2-12] to the storage nodes of standard/conventional SRAM/FF cells. The PS-MOSFET is a circuit for reproducing spin-transistor functions using a spin-transfer-torque magnetic tunnel junction (STT-MTJ) and an ordinary MOSFET [2,4-12]. Therefore, the NV-SRAM and NV-FF circuits can be implemented by application of the present STT-MRAM technology to the CMOS logic platform.

Pseudo-spin-transistors employing FinFETs [13-15] (hereafter, referred to as PS-FinFETs) are attractive for the NVPG architecture adaptable to FinFET-based low-voltage logic systems. In this paper, we computationally analyze near-threshold voltage operation and stability of the NV-SRAM cell using PS-FinFETs. The effects of the power switch on the cell operations are also clarified.

**Cell configuration and operation:** Figure 1(a) shows the circuit configuration of the NV-SRAM cell using PS-FinFETs, in which the power switch connected to the cell is also shown. In this study, virtual supply-voltage ($VV_{DD}$) architecture is employed, i.e., the power switch is installed between the cell and the power supply line. The cell consists of a cross-coupled inverter loop and two PS-FinFETs connected to the storage nodes of the inverter loop. Hereafter, the fin numbers of the load transistors, driver transistors, pass transistors, and PS-FinFETs are denoted by $N_{FL}$, $N_{FD}$, $N_{FP}$, and $N_{FPS}$, respectively. Also, the fin number of the power switch is denoted by $N_{FSW}$. Design of $N_{FL}$ and $N_{FD}$ is highly important, since it restricts the occupied area and static noise margins (SNMs) of the cell. The base design of $(N_{FL}, N_{FD}) = (1,1)$ is beneficial to minimize cell area, although the cell stability is lowered. However, various bias assist techniques are helpful to achieve sufficiently stable operations even for this aggressive design. In this study, near-threshold voltage operation of the NV-SRAM cell with the $(N_{FL}, N_{FD}, N_{FP}, N_{FPS}) = (1,1,1,1)$ design in conjunction with adequate bias-assist techniques is investigated.

In shutdown and wake-up modes for NVPG, the NV-SRAM cell executes store and restore operations, respectively. The store operation is divided into two steps, as shown in Fig. 1(b). In the first step, H-level data on the storage node (Q or QB) is stored into the STT-MTJ connected to this node (H-store operation), and in the second step, L-level data on the other storage node is stored into the other STT-MTJ (L-store operation) [12]. After these operations, the cell can be shut down without losing its data. At the initial stage of the restore operation, the PS-FinFETs are turned on, and the data stored in these STT-MTJs are restored to the storage nodes of the bistable circuit by pull-up of power supply voltage $V_{DD}$, as shown in Fig. 1(c).

All the simulations examined here were performed using HSPICE with a predictive technology model for a 20nm FinFET [16] and our developed STT-MTJ macromodel [10]. The device and circuit parameters of the FinFET and STT-MTJ models are shown in Table I. The STT-MTJ parameters were determined by reference to recently reported STT-MTJs [17-19]. Assuming usage of a device process for full-swing (0.9V) operation, we set the threshold voltages $V_{th}$ of the FinFETs for the low-voltage (0.3 - 0.5V) operations to the same as those for the full-swing operation. Although these $V_{th}$ are not optimized for the low-voltage operations, the usage of an already-developed process would yield a benefit for production cost.

**Operation and stability of the NV-SRAM cell:** Figure 2 shows $VV_{DD}$ as a function of $N_{FSW}$ during the hold, read, and write operations with $V_{DD} = 0.9, 0.5,$ and $0.3V$, in which the wordline underdrive (WUD) technique [20] is used for the read operation. The data retention during the hold operation does not degrade $VV_{DD}$, whereas $VV_{DD}$ for the read and write operations decrease with decreasing $N_{FSW}$. Although the worst-case degradation of $VV_{DD}$ is for the write operation, a sufficient noise margin can be obtained for the write operation with $V_{DD} = 0.3 - 0.9V$ even for $N_{FSW} = 1$, as shown later. The degradation effect of $VV_{DD}$ for the read operation is not so high compared with the write operation. However, this strongly affects the SNM for the read operation. The WUD technique is highly effective at achieving a sufficient read SNM. Figures 3(a)-(c) show butterfly curves for the hold, read, and write operations of the NV-SRAM cell with $N_{FSW} = 1$ for $V_{DD} = 0.9, 0.5,$ and $0.3V$, in which WUD is used for the read operations. Although the worst case is the read operation, a satisfactorily large read SNM (~ 20% of $V_{DD}$) can be achieved even for the 0.3V operation using an appropriate WUD. Note that the SNMs of the NV-SRAM cell are completely the same as those of the equivalent volatile FinFET-based 6T-SRAM cell, since the STT-MTJs can be electrically separated from the bistable circuit part of the cell by the PS-FinFETs during the normal SRAM operation mode. Figure 4 shows read SNM as a function of $N_{FSW}$ for the 0.9, 0.5, and 0.3V operations. The WUD technique is highly effective and the read SNMs are sufficiently large (15 - 20% of $V_{DD}$) even for $N_{FPW} = 1$. Figures 5(a) and (b) show write currents $I_{MTJ}^{P \to AP}$ and $I_{MTJ}^{AP \to P}$ for the H-store and L-store operations as a function of $V_{SR}$ and $V_{CTRL}$, respectively, for $N_{FSW} = 1$. To ensure write currents that satisfy a sufficient margin (e.g., $1.5 \times I_C$) [11], the pull-up of $V_{DD}$ is needed for the 0.5V and 0.3V operations. $V_{SR}$ and $V_{CTRL}$ so as to assure a required current margin can be optimized, as shown in Figs. 5(a) and (b). For instance, when $V_{DD} = 0.9V$, $V_{SR} = 0.65V$ and $V_{CTRL} = 0.5V$ can be chosen to ensure current margins of $1.5 \times I_C$. Figures 6(a) and (b) show butterfly curves for the H-store and L-store operations with $V_{DD} = 0.9V$, $V_{SR} = 0.65V$, and $V_{CTRL} = 0.5V$, in which the power switch with $N_{FPW} = 1$ is used. The H- and L-store operations with magnetization switching can be stably achieved using this bias condition. Figure 7(a) shows $VV_{DD}$ during the H- and L-store operations as a function of $N_{FSW}$, in which $V_{DD}$ is varied from 0.7 - 0.9V and $V_{SR}$ and $V_{CTRL}$ are set to 0.65 and 0.5V, respectively. $VV_{DD}$ for the H-store operation decreases with decreasing $N_{FSW}$, since the cell impedance is lowered owing to the electrical connection of the STT-MTJs to the storage node. Figure 7(b) shows SNMs for the H- and L-store operations as a function of $N_{FSW}$ for $V_{DD} = 0.7 - 0.9V$, in which $V_{SR}$ and $V_{CTRL}$ are set to 0.65 and 0.5V, respectively. To ensure sufficient SNMs for the H- and L-store operations, $V_{DD}$ needs to be set to 0.8V for $N_{FSW} = 1$. Figures 8(a)-(c) show butterfly curves for the restore operation with $V_{DD} = 0.9, 0.5,$ and $0.3V$. When $V_{SR}$ is set to the same as $V_{DD}$, the SNMs are deteriorated by the high currents passing through the STT-MTJs. The SNMs can be satisfactorily improved by regulating $V_{SR}$, as shown in Figs. 8(a)-(d).

**Conclusion:** The bias assist techniques and power switch design for stable near-threshold voltage operations of the NV-SRAM cell are investigated. Although the pull-up of $V_{DD}$ is necessary for the store operation, this executes only at the moment of the shutdown of the cell and the energy required for the store operation can be completely compensated by the shutdown during a period prescribed by break-even time [8]. The NV-SRAM cell using PS-FinFETs is promising for the NVPG architecture of near-threshold-voltage CMOS logic systems.

**References:** [1] S.Jain *et al., 2012 IEEE ISSCC*, paper 3.6. [2] Y.Shuto *et al., IEDM 2012*, paper 29.6. [3] S Sugahara, and J.Nitta, *Proc. IEEE*, **98**, 2124 (2010). [4] S.Yamamoto *et al., JJAP*, **51**, p. 11PB02 (2012). [5] S.Yamamoto *et al., IEEE ISCDG 2012*. [6] Y.Shuto *et al., IEEE SNW 2012*, paper 4-3. [7] Y.Shuto *et al., IEEE IMW 2012*, paper 16. [8] S.Yamamoto *et al., Electron. Lett.*, **47**, p. 1027 (2011). [9] S.Yamamoto and S.Sugahara, *JJAP*, **49**, p. 090204 (2010). [10] S.Yamamoto and S.Sugahara, *JJAP*, **48**, p. 043001 (2009). [11] Y.Shuto *et al., JJAP*, **51**, p. 040212 (2012). [12] Y.Shuto *et al., JAP*, **105**, p. 07C933 (2009). [13] Y.Shuto *et al., IEEE ISCDG 2013*. [14] Y. Shuto *et al., IEEE SNW 2014*, paper P2-13. [15] Y. Shuto *et al., SISPAD 2014*. [16] Predictive Technology Model (PTM), http://ptm.asu.edu/. [17] H.Yoda *et al., Curr. Appl. Phys.*, **10**, p. e87 (2010). [18] J.H.Park *et al., 2012 VLSI Symp.*, paper 7.1. [19] M.Gajek *et al., APL*, **100**, p. 132408 (2012). [20] K.Nii *et al., IEEE Symp. VLSI Circuits Dig.*, p.212 (2008).

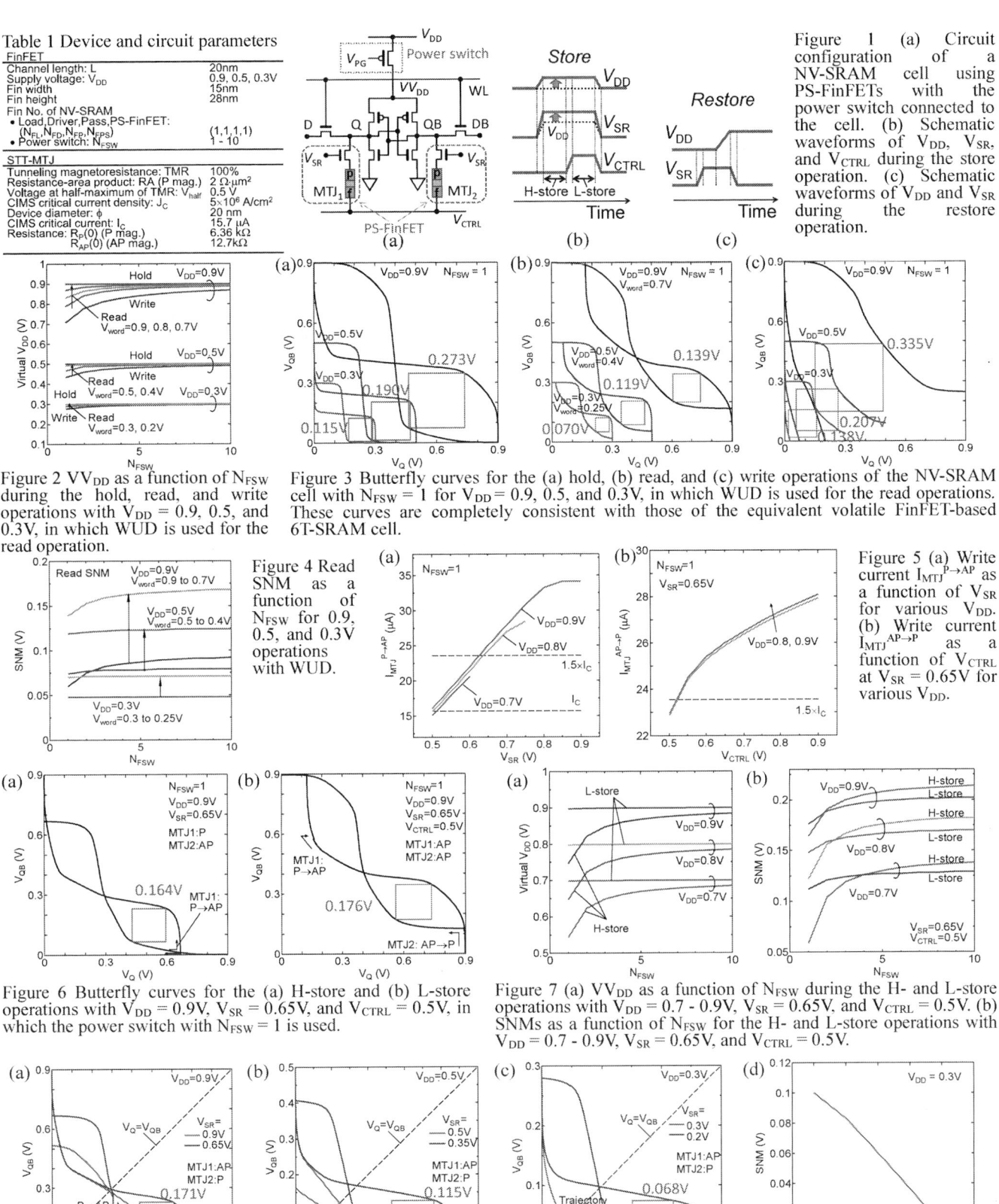

**Table 1 Device and circuit parameters**

| FinFET | |
|---|---|
| Channel length: L | 20nm |
| Supply voltage: $V_{DD}$ | 0.9, 0.5, 0.3V |
| Fin width | 15nm |
| Fin height | 28nm |
| Fin No. of NV-SRAM | |
| • Load,Driver,Pass,PS-FinFET: $(N_{FL}, N_{FD}, N_{FP}, N_{FPS})$ | (1,1,1,1) |
| • Power switch: $N_{FSW}$ | 1 - 10 |

| STT-MTJ | |
|---|---|
| Tunneling magnetoresistance: TMR | 100% |
| Resistance-area product: RA (P mag.) | 2 $\Omega \cdot \mu m^2$ |
| Voltage at half-maximum of TMR: $V_{half}$ | 0.5 V |
| CIMS critical current density: $J_C$ | $5 \times 10^6$ A/cm$^2$ |
| Device diameter: $\phi$ | 20 nm |
| CIMS critical current: $I_C$ | 15.7 $\mu$A |
| Resistance: $R_P(0)$ (P mag.) | 6.36 k$\Omega$ |
| $R_{AP}(0)$ (AP mag.) | 12.7 k$\Omega$ |

Figure 1 (a) Circuit configuration of a NV-SRAM cell using PS-FinFETs with the power switch connected to the cell. (b) Schematic waveforms of $V_{DD}$, $V_{SR}$, and $V_{CTRL}$ during the store operation. (c) Schematic waveforms of $V_{DD}$ and $V_{SR}$ during the restore operation.

Figure 2 $VV_{DD}$ as a function of $N_{FSW}$ during the hold, read, and write operations with $V_{DD}$ = 0.9, 0.5, and 0.3V, in which WUD is used for the read operation.

Figure 3 Butterfly curves for the (a) hold, (b) read, and (c) write operations of the NV-SRAM cell with $N_{FSW}$ = 1 for $V_{DD}$ = 0.9, 0.5, and 0.3V, in which WUD is used for the read operations. These curves are completely consistent with those of the equivalent volatile FinFET-based 6T-SRAM cell.

Figure 4 Read SNM as a function of $N_{FSW}$ for 0.9, 0.5, and 0.3V operations with WUD.

Figure 5 (a) Write current $I_{MTJ}^{P \rightarrow AP}$ as a function of $V_{SR}$ for various $V_{DD}$. (b) Write current $I_{MTJ}^{AP \rightarrow P}$ as a function of $V_{CTRL}$ at $V_{SR}$ = 0.65V for various $V_{DD}$.

Figure 6 Butterfly curves for the (a) H-store and (b) L-store operations with $V_{DD}$ = 0.9V, $V_{SR}$ = 0.65V, and $V_{CTRL}$ = 0.5V, in which the power switch with $N_{FSW}$ = 1 is used.

Figure 7 (a) $VV_{DD}$ as a function of $N_{FSW}$ during the H- and L-store operations with $V_{DD}$ = 0.7 - 0.9V, $V_{SR}$ = 0.65V, and $V_{CTRL}$ = 0.5V. (b) SNMs as a function of $N_{FSW}$ for the H- and L-store operations with $V_{DD}$ = 0.7 - 0.9V, $V_{SR}$ = 0.65V, and $V_{CTRL}$ = 0.5V.

Figure 8 Butterfly curves for the restore operation with (a) $V_{DD}$ = 0.9V, (b) 0.5V, and (c) 0.3V immediately after the pull-up of $V_{DD}$ under $V_{SR}$ = $V_{DD}$ and regulated-$V_{SR}$ conditions. (d) Restore SNM as a function of $V_{SR}$ for $V_{DD}$ = 0.3V.

# More than An Order of Magnitude Energy Improvement of FPGA by Combining 0.4V Operation and Multi-Vt Optimization of 20k Body Bias Domains

Hanpei Koike[1], Chao Ma[1,2], Masakazu Hioki[1], Yasuhiro Ogasahara[1],

Toshiyuki Tsutsumi[2], Tadashi Nakagawa[1] and Toshihiro Sekigawa[1]

1) National Institute of AIST, 2) Meiji University, Japan

Email: h.koike@aist.go.jp

## Abstract

In this paper, more than an order of magnitude (1/13) energy improvement of FPGA by combining low voltage operation and fine-grained body bias optimization is demonstrated from the measurement of the new SOTB implementation of Flex Power FPGA test chip. (keywords: minimum energy operation, FPGA, static power rediction, body biasing, SOTB and ET-SOI)

## Introduction

Sub-threshold or near-threshold minimum energy operation of logic circuits is a new design paradigm in the post-scaling era. By minimizing the energy per operation (power-delay product) of the circuit, we can fully *maximize the total amount of operations* squeezed from the limited amount of energy, such as a battery, as long as the operating frequency meets the requirement of the application such as a sensor node. The dynamic energy consumption per operation can be reduced by lowering VDD, while static energy consumption per cycle gradually increases because of the prolonged cycle time. This trade-off determines the *minimum energy point* (MEP) of the circuit, which is typically around 0.4V. Minimum energy operation of FPGA is particularly complicated because of the relatively high static power consumption by FPGAs. MEP of FPGA tends to be higher than ordinary logic circuits, and energy improvement is substantially limited.

Flex Power FPGA [1] uses body biasing technique to implement the fine-grained Vt programmability of the FPGA component circuits such as Look Up Table (LUT) and Multiplexer (MUX), so that Multi-Vt optimization technique can be also applied to FPGAs. Low-Vt state is assigned only to the components along the critical path of the user application design, while High-Vt state is assigned to the most part of the FPGA. As a result, drastic reduction of the static power consumption without speed degradation can be realized. The recent implementation of Flex Power FPGA using SOTB (Silicon On Thin BOX) transistors presented in this conference last year [2, 3] exhibits 1/50 static power reduction performance owing to the excellent Vt controllability of SOTB device.

This paper introduces a new version of the Flex Power FPGA test chip. The new chip can operate down to 0.4V this time, owing to the good SOTB performance, the careful circuit design and the new low voltage operation cell libraries. From the measurement results of the test chip, more than an order of magnitude energy improvement of FPGA by combining low voltage operation and fine-grained body bias optimization is demonstrated for the first time.

## Overview of the New Flex Power FPGA Test Chip

The new test chip (Fig. 1) developed is an advanced version of the first SOTB FPGA described in [2, 3]. FPGA tile numbers are extended from 11x11 to 20x20, and the chip is carefully redesigned to operate down to 0.4V with new low voltage operation cell libraries for SOTB design.

The test chip has a typical island-style FPGA structure with 20x20 FPGA tiles. Each tile includes 4 logic elements, and wire segments and MUXs are provided for connection between logic elements. The size of the configuration SRAM in one tile is 350 bits, which determine all the logic functions of LUTs and all the routing through the MUXs, as well as Vt configurations described later. The total size in the test chip is 140k bits.

For each FPGA components such as LUTs and MUXs, associated Vt configuration SRAM bit and body bias control circuit are provided. Four different bias voltages for LVT/HVT to NMOS/PMOS are supplied from the external sources. The Vt states (LVT or HVT) for NMOS and PMOS of the component circuits, are determined by the contents of the Vt configuration SRAM. New body bias control circuit is carefully redesigned for low voltage operation by cascading the critical level conversion circuits (Fig. 2).

Grain size of the Vt control domains is set to extremely fine, i.e. MUX level, in order to fully demonstrate the maximum static power reduction performance. The number of the Vt control domains in a tile is 49, and the total number in the test chip reaches up to 19,600. 20k bits of Vt configuration SRAM and body bias control circuits, as well as back gate separation areas, are the substantial overheads for the fine-grained Vt control, which will be optimized in the later versions.

Dedicated CAD tool flow for Flex Power FPGA includes a newly-introduced software called *Vt mapper*, in addition to the conventional FPGA tools such as placer and router. Vt mapper analyzes the critical path of the user application design, and assigns HVT state to as much Vt control domains as possible, while keeping the operation speed constant.

## Evaluation Results

In the previous paper [2, 3], we have shown the basic characteristics of 32-bit binary counter circuit mapped on the first SOTB Flex Power FPGA chip operating at VDD=0.8V. In this evaluation, we also use 32-bit binary counter circuit as a sample design to be easily compared with the previous result, and its behavior is investigated in the wider range of VDD form the wider point of view (i.e. power delay product evaluation).

First, maximum operating frequencies for several VDDs down to the minimum were measured with VBNL= VBNH=0V,

978-1-4799-7440-5/14 $31.00 © 2014 IEEE       138

i.e. normal FPGA without body biasing. Fig. 3 shows the relationship between maximum operating frequency and VDD. The result clearly shows that the test chip can work well down to 0.4V owing to the reduced SOTB device variation and the careful circuit redesign. The decrease in operating frequency is nothing other than what we expect theoretically, which also demonstrates the excellent SOTB device performance.

Then, the maximum operating frequency of the counter circuit, the dynamic current at that frequency and the static leakage current of the data-path of the FPGA were measured in various bias voltage conditions and VDD conditions. 15 combinations of the bias voltages for LVT (VBLVT) and for HVT (VBHVT), chosen from {-1.2, -0.8, -0.4, 0.0, 0.4} so that VBHVT <= VBLVT, are examined for each VDD from {0.4, 0.5, 0.6. 0.8, 1.2}. Vt mapper is re-calibrated and re-executed condition by condition, so that the correct Vt mapping result is produced based on the speed ratio of LVT and HVT specific to each bias voltage condition.

Fig. 4 shows the measured static leakage currents of the data-path part of the test chip at the lowest VDD. The dashed line shows leakage characteristics of a virtual single-Vt FPGA (i.e. VBLVT=VBHVT), which demonstrates excellent Vt controllability of SOTB device. Each line for a VBLVT gradually departs from the dashed line, and is saturated in the deep reverse biasing region, as VBHVT is lowered. This is because the leakage of the critical path part gradually dominates the total leakage. Despite of this saturation, static leakage can be drastically reduced by 1/31.

Both 2D and 3D graphs in Fig. 5 show how power-delay product (energy per cycle) calculated from the measured data changes as VDD is lowered, for several VBHVT conditions (from 0V to -1.2V) and fixed VBLVT condition (0V). The top solid line in the left-hand side 2D graph shows energy characteristics of a conventional single-Vt FPGA without any body biasing (i.e. VBLVT=VBHVT=0V). The minimum energy

point is observed at VDD= 0.6V and energy reduction from 1.2V is only 1/2. This is due to the dominating amount of static leakage (dashed line). As body bias optimization is applied by lowering VBHVT, the total energy is drastically reduced and the minimum energy point gradually shifts from 0.6V to 0.4V. Observed energy ratio from VDD=1.2V without body biasing to the minimum energy, attained by combining lowering VDD and body bias optimization, is 1/13. We can also see that 0.8V of LVT/HVT bias voltage difference is enough to attain minimum energy.

From the results shown above, we can conclude that, not only the low voltage operation, but also an effective leakage reduction technique, such as the fine-grained body bias optimization proposed in our Flex Power FPGA, plays a very important role in achieving substantial (more than an order of magnitude in this case) energy efficiency improvement in FPGAs.

## Acknowledgements

This work was performed in "Ultra-Low Voltage Device Project" of Low-power Electronics Association and Project (LEAP) funded and supported by the Ministry of Economy, Trade and Industry (METI) and the New Energy and Industrial Technology Development Organization (NEDO).

## References

[1] T. Kawanami et. al. "Preliminary Evaluation of Flex Power FPGA: A Power Reconfigurable Architecture with Fine Granularity," IEICE Trans. Inf. & Syst., Vol.E87-D, No.8, Aug. 2004.

[2] H. Koike et. al. "The First SOTB Implementation of Flex Power FPGA," IEEE S3S Conf. 2013, Oct. 2013

[3] M. Hioki et. al. "SOTB Implementation of Field Programmable Gate Array with Fine Grained Vt Programmability," *J. Low Power Electron. Appl.* 2014, 4, 188-200.

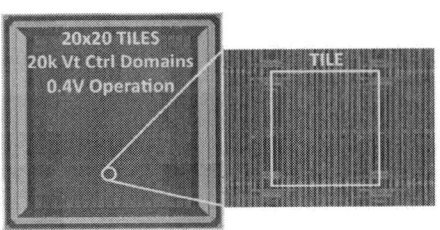

Fig. 1: Die photos of the new test chip and a tile

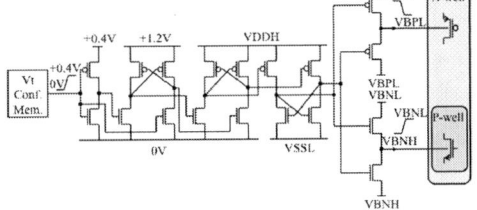

Fig. 2: New body bias control circuit designed to operate down to 0.4V

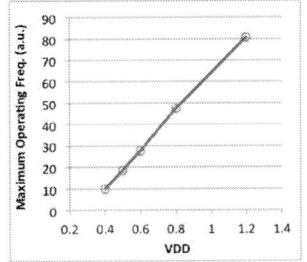

Fig. 3: Maximum operating frequency measurement results down to 0.4V (VBLVT=VBHVT=0V)

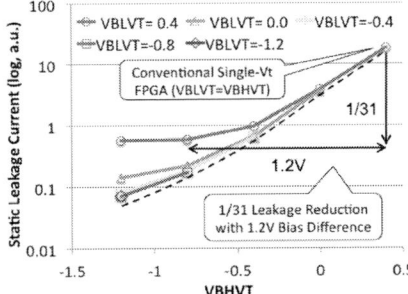

Fig. 4: Static leakage measurement results in the various body bias conditions (VDD=0.4V)

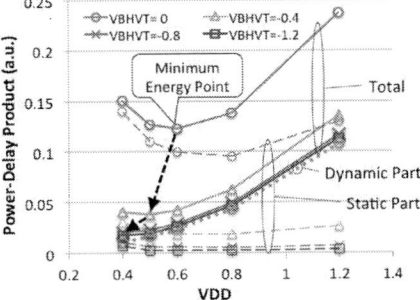

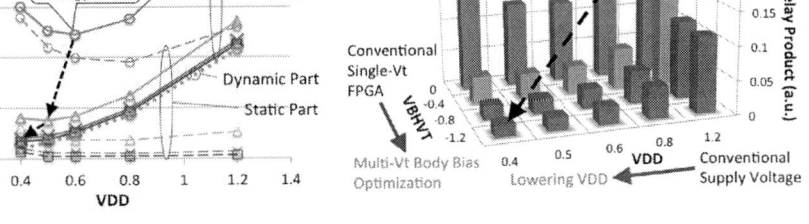

Fig. 5. Power-Delay Products (Energy per cycle) in 2D and 3D graphs calculated from the measurement results in various VDD and VBHVT conditions (VBLVT=0V)

978-1-4799-7440-5/14 $31.00 © 2014 IEEE

# Monolithic IC Integration
## Key Alignment Aspects for High Process Yield

Thomas Uhrmann, Thomas Wagenleitner, Thomas Glinsner, Markus Wimplinger, Paul Lindner
EV Group
St. Florian am Inn, Austria
t.uhrmann@evgroup.com

*Abstract*—Lithographic scaling has been the main growth driver to follow Moore's law of cost reduction and performance increase for several decades. However, the 22nm node appears to be a game changer, where other core processes besides lithography have to be taken into account. Monolithic integration is one solution, where lithographic scaling is replaced by integration in vertical direction. Stacking and electrically contacting several semiconductor layers is challenging, as multiple unit processes have to be solved and put together. One of the key processes for monolithic integration is aligned wafer-to-wafer bonding. Besides optimization of the alignment accuracy, particle cleaning or plasma activation, earlier processing steps have important influence to a high yield processing of monolithic integrated circuits.

*Keywords—monolithic integration; alignment accuracy; fusion bonding; hybrid bonding*

## I. INTRODUCTION

Scaling and Moore's law have been the economic drivers in the planar silicon arena for the last 30 years. During that period, major technology evolutions have been implemented in CMOS processing. The most recent of these evolutions have been extremely complex, including multiple-step lithographic patterning, new strain enhancing materials and metal oxide gate dielectrics. Despite these great feats of engineering and material science, the often predicted "red brick wall" is once again fast approaching and requires evasive action. In fact, several semiconductor suppliers have already shown that the "economic" brick wall has arrived at the 22nm node, where scaling can no longer decrease the cost per transistor [1]. Solutions are getting more difficult to track down in an industry driven by increasing performance at lower cost. 3D-IC integration provides a path to continue to meet the performance/cost demands of next-generation devices while avoiding the need for further lithographic scaling, which requires both increasingly complex and costly lithography equipment as well as more patterning steps. 3D-IC integration, on the other hand, allows the industry to increase chip performance while remaining at more relaxed gate lengths with less process complexity—without necessarily adding cost [1].

Minimizing through silicon via (TSV) dimension for via-last bonding, or TSV and bonding pad dimensions for hybrid bonding, are key requirements for bringing down the cost of 3D devices. Considering that the role of a TSV is essentially "only" for signal connection yet consumes valuable wafer real estate, further miniaturization has to be the logical consequence.

A first option for monolithic integration is hybrid bonding, whereby a dual damascene copper and silicon oxide hybrid interface serves as both the full-area bonding mechanism and the electrical connection. A second option is the transfer of a thin processed semiconductor layer (ranging from tens to a few hundred nanometers in thickness) using a full-area dielectric bond. In contrast to hybrid bonding, the electrical connection is introduced in a subsequent process between early interconnect metal levels on the bottom wafer and the second transferred transistor layer. Both hybrid bonding and full-area dielectric bonding can be achieved through aligned wafer-to-wafer fusion bonding. However, high-interconnect density along with small routing dimensions set a high bar for bond alignment precision, which is necessary for fusion bonding.

## II. ALIGNED WAFER-TO-WAFER FUSION BONDING

Fusion bonding is a process based on forming molecular bonds between two surfaces placed into contact which are then further thermally annealed in order to allow the weak bonds formed at room temperature to transform into strong, irreversible, covalent bonds. The thermal annealing temperatures required are far too high for CMOS technology, in which the main temperature limitation is imposed by the Al interconnects (400°C or 450°C for very short time, in the range of minutes) [2-4].

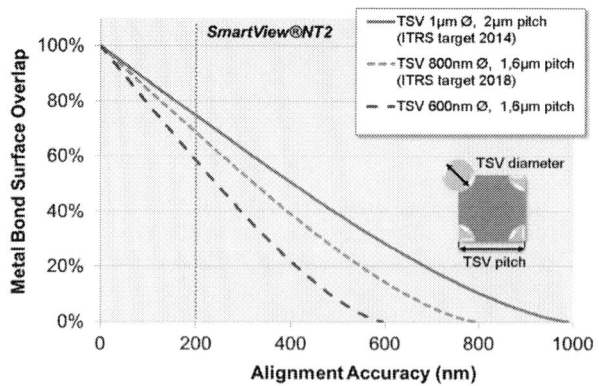

Fig. 1. Calculated surface overlap of metal TSVs for hybrid bonding as a function of wafer-to-wafer alignment accuracy. Comparison of ITRS roadmap relevant TSV pitches and diameters reveal, alignment accuracy of better than

200nm (3σ) is needed to achieve 60% and more TSV overlap for hybrid bonding.

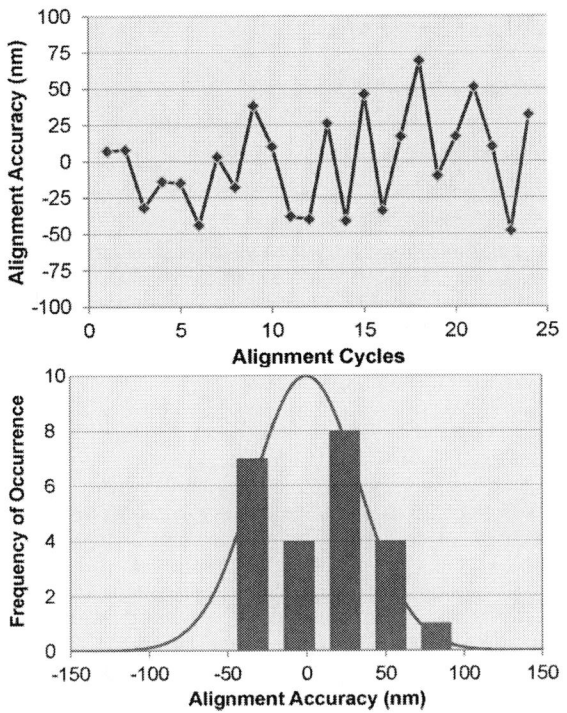

Fig. 2. EVG SmartView ®NT2 alignment data for consecutive alignments (top), revealing an alignment accuracy of 200nm (3σ) from the histogram and corresponding normal distribution (bottom).

Several factors contribute to the global alignment of the wafers besides the in-plane measurement and placement of the wafers relative to each other. In fusion bonding, both wafers are aligned and a pre-bond is initiated. When bringing the device wafers together, wafer stress and/or bow can influence the formation of a bond wave. The bond wave describes the front where hydrogen bridge bonds are formed to pre-bond the wafers. Controlling the continuous wave formation and controlling influencing parameters is key to achieving the tight alignment specifications noted above. In essence, optimizing a fusion bonding process means that one must optimize the force generated during the bonding. Any wafer strain manifests in distortion of the wafer, which leads to an additional alignment shift. Process and tool optimization can minimize strain and significantly reduce local stress patterns. Typically, distortion values in production are well below 50nm. Indeed, further optimization of distortion values is a combination of many factors, including not only the bonding process and equipment, but also previous manufacturing steps and the pattern design.

The 2011 edition of the International Technology Roadmap for Semiconductors (ITRS) roadmap specified that for high-density TSV applications, the diameter of vias will be in the range of 0.8-1.5 μm in 2015, which requires an alignment accuracy of 500nm (3σ) in order to establish a good electrical connection for the larger ones. This postulated alignment would lead to just 40% of areal TSV metal overlap. In order to enable high process yield for hybrid bonding, the areal overlap

has to be increased. For a 1μm diameter copper TSV the alignment spec quickly scales for 50%, 70% and 90% via overlap to 400nm, 240nm and 80nm alignment accuracy, as being depicted in Fig. 2.

The Gemini® FB XT is EVG's latest generation fusion bonding equipment dedicated for high-volume manufacturing. As it became clear from the above description, fusion bonding is moving up in the manufacturing value chain, with a strong pull-in from integrated device manufacturers and foundries. The integration of essential pre-processing steps, optimization of process flows and productivity has been central objectives for the Gemini FB XT. For this reason, two slots for pre- and post-processing modules have been added in the Gemini FB XT, extending the module spaces to six. In this way, throughput could be raised by approximately 50%. Integrated process steps consist of wafer cleaning and/or light chemical surface treatment, plasma activation and wafer bond alignment. Most important, the system introduced a new bond aligner to the market. The newly introduced SmartView®NT2 bond aligner has demonstrated the ability to achieve face-to-face overlay alignment within 200nm (3σ) in the current development stage with a clear goal to cross the 100nm (3σ) overlay barrier within the next few months, corresponding to a factor of 2-3 improvement from the well-established SmartView NT version. Furthermore, integrated metrology has been added into the tool. In this way, overlay after wafer bonding can be qualified and corrected within couple of wafer bonds.

## III. CONCLUSION

In summary, aligned wafer-to-wafer fusion bonding is a key process for the fabrication of monolithic integrated circuits. Bonding alignment accuracy is the most determining factor to arrive at high process yields. Recent developments in wafer bonding technology have demonstrated the ability to achieve bond alignment accuracy of 200nm (3σ) or less, which is needed to support the production of the next monolithic devices. The alignment within the bond aligner is only one aspect for a high yielding process. Other important aspects include the initiation, manipulation and control of a bonding wave formed at the interface. Meeting certain preprocessing steps, including die strain, oxide surface roughness and global wafer strain / warp, are major contributors to be controlled and analyzed

## REFERENCES

[1] Z. Or-Bach, "Is the Cost Reduction Associated with Scaling Over?", June 18, 2012, http://www.monolithic3d.com/2/post/2012/06/is-the-cost-reduction-associated-with-scaling-over.html

[2] Q.-Y. Tong and U. Gösele, Semiconductor Wafer Bonding: Science and Technology (Wiley Interscience, New York, 1999)

[3] T. Plach, et al., "Investigations on Bond Strength Development of Plasma Activated Direct Wafer Bonding with Annealing", ECS Transactions, 50 (7) 277-285 (2012)

[4] T. Plach, et al., "Mechanisms for room temperature direct wafer bonding", J. Appl. Phys. 113, 094905 (2013)

# New Precision Alignment Methodology for CMOS Wafer Bonding

Isao Sugaya[1], Hajime Mitsuishi[1], Hidehiro Maeda[1], Masashi Okada[1],
and Kazuya Okamoto[1,2]

[1]Nikon Corporation, Nagaodai 471, Sakae-ku, Yohohama, Kanagawa 244-8533, Japan
[2]Osaka University, Yamadaoka 2-1, Suita, Osaka 565-0871, Japan

*Abstract-* **A new precision alignment methodology suitable for distorted CMOS wafer bonding is proposed. Using multiaxis-interferometer and load-cell measurements with in situ fine stage-position adjustment combined with newly designed wafer holders, two wafers are flattened and aligned precisely. The alignment capability is 250 nm or better without any damage to low-k materials, making the methodology suitable for future 3D integration.**

## I. INTRODUCTION

The increasingly high speed and performance of CMOS devices, the basic component in current semiconductor devices, have been realized through hyper-miniaturization technologies. Beyond the 1x nm node range, however, the signal delay in global wiring used to couple individual IPs within a chip will become a crucial issue. The conventional countermeasures taken so far against signal delay will no longer be sufficiently effective and it will be necessary to use an additional compensating circuit (repeaters). The repeaters will inevitably increase the chip size and also the power consumption. At the same time, to keep pace with the increasing popularity of mobile devices, even smaller ICs with more diverse functions will be required. The use of three-dimensional integrated circuits (3DICs), employing TSV technology formed by vertically stacking chips with two-dimensional circuit patterns, is becoming increasingly common as a solution to meet these requirements [1-2]. 3D chips can be formed through stacking by adopting any of the following three modes: die-to-die (D2D), die-to-wafer (D2W), and wafer-to-wafer (W2W). Because of the reduced cost of chip fabrication and future 3DIC formation, the use of W2W mode is expected to be inevitable [3]. The precision wafer bonding of Cu-Cu interconnects is expected to be the key to fabricating the 3DICs, although many challenges must be overcome to achieve this such as the compensation of distorted wafers, low-force wafer contact, and precision alignment capability. On the basis of our long experience of CMOS lithography tools, we propose a methodology for alignment with a precision that cannot be achieved by conventional assembly tool.

## II. PRECISION ALIGNMENT METHODOLGY

This methodology includes two critical functions: the compensation of wafer distortion, and in situ monitoring and fine adjustment during the alignment and clamping procedure.

### A. Compensation of wafer distortion

A wafer has a low thickness relative to its surface area, thus does not have high mechanical strength and is prone to brittle failure. In addition, a CMOS wafer itself has local and global distortion including bow and warpage. Therefore, to handle a wafer without damaging it and to achieve higher alignment accuracy, the wafer should be fixed to wafer holders using an electrostatic chuck (ESC) with a flat adhesion surface, so that the wafer is handled together with the wafer holders. When bonding wafers, the bonding process can be easily achieved via the wafer holders by holding the wafers between the wafer holders as shown in Fig. 1. Furthermore, each ceramic low-thermal-expansion wafer holder includes three peripheral clamping mechanisms that enable very soft contact with a force of less than 20 N between wafers without coming in direct mechanical contact with the wafers. Moreover, a well-controlled shockless magnetic clamp operation is used to avoid the generation of extra particles along with chemical contamination (class 1 operation). This clamping function should be performed after the wafer alignment process described below.

### B. Precision alignment procedure

Alignment methods involve the visualization of alignment marks from each of the substrates and the mechanical positioning of the wafers to bring the marks into registry. Here we define the precision bonding technology required for stricter alignment tolerances to facilitate higher-level 3DICs. The stable alignment accuracy for all types of CMOS wafers must be 250 nm or better over a 300 mm wafer. Therefore, a new technology of precision bonding technology is required. Classical alignment includes front-side (through-wafer including IR light transparency), back-side, and intersubstrate alignment strategies. This is conventionally achieved with two microscopes and stages to align the wafers in the *X/Y/Z* and $\theta$ directions. To obtain higher alignment accuracy, we consider that enhanced wafer global alignment (EGA) is required because of the distortion characteristics of a 300mm diffused wafer. The basic EGA procedure for bonding wafers is based on the standard coordinate system, determined by fiducial marks. Here we consider the bonding of two wafers, wafer-1 and wafer-2. The coordinate system of wafer-1 is determined on the basis of fiducial coordinate system. After obtaining the relation between the coordinate systems of wafers-1 and -2, the alignment unit is adjusted and aligned using *X/Y/Z/θ* stage-driving units. The translation and rotation of wafer-2 are described as follows:

$$\begin{pmatrix} x_c \\ y_c \end{pmatrix} = \begin{pmatrix} \cos\theta & \sin\theta \\ -\sin\theta & \cos\theta \end{pmatrix} \begin{pmatrix} x_m \\ y_m \end{pmatrix} + \begin{pmatrix} T_x \\ T_y \end{pmatrix}. \quad (1)$$

where $(x_c, y_c)$ are coordinates in the converted coordinate system, $(x_m, y_m)$ are the measured coordinates of alignment marks, and $(T_x, T_y)$ are translation distances. As described above, however, actual diffused wafers are distorted; thus, to precisely align the two wafers, the optimization requires EGA with multiple alignment marks. For instance, when the alignment mark positions of wafer-2 in the fiducial coordinate system are at $(A_{xi}, A_{yi})$, where $i$ is the mark number, after translation and rotation using the determined values, the positions in the converted coordinate system $(M_{xi}, M_{yi})$ are

$$\begin{pmatrix} M_{xi} \\ M_{yi} \end{pmatrix} = \begin{pmatrix} \cos\theta & \sin\theta \\ -\sin\theta & \cos\theta \end{pmatrix} \begin{pmatrix} A_{xi} \\ A_{yi} \end{pmatrix} + \begin{pmatrix} T_x \\ T_y \end{pmatrix}. \quad (2)$$

Next, denoting the coordinates of the alignment marks of wafer-1 in the fiducial coordinate system as $(D_{xi}, D_{yi})$, we can determine the optimized shift $(T_x, T_y)$ and rotation $(\theta)$ by minimizing the function $F(\theta, T_x, T_y)$ given by

$$F(\theta, T_x, T_y) = \sum \left\{ (D_{xi} - M_{xi})^2 + (D_{yi} - M_{yi})^2 \right\}. \quad (3)$$

The alignment equipment should be set up in accordance with the above theory to obtain higher alignment accuracy between two wafers. Equipment should also be developed to increase the precision of the stage device used for alignment mechanism and the DSP control system. Note that the device used for alignment should be equipped with an alignment microscope and multiple laser interferometers to measure the distance traveled by the wafers along two mutually perpendicular directions. To ensure the soft contact between both wafers, multiple precision load cells should also be implemented. We developed a new system based on the above methodology.

*C.  IR metrology tool*

We also developed a tool for measuring the overlay alignment accuracy that employs an IR microscope. The temperature of the tool is controlled at 23 ± 0.1 °C. The alignment accuracy is determined by measuring the deviation between the line and space marks patterned on the two wafers. Fig. 2 shows a representative IR image of overlay accuracy measurement marks on a pair of 300 mm wafers. A measurement repeatability of better than 16 nm was achieved at 18 measurement points as shown in Fig. 3.

## III. EXPERIMENTAL RESULTS

Fig. 4 shows typical position- and force-control time axis data obtained using Cu bump wafers with a Cu size of micron order. After the EGA procedure, the wafers are in soft contact with a force of less than 20 N per load cell using our newly developed impedance-control algorithm engine. During the application of load and the clamping of the wafers, no position shift exceeding 100 nm was detected by interferometers. An overlay deviation map of bonded wafers in five times continuous pre-bonding is shown in Fig. 5. Fig. 6 shows five results in the case of continuous pre-bonding over 300 mm wafers. The bonding process was clearly well controlled. Deviations in the X and Y directions of less than 250 nm were confirmed, which can be obtained to be reduced less than 100 nm over time.

## IV. CONCLUSIONS

A new EGA precision wafer bonding methodology involving in situ monitoring was proposed. To meet the demand for more precise wafer bonding, we theoretically and experimentally obtained stable and higher alignment accuracy of better than 250 nm. This will contribute to the fabrication of future 3DICs such as DRAM, MPUs and image sensors.

## ACKNOWLEDGMENT

We would like to acknowledge the continuous encouragement of Dr. H. Ohki, and are also grateful for the engineers in Nikon Precision Equipment Company.

## REFERENCES

[1] M. Koyanagi, et al., "Future system-on-silicon LSI chips," *IEEE Micro*, vol. 18, no. 4, pp. 17-22, July/August 1998.

[2] Y. Xie, G. Loh, B. Black, K. Bernstein, "Design Space Exploration for 3D Architectures," *J. Emerging Technologies in Computing Systems*, vol. 2, no. 2, pp. 65-103, April 2006.

[3] K. Okamoto, "Importance of wafer bonding for the future hyper-miniaturized CMOS devices," *ECS Trans.*, vol. 16, no. 8, pp. 15–29, 2008.

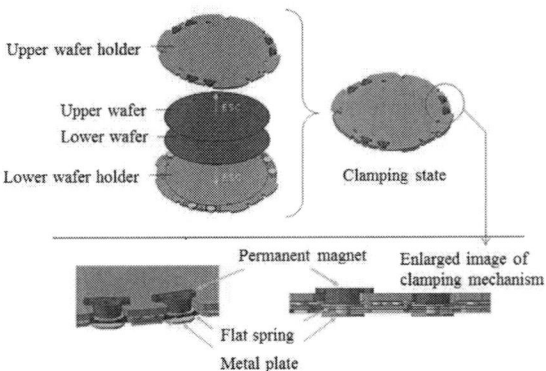

Fig. 1. Wafer holders and new peripheral clamping mechanism.

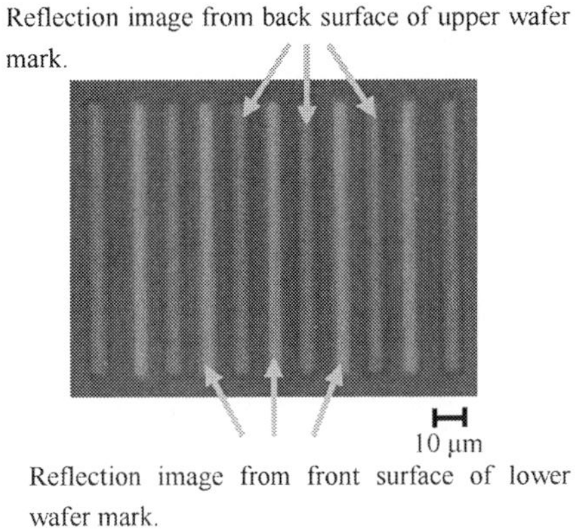

Reflection image from back surface of upper wafer mark.

10 µm

Reflection image from front surface of lower wafer mark.

Fig. 2. Representative IR image of overlay measurement marks on a Cu pre-bonded wafer pair.

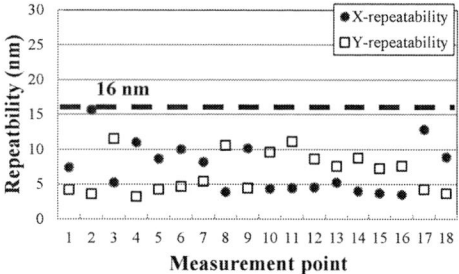

Fig. 3. Measurement repeatability results of IR metrology system. The repeatability is 3σ of 11 times measurements at each measurement point.

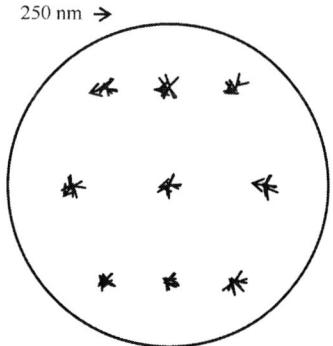

Fig. 5. Overlay deviation map of 9 measurement points on pre-bonded 300 mm wafer measured by IR metrology tool in five times continuous pre-bonding.

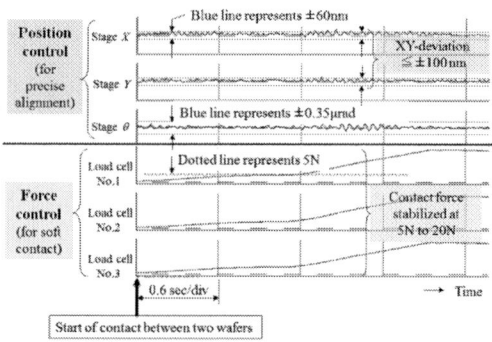

Fig. 4. Results of wafer bonding process using in situ position/load monitoring.

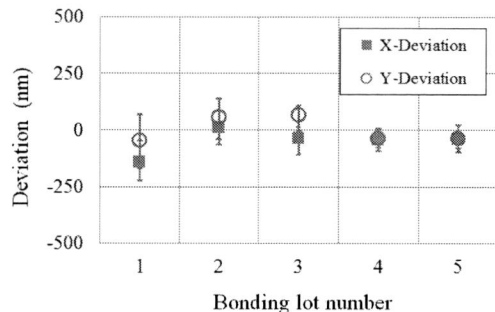

Fig. 6. Results of wafer bonding obtained using IR measurement over 300 mm wafers.

# Precision Bonders - A Game Changer for Monolithic 3D

Zvi Or-Bach, Brian Cronquist, Zeev Wurman, Israel Beinglass, and Albert Henning

MonolithIC 3D Inc., San Jose, CA 95124, USA E-mail:zvi@monolithic3d.com

*Abstract* — It is well recognized that dimensional scaling has reached its diminishing return phase and the industry is now looking to monolithic 3D to be the future technology driver. This was clearly voiced in the Qualcomm keynote at DAC 2014. This paper will present the game changing impact of the emerging precision bonders, such as EVG's Gemini® XT FB. Innovative process flows combined with a 'Smart Alignment' technique could enable any semiconductor vendor to integrate monolithic 3D into its existing manufacturing line and existing process flows with minimal technology challenge. This provides a natural path for product innovation and an unparalleled competitive edge. In sharp contrast, prior published works on monolithic 3D were conditioned on new process flows and new transistor formation recipes. In addition, this game-changer breakthrough offers a very cost competitive flow.

## Introduction

For many years monolithic 3D was considered untenable due to the strict 400 °C temperature limit imposed by the aluminum or copper interconnect. This led to the focus on TSV technology as the only viable path for 3D ICs. Unfortunately, it is now clear that the TSV flow in intrinsically expensive and accordingly being perpetually pushed to the future. In the recent years pioneering efforts were published providing practical paths for monolithic 3D logic devices [1-7]. But each and every one of those presented new transistor formation flows and comes along with some other non trivial process development challenges. Recently a new wafer bonder has been introduced to the market by EVG [8]. While prior wafer bonders had about 1 micron alignment accuracy, the newly introduced Fusion Bonder has an alignment precision of 200nm (3σ). This paper proposes a process flow that leverages such precise bonder to provide a true monolithic 3D IC without the need for a new recipe for transistor formation. The process could be adapted by any current fab providing very competitive costs for a range of product enhancements and offer a long term road map for better offerings by scaling up.

## Ion-Cut - A Layer Transfer Technology

The proposed flow utilizes a well-known process for single crystal thin layer transfer known as ion-cut technology. It involves hydrogen implantation, wafer bonding, and cleaving (Fig. 1). Ion-cut is a volume production qualified process that has been used for two decades in SOI wafer manufacturing. The technology was owned exclusively by Soitec for many years, which named it smart-cut®, but in late 2012 Soitec's fundamental patent expired and the technology is now widely available. Our estimates suggest that with re-use of substrates, ion-cut would cost less than $60 per transferred layer.

## Monolithic 3D IC

The following flow (Fig. 2) is built on what we call 'gate replacement' [9] and leverages the precision bonder

alignment accuracy. Step 1 - a 'donor' wafer will be used to process a transistor layer labeled Stratum 3. The existing front end process could be used. Alternatively for a gate-last flow, the process will hold before the gate replacement phase. Then H+ would be implanted at the desired depth (~100nm) in preparation for the layer transfer step. Step 2 - the donor wafer is bonded (oxide to oxide) to a 'carrier wafer' and ion-cut off. This bonding step does not require precise alignment. Step 3 - the carrier wafer could be now annealed to repair the potential H+ implant damage. Step 4 - the donor wafer is now processed to form Stratum 2. The existing front line process could be used including FinFET or any other available front line process. The choice of the transistor and the architecture for Strata 2 and 3 should consider the need for vertical isolation in-between them. Note that between the transferred layer and the carrier wafer there is an oxide layer which would be an excellent etch stop allowing the transfer onto the target layer without the need for ion-cut. A preferred strategy is to use Stratum 2 for the high performance circuits while Stratum 3 would be used for support of less sensitive circuits. All high temperature should be completed at this point, as in the following step interconnects are added. Step 5 - add contacts and at least one metal layer. Step 6 - bond (oxide to oxide or metal to metal) to the target wafer using the precise bonder alignment with less than 200nm misalignment. Now grind and etch off the carrier wafer. (Not presented here are options to remove the carrier wafer for reuse.) Step 7 - the dummy gate and the gate oxide of Stratum 3 can be now replaced, and connections could be made between Stratum 2, Stratum 3 and the underneath target wafer. Alignment and via processing are just as if between conventional BEOL metal layers, as the transferred layer is very thin (~100nm).

## Smart Alignment

Having a thin transferred layer allows the through layer via to be as small as a conventional BEOL interconnection via (~50 nm). Yet the 200 nm bonding alignment window would appear to require a landing pad of 200 nm by 200nm for each vertical connection. With Smart Alignment the connection is made by two perpendicular 200 nm long strips as seen in Fig 3. The vertical strip is part of the top layer of the target (bottom) wafer. After bonding, the through layer via would be aligned to the target wafer in the X direction and to the transferred layer in the Y direction as seen in Fig 4. The top connection strip could be then processed aligned to the transferred (top) layer. This alignment scheme reduces the vertical connection overhead to minimum and allows for multiple vertical connections per unit area of 200 nm x 200 nm.

## Strata 2, 3 - Examples

Fig. 5 illustrates one example for circuit allocation for Stratum 2 and Stratum 3 with an intrinsic vertical isolation.

For Stratum 2, most advanced devices could be used such as FinFET transistors and forming high speed logic. The SRAM for the high speed logic circuit could be placed onto the close by Stratum 3. A compelling option for the SRAM would be the use of Zeno technology [10] where a two stable states one transistor SRAM are enabled by a deep implanted back-bias. The vertical isolation is achieved by the back-bias. The FinFET transistor by design is also isolated from the substrate. This use of Stratum 2 and Stratum 3 is compelling as there is no obstruction to the memory blockages and creates a very short path for memory access. Such dual functional layer (Stratum 2 + Stratum 3) could be a product by itself offered as an add-on to many designs and 1-chip systems.

Such flows with a dual functional layer could enable many new innovative devices such as:
* An image sensor on Stratum 3 with pixel electronics on Stratum 2 could provide an unparalleled dynamic range to cameras.
* A full redundancy layer [11] as Stratum 3 provides redundancy to Stratum 2, allowing almost unlimited logic integration on huge dies, essentially a server farm on a device.
* A configurable logic fabric as an add-on...

## Monolithic 3D Cost Estimates

It is well known that high cost is the number one issue which slows down the adoption of 3D ICs based on TSV. The proposed monolithic 3D flow has the potential to overcome these barriers as it avoids use of a thick layer with lengthy etch and deposition processes. In fact, it can provide circuit fabrics for two strata for a cost that is less than one wafer substrate. The donor wafer is reusable and the cost of the first ion-cut is estimated to be less than $60 [12]. The carrier wafer could be reusable or utilize an inexpensive test wafer costing about $30. The estimated per wafer cost of precision bonding is less than $20. Other steps involved in layer transfer, cleaning, etch, etc., are estimated at about $30 total. The costs for transistor formation for Strata 2 and 3 and their associated interconnects are no different than any other circuit fabrication costs. Accordingly we estimate that the cost structure is comparable with the fabrication cost of 2D devices. Yet having the overall design built in a 3 strata fabric provides huge power, performance, and cost benefits.

## Heat Removal

A point of concern is always the heat removal aspect for 3D IC. The first question relates to having more transistors in a smaller space. While more complex circuits present an ever increasing power challenge, having it built in monolithic 3D is an important part of the solution as it is well documented [13] that 80% of the power consumption is due to on-chip connectivity. The more interesting question relates to the fact that Strata 2 and 3 transistors are thermally isolated (surrounded by oxide) and without direct access to the silicon bulk for heat removal. Fig. 6 illustrates the solution of using the power delivery network for heat removal. This work was reported in IEDM 2012 [14]. Having Stratum 2 only about 1 micron away from the bulk

allows a very effective heat removal path through the power delivery network. Additional supporting techniques such as heat spreaders and thermally conducting but electrically isolated contacts could also be implemented [15].

## Summary

Precision Bonders combined with innovative layer transfer and alignment techniques enable a simple path to 3D IC providing the best of all worlds:

- Vertical connectivity density comparable with the horizontal one
- Use of existing transistor and interconnect flows
- Compatible with advanced and older fabs
- Low cost competitive with 2D IC cost structure
- Heterogonous integration - Fab lines, process nodes, device materials, processes
- Parallel and Sequential process (short TAT)
- Enables many new classes of device and systems untenable with 2D IC
- Multiple paths for cost reduction that were untenable with 2D IC

It seems that the new form of 3D IC combines the best of TSV with monolithic 3D IC to offer the most attractive path to keep Moore's Law, while opening an unparalleled path for all fabs to keep enhancing their product range using their existing equipment and flows.

And yes, this opens a new horizon for the semiconductor industry.

## References

[1] Z. Or-Bach, IEEE 3DIC Conference, 2013.
[2] P. Batude et al., Proceedings of the Electro-Chemical Society (ECS) spring meeting, Vol. 16, pp.47 (2008)
[3] D. C. Sekar, IEEE 3DIC Conference, 2012.
[4] B. Rajendran, IEEE 3DIC Conference, 2013.
[5] Chih-Chao Yang 29.6 IEDM 2013
[6] Chang-Hong Shen pp. 262 IEDM 2013
[7] Sang-Yun Lee, US Patent 7,470,142
[8] Thomas Uhrmann et al., Solid State Technology, pp.14, July 2014
[9] http://www.monolithic3d.com /hkmg.html
[10] Yuniarto Widjaja US Patent 8,514,623
[11] http://www.monolithic3d.com/ultra-large-integration---redundancy-and-repair.html
[12] http://www.monolithic3d.com/blog/how-much-does-ion-cut-cost1
[13] L. Chang, IBM Short Course, IEDM 2012
[14] Hai Wei, et al., IEDM 2012
[15] D. C. Sekar US Patent 8,686,428.

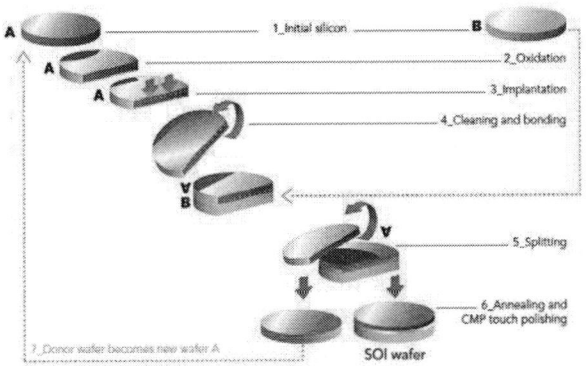

Fig. 1: Ion-Cut (smart-cut®) process, stacking single crystal silicon; Soitec

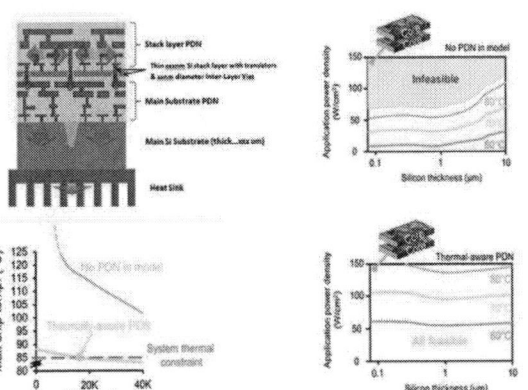

Fig. 6. Heat removal by the power delivery network (PDN) [14]

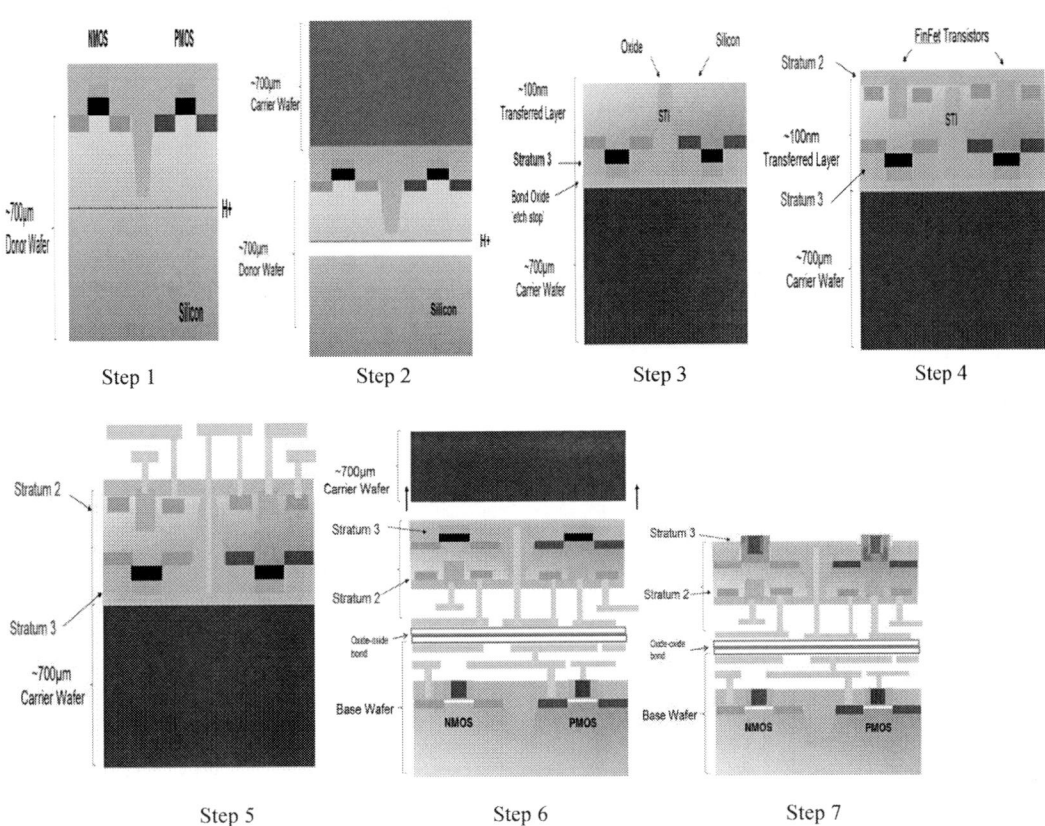

Fig. 2: Process Flow for Gate Replacement Process and precise bonding.

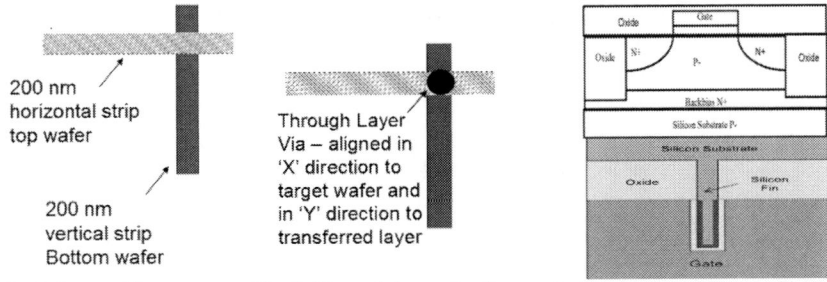

Fig. 3: Smart-Alignment    Fig. 4: Through layer via alignment    Fig 5. 1T SRAM over FinFET

978-1-4799-7440-5/14 $31.00 © 2014 IEEE        147

# Fully Functional Fine-grain Vertically Integrated 3D Focal Plane Neuromorphic Processor

M. Di Federico*[†], P. Julián*[†], A. G. Andreou[‡], P. S. Mandolesi*[§]

*Departamento de Ingeniería Eléctrica y de Computadoras, Universidad Nacional del Sur
Av. Alem 1253, Bahía Blanca - Argentina,
Email: mdife@uns.edu.ar
[†] CONICET - Argentina
[‡] Electrical and Computer Engineering Department, Johns Hopkins University - USA
[§] CIC - Pcia. Bs. As., Argentina

*Abstract*—**This paper presents the first fully functional fine-grain 3D vertically integrated focal plane system processor, implemented in the 3D interconnection technology of Tezzaron Semiconductors. The processor consists of an array of cells, auxiliar structures and general purpose blocks. The chip can acquire video and apply a series of image processing tasks to each frame employing local computation among cells. The chip has been successfully tested at 50Mhz.**

*Index Terms*—**3D integration, focal plane processor**

## I. INTRODUCTION

Event though the benefits of vertical 3D integration have been shown in the literature [1], few functional experimental prototypes have been reported. [2] and [3] report imagers built with two tiers, where the photodiodes are in one tier and the acquisition electronics are on the other tier. [4] reports a fine-grain DSP processing unit, with electronics in both levels, and [1] reports one wafer with 64 processor cores connected to another SRAM memory wafer; both chips were fabricated using a 130nm process and Tezzaron interconnection technology.

This paper presents the first fully functional fine-grain 3D vertically integrated focal plane system processor, based on the simplicial CNN architecture [5], [6]. The 3D IC has two tiers implemented on the Chartered 130nm technology and Tezzaron interconnection technology. The integrated circuit (IC) consists of an array of cells, where every cell is distributed in the two tiers, together with auxiliar and general purpose computation blocks (see Fig. 1) so that the chip can perform low and medium level image processing operations. The chip has been successfully tested at 50MHz.

## II. ARCHITECTURE

The array is composed by $48 \times 32$ elements. The auxiliary computation structures are located on one side of the array, and include: a) a core for nonlinear (piecewise linear) computation; b) a BUS manager; c) a control and decoder control unit; d) a program memory; e) a column adder and an integrator; f) a multiplier and a divider; g) a chain code generator; h) a correlator. The column adder and the integrator are used to calculate the sum of the value of all the pixels in the image. The multiplier and divider, together with the sum of all pixels can be used to calculate several image descriptors. The general purpose structures are an 8051 micro controller and a UART unit.

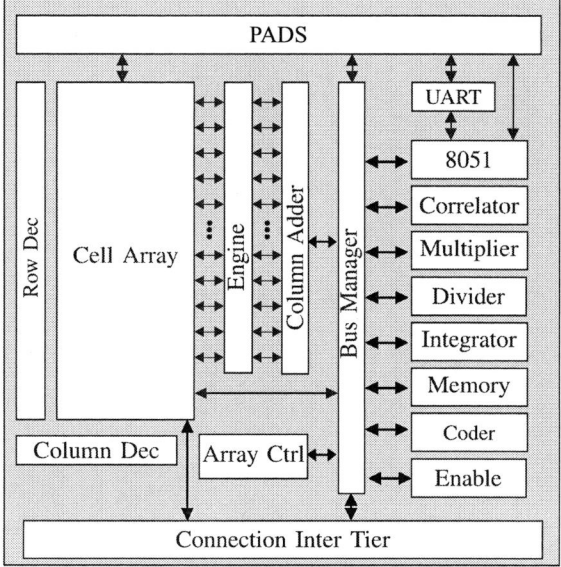

Fig. 1: Block Diagram.

Every cell in the array is composed of the blocks illustrated in Fig. 2, namely, two 7-bit registers, one 7-bit counter, two latches and one multiplexer. There are 9 3D bias per cell. The array computes one column at a time. The pixel value is acquired and converted into a 7-bit digital word in the top tier. Then, it is stored in U register in the bottom tier. During computation, X and U digital words are placed on the bus and compared with a digital ramp running on the internal bus. This is done using the column comparators, and the obtained results are one-bit time-coded (PWM) signals. These value are latched into the F and G latches in the cell bottom tier. Then these two latch values are placed on the bus, and grouped together with the corresponding signals of the 4 local neighbour cells. These 5 values access the $32 \times 1$ parameter memory, located on the column array, which are digitally operated by the programmable *FoG* block, and subsequently added by the counter located in the top tier of the cell. This process is repeated until all columns have been computed.

The IC can work with different pixel resolutions: 7-bit grayscale or 1-bit black/white. The processing time depends on the resolution: black and white images can be processed in 1 or

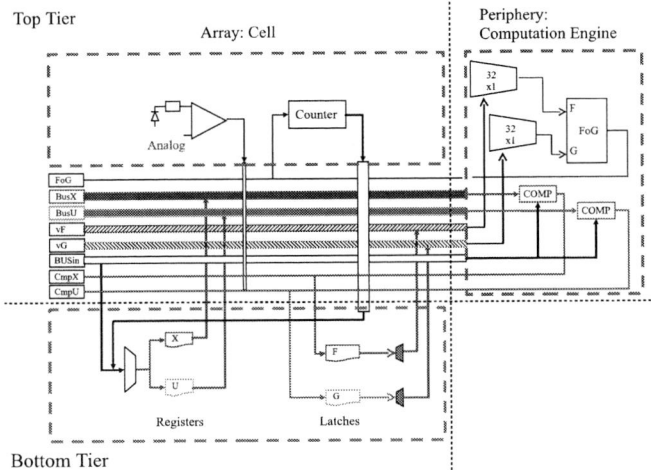

Fig. 2: Cell structure.

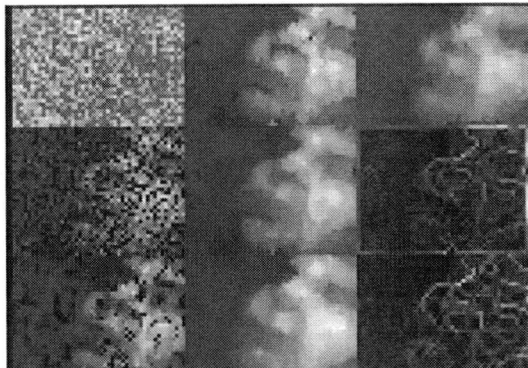

Fig. 4: Image of a face acquired by the chip (Top Left) and outputs from different processing algorithms executed by the chip.

2 clock cycles, using one or two registers. Similarly, images of 7-bit pixel resolutions are processed in 64 or 128 cycles. With a clock running at 50Mhz, this architecture can process 375 thousand grayscale images or 25 million black/white images per second.

### III. VLSI Implementation and Results

The full chip size is 2mmx2.5mm, the working voltage for the core is 1.5V, and the I/O works at 3.3V. The chip has 70 pads, encapsulated in a PGA85 package.

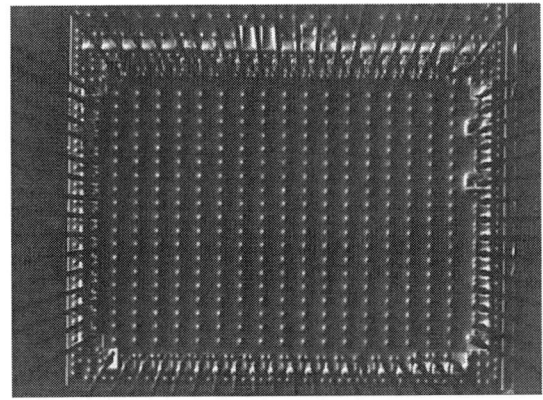

Fig. 3: IC microphotograph.

A test board to control and test the prototype was fabricated. This board is controlled by an FPGA with a microprocessor that communicates with the PC. This test board was designed to send and receive a large number of signals: it can manage 256 8-bit digital input/output ports. The PC send commands to the microcontroller in the FPGA through the UART. The clock frequency in this test was 50Mhz, mainly limited by the test board. Fig. 4 shows a snapshot of an image acquired by the chip and the result of several processing operations applied on the fly by the chip. The left column, from top to bottom, shows: the original image captured by photodiodes; the fixed pattern noise removal with Correlated Double Sampling; a Median filter

applied to the Center Left Image. The center column, from top to bottom, shows: the image closure, a median filter applied to the closed image; the dilation of the filtered image. The right column, from top to bottom, shows: opening and closing; edge recognition using the difference between the eroded image and the original; edge recognition using the difference between the dilated image and the original. The video captured and processed by the chip is available at [7].

### IV. Conclusions

3D technology can be used at present to build complex systems. We have provided experimental evidence of a full working prototype of a system that acquires images, and performs A/D conversion plus local nonlinear computation at the pixel level, with a fine-grain design methodology. In addition, the system has blocks that can be used to perform mid and high level processing tasks.

### Acknowledgment

This project was partially funded by the International Project PICT 2010 No. 2657 3D Gigascale Integrated Circuits for Nonlinear Computation, Filter and Fusion with Applications in Industrial Field Robotics, ANPCyT, Ministry of Science and Technology of Argentina.

### References

[1] D. H. Kim et al. "3D-maps: 3D massively parallel processor with stacked memory," *2012 IEEE International Solid-State Circuits Conference, Digest of Technical Papers* , pp. 188–190, 2012.

[2] V. Suntharalingam et al. "A 4-side tileable back illuminated 3D-integrated Mpixel CMOS image sensor," *2009 IEEE International Solid-State Circuits Conference, Digest of Technical Papers*, pp. 38 – 39, 2009.

[3] J. Burns et al. "Three-dimensional integrated circuits for low-power, high-bandwidth systems on a chip," *2001 IEEE International Solid-State Circuits Conference.*, pp. 268–269, 2001.

[4] T. Thorolfsson, S. Lipa, and P. Franzon, "A 10.35 mw/Gflop stacked SAR DSP unit using fine-grain partitioned 3D integration," *2012 IEEE Custom Integrated Circuits Conference*, pp. 1–4, 2012.

[5] M. D. Federico, P. Julian, and P. Mandolesi, "SCDVP: A simplicial CNN digital visual processor," *IEEE Transactions on Circuits and Systems I: Regular Papers*, vol. PP, no. 99, pp. 1–8, 2014.

[6] P. Mandolesi, P. Julian, and A. G. Andreou, "A scalable and programmable simplicial CNN digital pixel processor architecture," *IEEE Transactions on Circuits and Systems I: Regular Papers*, vol. 51, no. 5, pp. 988 – 996, 2004.

[7] M. D. Federico, "3D-IC video processor," 2013. [Online]. Available: https://www.youtube.com/watch?v=88br5L9WMBo

978-1-4799-7440-5/14 $31.00 © 2014 IEEE

# Smart Co-Integration of Light Sensitive Layers with FDSOI Transistors for More than Moore Applications

L. Grenouillet, B. De Salvo, L. Brunet, J. Coignus, C. Tabone, J. Mazurier, C. Le Royer,
P. Grosse, M.A. Jaud, P. Rivallin, Z. Chalupa, O. Rozeau, O. Faynot, M. Vinet.

CEA, LETI, Minatec Campus, 17 avenue des Martyrs, Grenoble, France
Email : laurent.grenouillet@cea.fr, Phone : +33 4 38 78 99 23

## Abstract

We demonstrate for the first time that FDSOI technology can be turned into an integrated light-sensitive device technology capable of detecting or interacting with light. By designing a dedicated diode below the BOX, light absorption induced $V_T$ shifts as large as 100mV and saturation drain current modulation of 70% are measured for the transistors above the BOX. Those experimental results are supported by TCAD simulations and pave the way to More than Moore applications for FDSOI technology.

## Introduction

Fully Depleted Silicon-On-Insulator (FDSOI) technology is considered one of the most promising solutions to pursue the CMOS scaling in the More Moore industry [1,2], as planar bulk technology has several intrinsic limitations and 3D FinFET devices still present technological challenges extremely difficult to tackle. One of the main differentiating features of FDSOI technology is the possibility to tune, either statically or dynamically, the transistor electrical properties by applying a potential on the back plane below the Buried Oxide (BOX) (cf Fig.1), which acts as a second gate [3, 4].

In other words it means that FDSOI transistors can be considered as very small footprint *probes* which are *sensitive* to what happens below the BOX. In this paper we demonstrate that FDSOI transistor electrical properties can be sensitive to light illumination providing a diode is implemented below the BOX. The light-induced transistor performance modifications are quantitatively explained by TCAD simulations. Finally the transistor parameters are modified to amplify the transistor sensitivity to light absorption below the BOX.

## FDSOI transistor performance modification under illumination

NFET and PFET transistors featuring 25nm BOX, 6nm Si channel, and gate lengths $L_G$ down to 30nm were fabricated (Fig. 1). The backside of the BOX consists of either a unipolar P-type region, or a N/P diode (Fig.2). The diode is obtained by means of ion implantation of donors and acceptors through the BOX before gate stack deposition. The doping levels are in the $1 \times 10^{18}$ cm$^{-3}$ range, and the junction depth lies between 50nm and 150nm below the BOX/substrate interface.

For both backside configurations, NFET and PFET threshold voltage $V_T$ is measured either in the dark or when the wafer is exposed to a wide wavelength spectrum, the substrate being grounded. The light source comes from the top of the wafer. As reported in Fig.3, while light has no impact on NFET $V_T$ and PFET $V_T$ for the P-type backside reference configuration, NFETs shift towards high $V_T$ and PFETs shift towards low $V_T$ under illumination when the N/P diode is implemented below the BOX. The Light-Induced $V_T$ shift (LIVS), defined as $V_{T\ LIGHT} - V_{T\ DARK}$ is therefore attributed to light absorption in the N/P diode below the BOX.

## Light Induced $V_T$ Shift explanation and TCAD simulations

TCAD simulations taking into account light/matter interaction were performed to quantify the impact of light absorption on the transistor characteristics. The structure is exposed from the top to a light source matching the Air Mass 1.5 (AM1.5) spectrum.

Under illumination, photo-generated carriers tend to reduce the built-in electric field in the diode, therefore forward biasing it. In the BOX/N/P configuration case, as the P-type region of the diode is clamped to a given voltage (here 0V), it implies that the floating N-type region band structure lifts upwards under illumination to reduce the diode internal electric field. This in turn lifts upwards the BOX band structure as if a negative *equivalent* voltage $-V_{eq}$ was applied directly at the BOX/substrate interface (Fig.4), therefore reverse

biasing NFET devices and forward biasing PFET devices, as observed experimentally (Fig.3). Similarly, an opposite effect in terms of LIVS would be seen for a BOX/P/N configuration. Table 1 summarizes the expected LIVS effect for all FET and backside combinations. The impact of light source intensity on the $I_D$-$V_G$ curve is then investigated. Increasing the light flux intensity within 4 orders of magnitude continuously modifies the band curvature (inset of Fig.5), which in turns shifts the $I_D$-$V_G$ curve (Fig.5). LIVS values around 30mV are derived under regular illumination (Fig.7), in very good agreement with experimental results reported in Fig.3.

## Body Factor increase and substrate biasing to enhance light detection – transistor interaction

In order to enhance FDSOI transistor sensitivity to light absorption below the BOX, the Body Factor (BF) [5] was experimentally increased to improve LIVS magnitude. To do so a thicker Si channel (25nm instead of 6nm) was implemented. According to TCAD simulations shown in Fig.7, the expected BF should raise from 60mV/V to more than 200mV/V with a 25nm Si channel thickness. Thick Si channel thickness being detrimental for short channel effect control, only long channel ($L_G$>200nm) devices are characterized hereafter. Another way to increase the BF would have been either to decrease the BOX thickness (from 25nm to 10nm) or to increase the gate oxide thickness, the advantage of the former being to preserve the short channel effect control.

NFET $I_D$-$V_G$ curves recorded with or without illumination for different substrate bias are plotted in Fig.8 with a similar BOX/N/P backside diode configuration. BF values up to 225mV/V are extracted from Fig.9, in very good agreement with TCAD (Fig.7). The derived LIVS values (Fig.10) are proportional to the BF, reaching 100mV when the substrate is biased at +2V. This corresponds to a significant x20 light absorption induced modulation of the NFET drain current in the sub-threshold regime. LIVS is not impacted by the transistor regime (linear or saturation mode), as shown in Fig.10. The predicted continuous $V_T$ shift with light illumination (Fig.5) is experimentally observed in Fig.11, without any degradation in the sub-threshold slope. It is worth mentioning that the LIVS effect is reversible. So it is possible to switch from one $V_T$ to the other by turning on/off the light source, as illustrated in Fig.12 and Fig.13, respectively. Fig.13 clearly suggests that both steeper slope and lower $I_{OFF}$ FET devices could be fabricated provided the illumination switching mechanism is controlled by $V_G$ around the $V_T$ value.

NFET $I_D$-$V_D$ curves recorded with or without illumination (Fig.14) show that light absorption in the diode has also a strong impact in the saturation current, which can be modulated by up to 70% at $V_G$=0.3V with light illumination (Fig.15).

## Conclusion

We demonstrate experimentally that the implementation of a photodiode below the BOX of FDSOI transistors modifies $V_T$ under illumination (up to 100mV $V_T$ shift), therefore transforming the transistor into a light intensity probe. Significant saturation drain current modulation (70%) is induced by light absorption in the diode, and technological parameters (BOX and gate oxide thickness) are identified to enhance further the light detection – transistor interaction.

The described results suggest that a light detection mechanism co-integrated with FDSOI technology can benefit device performance (lower $I_{OFF}$, steeper slope) and also has potential to address More than Moore analog or logic applications like imaging or light-controlled logic.

978-1-4799-7440-5/14 $31.00 © 2014 IEEE

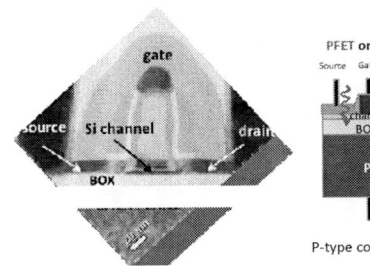

Fig.1: TEM picture of a typical FDSOI transistor featuring a 25nm BOX, a 6nm Si channel, a 30nm gate length, and the back plane acting as a second gate.

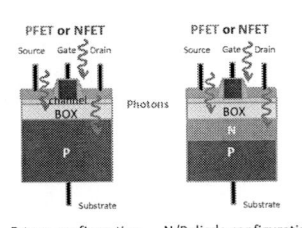

Fig.2: Schematics representing the different configurations below the BOX tested either in the dark or under illumination.

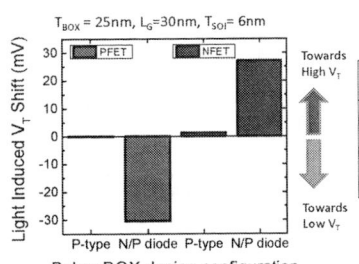

Fig.3: Experimental Light-Induced $V_T$ Shift (LIVS) of PFET and NFET FDSOI transistors for different doping configurations below the BOX (cf Fig.2).

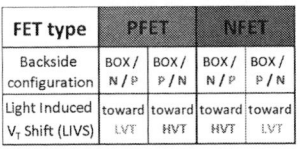

Table 1: Effect of light illumination on PFET and NFET $V_T$ for different diode configurations below the BOX. (LVT=Low $V_T$, HVT=High $V_T$).

| FET type | PFET | | NFET | |
|---|---|---|---|---|
| Backside configuration | BOX/N/P | BOX/P/N | BOX/N/P | BOX/P/N |
| Light Induced $V_T$ Shift (LIVS) | toward LVT | toward HVT | toward HVT | toward LVT |

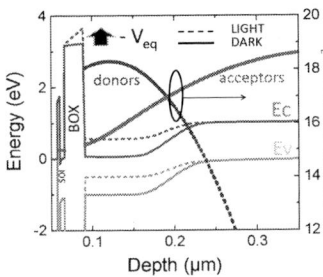

Fig.4: TCAD simulation showing the band structure of an NFET FDSOI device with an N/P diode implanted at the backside of the 25nm BOX. $V_G$=0.9V, $V_{DS}$=50mV, $V_{SUB}$=0V.

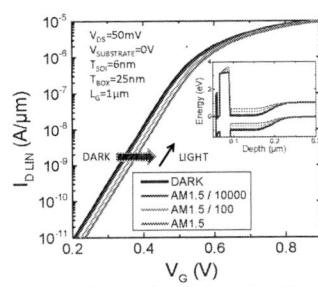

Fig.5: TCAD simulation showing NFET $I_{DLIN}$-$V_G$ curves for different illumination conditions on the N/P backside diode (corresponding band structure is plotted in the inset).

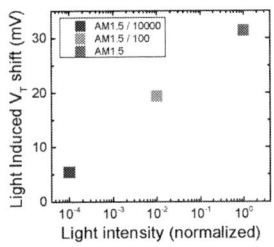

Fig.6: NFET LIVS values derived from Fig.5 showing a good agreement with experimental ones (Fig.3) under regular illumination.

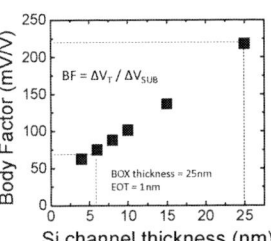

Fig.7: Body Factor as a function of channel thickness for a BOX thickness of 25nm and an EOT of 1nm (TCAD simulation).

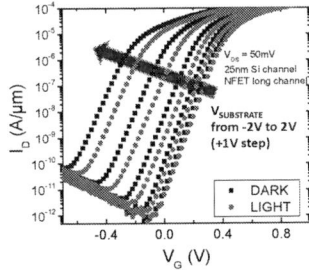

Fig.8: Experimental NFET $I_D$-$V_G$ curves recorded for different substrate voltages, without (black squares) or under (red circles) illumination. The backside configuration is BOX/N/P.

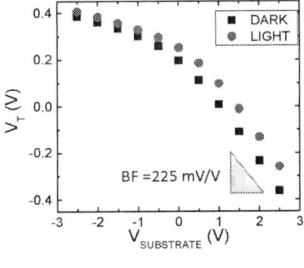

Fig.9: NFET $V_T$ derived from Fig.8 for different substrate voltage (with and without illumination). The BF reaches 225mV/V when substrate is biased at +2V.

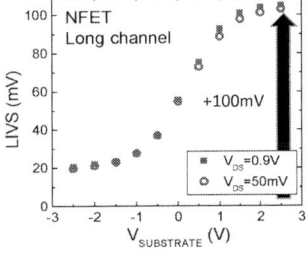

Fig.10: NFET LIVS derived from Fig.9 for different substrate voltages. LIVS reaches 100mV when substrate is biased at +2V.

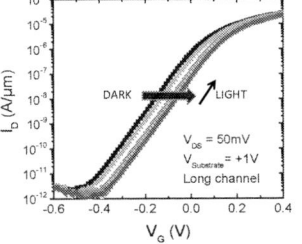

Fig.11: Experimental NFET $I_D$-$V_G$ curves recorded for different illumination conditions with an improved BF. The backside configuration is BOX/N/P.

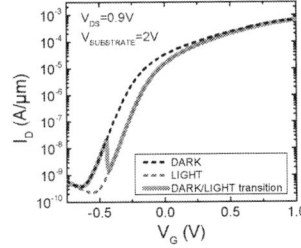

Fig.12: $I_D$-$V_G$ curve measured on a NFET transistor when light is turned ON during acquisition (from negative to positive $V_G$). $I_D$-$V_G$ curves with and without illumination correspond to the dashed lines as a guide for the eyes. The backside configuration is BOX/N/P.

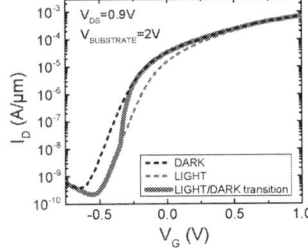

Fig.13: $I_D$-$V_G$ curve measured on a NFET transistor when light is turned OFF during acquisition (from negative to positive $V_G$). $I_D$-$V_G$ curves with and without illumination correspond to the dashed line as a guide for the eyes. The backside configuration is BOX/N/P.

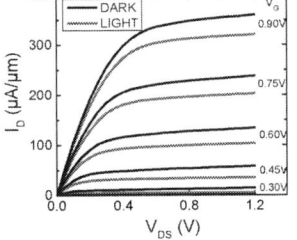

Fig.14: Experimental NFET $I_D$-$V_D$ curves recorded in the dark (black) or under illumination (red). The backside configuration is BOX/N/P and the substrate voltage is 0V.

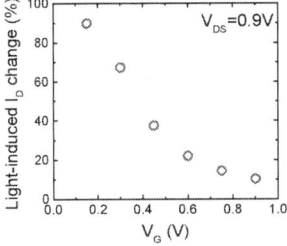

Fig.15: Experimental NFET light-induced drain current modulation at $V_{DS}$=0.9V for different gate voltages (derived from Fig.14).

**References**

[1] N. Planes *et al.*, VLSI 2012
[2] Press release, EETimes, Feb 2013
[3] L. Grenouillet *et al.*, IEDM 2012
[4] Q. Liu *et al.*, IEDM 2013
[5] M.A. Jaud *et al.*, SOICONF 2011

# Prototype of Multi-Stacked Memory Wafers Using Low-Temperature Oxide Bonding and Ultra-Fine-Dimension Copper Through-Silicon Via Interconnects

Wei Lin[1,*], Johnathan Faltermeier[1], Kevin Winstel[1], Spyridon Skordas[1], Troy Graves-Abe[2], Pooja Batra[2], Kenneth Herman2, John Golz[2], Toshiaki Kirihata[2], John Garant[2], Alex Hubbard[1], Kris Cauffman[1], Theodore Levine[1], James Kelly[1], Deepika Priyadarshini[1], Brown Peethala[1], Raghuveer Patlolla[1], Matthew Shoudy[1], James J Demarest[1], Jean Wynne[1], Donald Canaperi[1], Dale McHerron[1], Dan Berger[2], Subramanian Iyer[2,*]

IBM Corporation Systems and Technology Group, [1]Albany,NY, [2]Hopewell Junction, NY
*Correspondence should be addressed to S. Iyer at ssiyer@us.ibm.com and W. Lin at linw@us.ibm.com

*Abstract*—Reported for the first time is proof-of-concept multi-stacking of memory wafers based on low-temperature oxide wafer bonding using novel design and integration of two types of ultra-fine-dimension copper TSV interconnects. The combined via-middle (intra-via) and via-last (inter-via) strategy allows for the greatest degree of interconnectivity with the tightest allowable pitches and permits a highly integrated interconnect system across the stack. In combination with the successful metallization of the ultra-fine-dimension TSVs, the present work has shown the viability to extend the perceived TSV technology beyond the ITRS roadmap.

*Keywords—3D; DRAM; wafer stacking; TSV; oxide bonding*

## I. INTRODUCTION

3D IC integration (3DI) is one of the most promising semiconductor technologies beyond Moore's law for transistor/lithography scaling.[1] 3DI has the advantages in small form factor, high device density, and short interconnect path.[1,2] For example, 3D NAND by monolithic is being heavily invested due to its capability of multiplying the current device density by several times.[2] Implementation of 3DI technology into backside image sensor fabrication has shown a 100% fill factor of detector arrays.[3] Besides, 3DI also enables heterogeneous integration of different semiconductor materials and types of devices, which facilitates the realization of multi-functionality for future devices.[4]

3DI is the key to 3D DRAM stack for future memory-intensive computation. Fig. 1 illustrates a high-performance system based on 3DI technologies, where the two key elements are: a high-performance micro-processor (µP) chip stacked over a high-density cache memory, and the high-density low-power DRAM stacks supporting the µP.[5] Two early prototypes[6,7], were developed to demonstrate 3D cache feasibility and evaluate TSV processing[8] by stacking two embedded DRAM (eDRAM) chips. Compared with chip-to-chip 3DI, wafer-scale 3DI offers a high-volume low-cost production solution to achieve these benefits. In particular, low-temperature oxide wafer bonding renders wafer-scale 3DI a high throughput, high alignment accuracy, and a reliable (hermetic) bonding interface.[9,10] Furthermore, the associated bumpless interconnects enables a substantially high interconnect density for power distribution, signal transmission, and thermal dissipation. Prior work has demonstrated integration of two high-performance 45-nm SOI–CMOS embedded-DRAM wafers based on low-temperature oxide bonding to fit in the µP-on-3Dcache scheme in Fig. 1.[11] The present paper reports the integration and interconnection of four 300-mm wafers as a proof-of-concept technology for multi-stacked DRAMs to fit in the main memory stack scheme in Fig. 1. This technology is characteristic with multi-wafer stacking and two types of copper TSVs that interconnect the front-side and back-side wires of each stratum and the back-side wires between different strata through a combined via-middle (intra-via) and via-last (inter-via) strategy. The key enabling modules include high-quality low-temperature oxide wafer bonding, recent breakthrough in ultra-fine-dimension TSV metallization, and wafer thinning. To the authors' knowledge, this is the first report on multi-wafer stacking based on low-temperature oxide bonding with ultra-fine intra- and inter-strata copper TSV interconnects.

## II. INTEGRATION PROCESS

Fig. 2 illustrates the prototype structure of wafer stacking, including (from bottom up) a handle Si wafer, strata 1, 2, 3, and 4 wafers. Each stratum (x) starts with a full thickness and fabricated with the intra-vias ($A_x$, 0.25 µm in diameter) from the front side. After connecting the $A_x$ vias to the $F_x$ thin wires on the front side of the wafer, bonding dielectrics are deposited, planarized, and activated for low-temperature oxide bonding (flipped) to the lower wafer (handle or x-1).[12] The bonded wafer is thinned to <7 µm to reveal the $A_x$ via bottom from the backside. For x>1 strata, inter-vias ($E_x$, 1 µm in diameter) are fabricated from the backside using a process designed specifically to etch the remaining Si of the x strata and its complex FEOL/BEOL dielectric stack on the front side, the bonding dielectrics, and the interlayer dielectric stack on the back side of the x-1 strata. After backside wiring ($B_x$), $B_x$–$A_x$–$F_x$ and $B_{x-1}$–$E_x$–$B_x$–$A_x$–$E_x$ chains are formed for the intra- and inter-strata connection, respectively. The basic process flow for $A_x$ and $E_x$ via fabrications both include reactive ion etch, dielectric deposition for insulation, barrier/seed deposition, and electroplating of copper, although the materials and process windows for $A_x$ and $E_x$ vias are very different. The low resistivity of Cu is essential for TSVs of such dimensions.

978-1-4799-7440-5/14 $31.00 © 2014 IEEE

## III. HARDWARE RESULTS

Fig. 3 shows the cross-section SEM image of the four strata stacked on a handle wafer. The final thickness of each stacked stratum falls between 5 and 6.5 µm due to grinding endpoint variation and across-wafer thickness variation. Wafer bonding misalignment is consistently <1.25 µm (6σ), with bonding energy >1.6 J/m² as measured using the Maszara method.[12] Figs. 4 and 5 show the well-connected $B_1$–$A_1$–$F_1$ and $B_1$–$E_2$–$B_2$–$A_2$–$F_2$ chains, respectively, which play the most critical roles in the present multi-stacking interconnect technology. In-line test results confirm the robust connection of the TSV chains with reasonable resistances.

The unique and most challenging process module in the present technology is the metallization of the ultra-small intra-vias, which play a critical role in both the intra- and inter-strata interconnects. In Fig. 6, we compare the $A_x$ fill results from four different barrier/seed and plating processes, where process #3 has a poor barrier/seed coverage in the TSV, process #1 has a poor seed coverage at the TSV bottom, process #4 falls in the conformal plating regime, while process #2 has a good barrier/seed coverage and bottom-up plating waveform. The impact of $A_1$ metallization on the $B_1$–$A_1$–$F_1$ chain resistance is seen in Fig. 7. Process #2 achieves almost void-free fill of A1 vias and therefore dramatically reduced the resistance compared with process 1. The copper electroplating for TSVs of such a fine dimension, i.e., ~250 nm in diameter with > 25 aspect ratio, has never been reported before. The ITRS predicts minimum TSV diameter around 1 µm between 2009 and 2015, indicating difficulties in scaling to small TSV sizes needed for many applications. In this sense, the present TSV dimension and metallization success has extended the conventional TSV technology into a new era.

## IV. CONCLUSIONS

The concept of multi-stacking of memory wafers has been proved in the present study, based on the high quality of the low-temperature oxide wafer bonding and successful metallization of the 0.25-µm-diameter and 1-µm-diameter TSVs, the former interconnecting the front-side and back-side wires of each stratum, and the latter interconnecting the back-side wires between the different strata. This via system allows for the greatest degree of interconnectivity with the tightest allowable pitches and permits a highly integrated interconnect system across the stack. The present work opens a novel integration route toward high-volume production of high-density and low-power 3D DRAMs. Device readout from stacked fully-integrated memory wafers will be published in the near future.

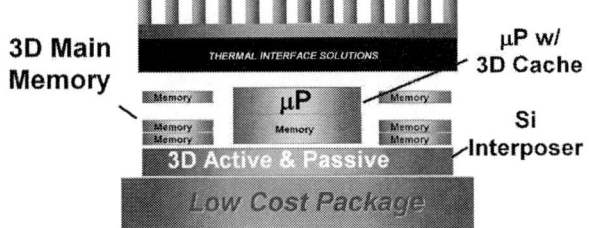

Fig. 1 A high-performance system using µP-on-3Dcache and 3D main memories.

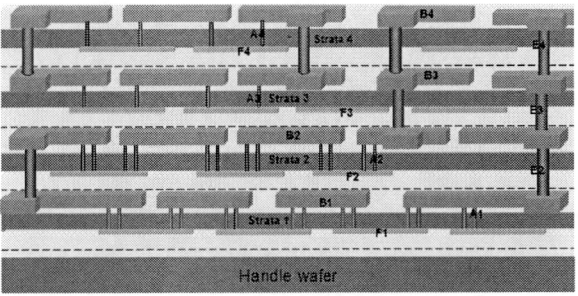

Fig. 2 Illustration of the prototype of a memory wafer stack.

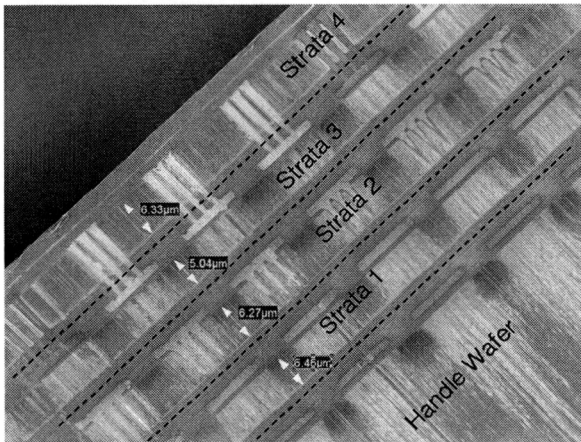

Fig. 3 Cross-section SEM image of the four wafers stacked on a handle Si wafer by oxide bonding and interconnection with copper TSVs. The dashed lines represent the bonding interfaces.

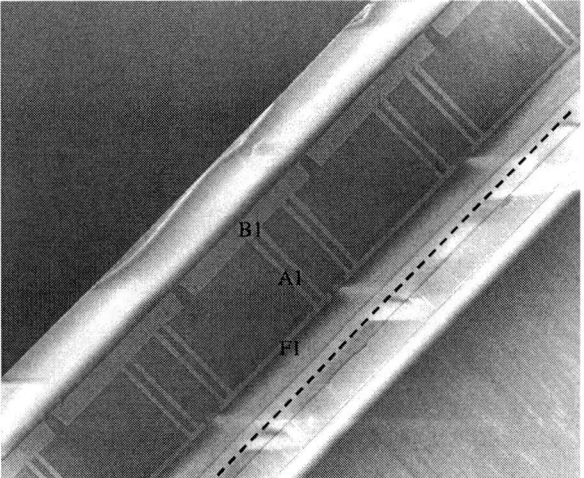

Fig. 4 Cross-section SEM image of the $B_1$–$A_1$–$F_1$ chain. The dashed line represents the bonding interface.

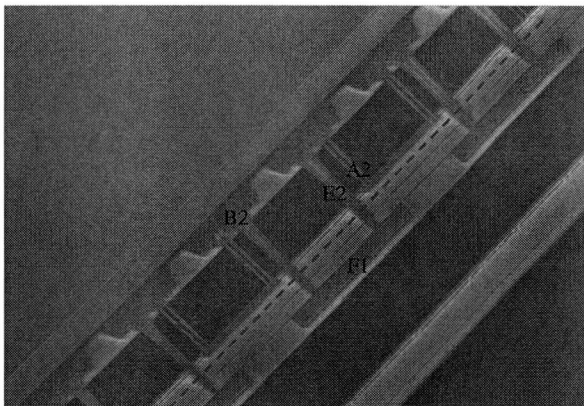

Fig. 5 Cross-section SEM image of the $B_1$–$E_2$–$B_2$–$A_2$–$F_2$ chain. The dashed line represents the bonding interface.

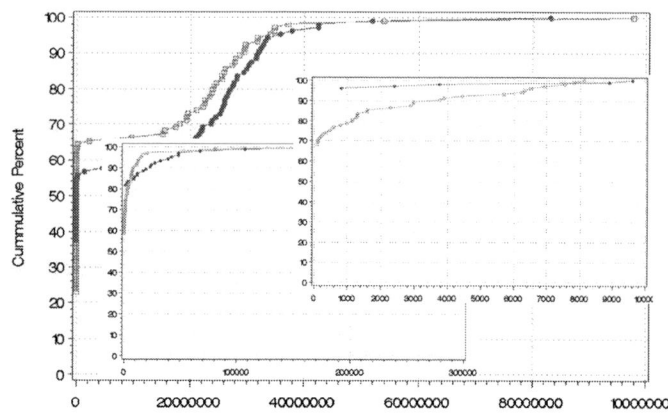

Fig. 7 Cumulative Rs result comparison based on the corresponding #1 (blue) and #2 (purple) processes in Fig. 6.

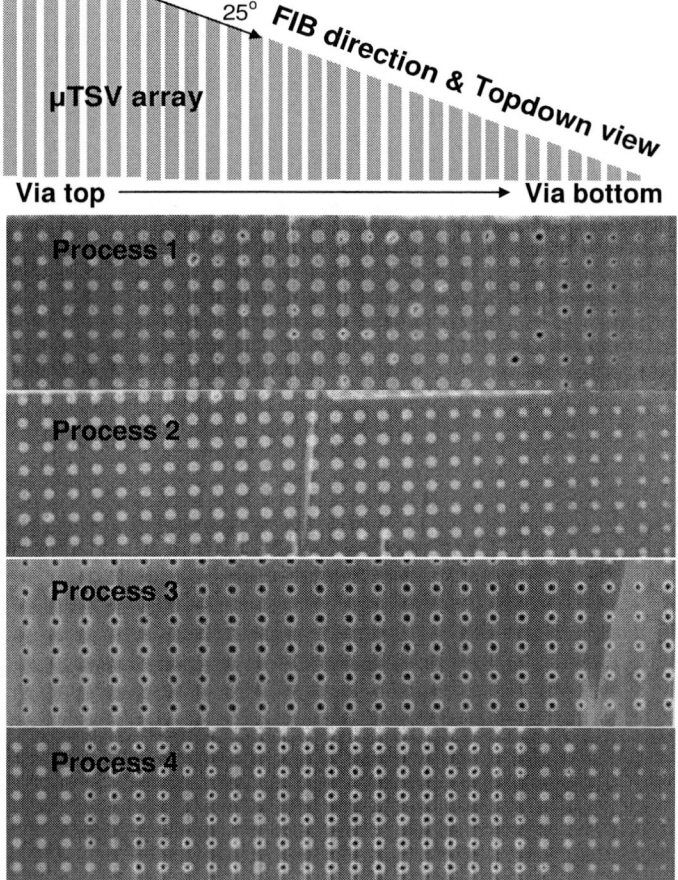

Fig. 6 Illustration of the FIB preparation of the T via array and comparison of the via fill based on four different barrier/seed and plating processes.

## References

[1] R. S. Patti, "Three-Dimensional Integrated Circuits and the Future of System-on-Chip Designs", Proc. IEEE, 2006, vol. 94, issue 6, pp. 11214-1224.

[2] D. C. Sekar, Z. Or-Bach, "Monolithic 3D-ICs with single crystal silicon layers", IEEE 3DIC, Osaka, Japan, Jan.-Feb. 2011.

[3] C. L, Chen, et al., "3D-enabled heterogeneous integrated circuits", IEEE S3S, Monterey, CA, Oct. 2013.

[4] C. K. Chen, et al., "Wafer-Scale 3D Integration of InGaAs Image Sensors with Si Readout Circuits", IEEE 3DIC, San Francisco, CA, Sept. 2009.

[5] T. Kirihata, "Three Dimensional Dynamic Random Access Memory", 4th Low-temperature Bonding Workshop, Japan, July 2014.

[6] J. Golz et. al., "3D Stackable 32nm High-K/Metal Gate SOI Embedded DRAM Prototype", Sump. on VLSI Circuits, 2011.

[7] M. Wordeman, et al. "A 3D System Prototype of an eDRAM Cache Stacked Over Processor-Like Logic Using Through-Silicon Vias", ISSCC Dig. Feb. 2012.

[8] M. Farooq, et. al, "3D copper TSV Integration,testing and reliability", IEDM, 2011.

[9] S. Skordas, et al, "Wafer-scale oxide fusion bonding and wafer thinning development for 3D systems integration: Oxide fusion wafer bonding and wafer thinning development for TSV-last integration", 3th Low-temperature Bonding Workshop, Japan, July 2012.

[10] W. Lin, et al, "Copper-to-Dielectric Heterogeneous Bonding for 3D Integration", 4th Low-temperature Bonding Workshop, Japan, July 2014.

[11] P. Batra, et al., "Three-Dimensional Wafer Stacking Using Cu TSV Integrated with 45 nm High Performance SOI-CMOS Embedded DRAM Technology", IEEE S3S, Monterey, CA, Oct. 2013.

[12] W. Lin, et al., "Low-temperature Oxide Wafer Bonding for 3D Integration: Chemistry of Bulk Oxide Matters", IEEE Trans. Semicon. Manufact., in press.

# A 262nW Analog Front End with a Digitally-Assisted Low Noise Amplifier for Batteryless EEG Acquisition

Pavan Bhargava[1], W. David Hairston[2], Robert M. Proie[1]

*U.S. Army Research Laboratory Sensors and Electron Devices Directorate[1], Human Research and Engineering Directorate[2]*

Email: pvnbhargava@berkeley.edu

*Abstract* – **This paper presents an analog front end (AFE) for a deploy-and-forget EEG monitoring system capable of indefinitely capturing and wirelessly transmitting a wide variety of neural measurements without battery power. This AFE design demonstrates a power consumption of 262nW per channel, lower than any other EEG acquisition system currently described in literature. In order to better utilize the smaller dynamic range of low power subthreshold ADCs, this architecture employs variable gain and a novel digital feedback loop to cancel electrode offset and other extraneous signals prior to amplification.**

## I. Introduction

The demand for mobility in electroencephalogram (EEG) systems has grown in recent years as new applications in neuroscience research and brain computer interface technologies require *in situ* measurements of brain activity [1]. While great strides have been made in the development of easy-application sensors and on-board signal processing [2], our goal of a true "deploy and forget" system requires a design capable of operation solely on locally harvested power (e.g. thermoelectric generators). However the development of long-term EEG measurement systems of this style poses multiple challenges to the design of the analog front end (AFE): namely, need for reduced power consumption while maintaining high-fidelity acquisition across varying environmental conditions and testing procedures [3]. Previous amplifier designs have shown some of the low power consumption [4], low noise [5], and electrode offset rejection [6] characteristics desired for mobile EEG measurement. However, to date these systems have been targeted towards short-term measurements or narrow applications.

Flexibility in EEG measurement is critical to developing a comprehensive research and detection platform. Although the ensemble EEG signal (between .1-100Hz) has a relatively wide bandwidth and large dynamic range, many specific neural measurements have narrower bands of interest. Traditional approaches fall into two categories - lower power architectures with pre-set bandpass analog signal processing [5], or AFEs with wide dynamic range and flexible digital feature extraction at the cost of higher power consumption.

This work's goal is to balance the flexibility of the latter within an ultra-low power solution. In order to do this, a subthreshold AFE with variable gain and digitally assisted low noise amplifier (LNA) has been developed.

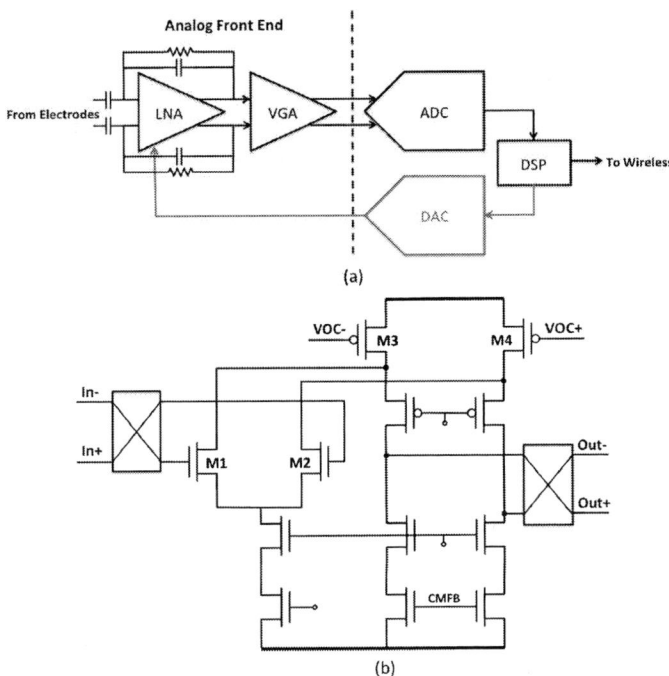

(a)

(b)

Figure 1 (a) Block diagram of EEG acquisition system. Digital feedback path is highlighted in red (b) Fully differential chopper-stabilized folded cascode amplifier with VOC through M3/M4

## II. Analog Front End Design

### A. Chopper-Stabilization Amplifier

Chopper-stabilization is an established technique used to isolate and remove 1/f noise in low-frequency amplifiers [4-6]. By modulating the input EEG signal to a frequency much higher than the 1/f noise corner prior to amplification, the noise can easily be suppressed via demodulation and low pass filtering. Figure 1(b) shows the implementation of the folded-cascode chopper amplifier. This design uses a structure similar to [6]; by placing blocking capacitors and feedback resistors prior to chopper modulation, the unregulated DC level of the scalp is effectively isolated from all of the on-chip components. In order to support research that requires very low frequency signals, this high-pass corner has been set to 0.2Hz via off-chip 500GΩ resistors. This design also employs a fully differential structure. This isolates power supply fluctuation from the energy-harvesting element from affecting the signal path.

978-1-4799-7440-5/14 $31.00 © 2014 IEEE

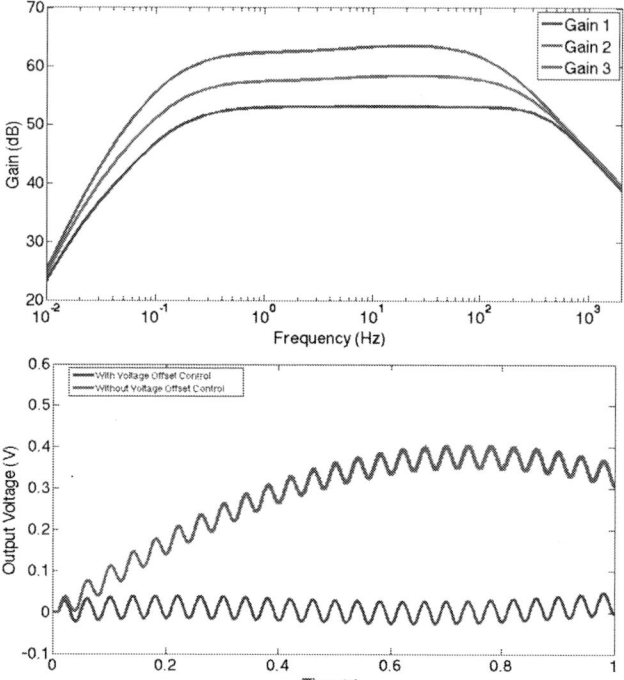

Figure 2 (a) Gain plots of LNA and VGA (b) Raw and offset cancelled outputs of AFE with 250μV .33Hz background signal

Table 1: Comparison with reported designs

| | Supply (V) | Power Consumption (μW) | Input Referred Noise (μVrms) | Input Impedance (MΩ) | Gain (dB) | CMRR (dB) | PSRR (dB) |
|---|---|---|---|---|---|---|---|
| This Work | 0.75 | 0.26 | 1.77 | >8 | 53-63 | 120 | 100 |
| [6] | 1 | 3.5 | 1.3 | >700 | 60 | 60 | - |
| [7] | 1.8-3.3 | 1.8-3.3 | .95 | >8 | 41, 51.5 | 80 | 100 |
| [5] | 1.2 | 6.4 | 0.46 | >4 | 40-74 | 80 | 60 |
| [4] | +-2.5 | 0.9 | 1.6 | - | 39.8 | >86 | >80 |

FFT. Depending on the type of measurement required, the extraneous spectral content is then converted back into the analog domain, and fed into the LNA as shown in Figure 1(a). The LNA receives the cancellation waveform at M3/M4 and acts opposite to main differential pair M1/M2, removing any offsets at the low-impedance node prior to amplification (see Figure 1(b)). As a result, brick-wall filtering of arbitrary frequency bands can be achieved with digital control of the cancellation waveform. By isolating frequency bands of interest for specific measurements, high gain can be used in the AFE without saturating the system with background signals.

## III. Results and Conclusions

The implemented front-end design shows an input-referred noise of 1.77μVrms, comparable to other EEG acquisition systems (Table 1), while using at least 3x less power. The use of a VGA in the front end allows for adjustable gain between 53-63dB (Figure 2(a)), while expanding the dynamic range of the system. The effect of the VOC is demonstrated in Figure 2(b). The application of a .33Hz signal of amplitude 250μV would normally consume a majority of the range of the subthreshold ADC after amplification. By adding a simulated cancellation waveform into the LNA, the offset is attenuated by 35dB, thereby enabling high gain without saturating the ADC. Given the software-controlled nature of the cancellation in this system, a wide range of EEG measurements can be taken with high resolution and greatly relaxed ADC requirements, enabling the use of ultra-low power ADCs.

## IV. References

[1] McDowell, K, et Al. "Real-World Neuroimaging Technologies," *Access, IEEE* , vol.1, no., pp.131,149, 2013

[2] Lance, B.J.; Kerick, S.E.; Ries, AJ.; Oie, K.S.; McDowell, K., "Brain–Computer Interface Technologies in the Coming Decades," *Proceedings of the IEEE* , vol.100, no.Special Centennial Issue, pp.1585,1599, May 13 2012

[3] Lun-De Liao, et Al., "Biosensor Technologies for Augmented Brain–Computer Interfaces in the Next Decades,"*Proceedings of the IEEE* , vol.100, no.Special Centennial Issue, pp.1553,1566, May 13 2012

[4] Harrison, R.R., "A low-power, low-noise CMOS amplifier for neural recording applications," *Circuits and Systems, 2002. ISCAS 2002. IEEE International Symposium on* , vol.5, no., pp.V-197,V-200 vol.5, 2002

[5] Fan Zhang; Mishra, A; Richardson, AG.; Otis, B., "A Low-Power ECoG/EEG Processing IC With Integrated Multiband Energy Extractor," *Circuits and Systems I: Regular Papers, IEEE Transactions on* , vol.58, no.9, pp.2069,2082, Sept. 2011

[6] Verma, N, et Al. "A Micro-Power EEG Acquisition SoC With Integrated Feature Extraction Processor for a Chronic Seizure Detection System," *Solid-State Circuits, IEEE Journal of* , vol.45, no.4, pp.804,816, April 2010

[7] Denison, T.; Consoer, K.; Santa, W.; Avestruz, A-T.; Cooley, J.; Kelly, A, "A 2 μW 100 nV/rtHz Chopper-Stabilized Instrumentation Amplifier for Chronic Measurement of Neural Field Potentials," *Solid-State Circuits, IEEE Journal of* , vol.42, no.12, pp.2934,2945, Dec. 2007

A large reduction in power was achieved by lowering the LNA bias current to 100nA. Since 1/f noise current is proportional to the bias current, this has the additional benefit of lowering the amount of generated 1/f noise. Due to the lowered 1/f noise, the required bandwidth of the chopper amplifier also decreases. The interplay between these factors enables 1/f noise cancellation to work more efficiently at ultra-low power levels. However, the lowered bias current results in a small $g_m$ for the main differential pair. This causes an increased thermal noise relative to higher power designs. Thermal noise reduction was achieved by using large W/L ratios in transistors M1/M2 for the high $g_m/I_d$ ratio in deep subthreshold operation. The rest of the transistors were biased in near-threshold mode with small W/L ratios to lower their $g_m/I_d$ ratio relative to M1/M2 [4]. Near-threshold biasing at low current values was achieved by using low-threshold voltage transistors available in the implemented 65nm process.

### B. Voltage Offset Control (VOC)

EEG electrode contact characteristics are not constant over time. As the electrode contact impedance changes and subject movement causes capacitive discharge, the voltage at the LNA input can drift over time. These effects are not uniform across all of the electrodes, and can result in large differential-mode input signals that saturate the AFE. This is especially a concern due to the low dynamic range available in the subthreshold ADC design.

We propose a digitally controlled filter to remove offset that could saturate the ADC prior to amplification while passing critical EEG data, without any added analog components. After the ADC, the signal is sent through a digital filter or

## A

| | |
|---|---|
| Aga, H. | 9a.1 |
| Agrawal, V.D. | 5.11 |
| Ahlbin, J.R. | 9b.2 |
| Akyel, K. | 7a.3 |
| Allain, F. | 3a.2 |
| Alles, M. | 9a.2 |
| Allibert, F. | 3a.1 |
| Allibert, F. | 3a.4 |
| Alper, C. | 8a.1 |
| Amir, M.F. | 7b.2 |
| Andreou, A.G. | 11.4 |
| Andrieu, F. | 3a.2 |
| Aoulaiche, M. | 6a.5 |
| Atias, L. | 9b.1 |
| Autran, J.-L. | 7.a2 |

## B

| | |
|---|---|
| Balakrishnan, K. | 3a.1 |
| Barba, M. | 8b.3 |
| Barral, V. | 5.6 |
| Barraud, S. | 3a.2 |
| Batra, P. | 12.2 |
| Batude, P. | 4.1 |
| Batude, P. | 6a.3 |
| Bawedin, M. | 5.8 |
| Bawedin, M. | 8a.2 |
| Beigné, E. | 7.a1 |
| Beinglass, I. | 11.3 |
| Belot, D. | 9a.3 |
| Berger, D. | 12.2 |
| Bernier, N. | 3a.2 |
| Berthelon, R. | 6a.2. |
| Bhargava, P. | 12.3 |
| Billoint, O. | 4.1 |
| Biswas, A. | 5.13 |
| Bol, D. | 5.10 |
| Bol, D. | 9b.3 |
| Boon, C.C. | 4.4 |
| Brito, R. | 8b.3 |
| Brunet, L. | 4.1 |
| Brunet, L. | 12.1 |
| Brunet, L. | 6a.3 |
| Bu, H. | 3a.3 |
| Buss, D. | 8b.2 |

## C

| | |
|---|---|
| Calhoun, B.H. | 3b.4 |
| Canaperi, D. | 12.2 |
| Cardoso, A.S. | 5.2 |
| Cassé, M. | 3a.2 |
| Cassé, M. | 6a.2. |
| Cassé, M. | 6a.3 |
| Cauffman, K. | 12.2 |
| Cauwenberghs, G. | 2b.1 |
| Chakraborty, P.S. | 5.2 |
| Chalupa, Z. | 12.1 |
| Chen, C.-J. | 7b.3 |
| Chen, S. | 6b.3 |
| Chen, S. | 6b.4 |
| Chen, Y.-N. | 7b.3 |
| Chen, Y.-Y. | 9a.4 |
| Cheng, K. | 3a.3 |
| Chertkow, O. | 9b.1 |
| Chiang, M.-H. | 5.4 |
| Choi, P. | 4.4 |
| Chuang, C.-T. | 7b.3 |
| Ciampolini, L. | 7a.3 |
| Ciampolini, L. | 7a.4 |
| Cibrario, G. | 4.1 |
| Claeys, C. | 6a.5 |
| Clerc, S. | 7.a2 |
| Clermidy, F. | 4.1 |
| Cochet, M. | 7.a2 |
| Coignus, J. | 12.1 |
| Cornell, A. | 10.3 |
| Cressler, J.D. | 5.2 |
| Cristoloveanu, S. | 5.8 |
| Cristoloveanu, S. | 8a.2 |
| Cristoloveanu, S. | 8a.3 |
| Cronquist, B. | 11.3 |
| Cros, A. | 6a.4 |

## D

| | |
|---|---|
| De Salvo, B. | 12.1 |
| de Souza, M. | 5.3 |
| de Streel, G. | 5.10 |
| De Vos, J. | 9b.3 |
| Delprat, D. | 3a.2 |
| Demarest, J.J. | 12.2 |
| Deprat, F. | 4.1 |
| DeSalvo, B. | 3a.4 |
| Deval, Y. | 9a.3 |
| Dhawan, V. | 5.5 |
| Di Federico, M. | 11.4 |
| Diab, A. | 5.9 |
| Doria, R.T. | 5.3 |
| Doris, B. | 3a.1 |
| Doris, B. | 3a.3 |
| Doris, B. | 3a.4 |

## E

| | |
|---|---|
| Ebrahimi, M.S. | 4.5 |
| Edelstone, M. | 1.3 |
| Esfeh, B.K. | 5.6 |

## F

| | |
|---|---|
| Faltermeier, J. | 12.2 |
| Fan, M.-L. | 7b.3 |
| Faynot, O. | 4.1 |
| Faynot, O. | 12.1 |
| Faynot, O. | 3a.2 |
| Fenouillet-Beranger, C. | 4.1 |
| Fenouillet-Beranger, C. | 6a.3 |
| Fenouillet-Beranger, C. | 8a.2 |
| Ferrari, P. | 8a.2 |
| Fish, A. | 9b.1 |
| Fitzgerald, E.A. | 4.4 |
| Flandre, D. | 5.6 |
| Flandre, D. | 6b.5 |
| Flandre, D. | 9b.3 |
| Flatresse, P. | 2a.2 |

Fonteneau, P.    8a.2
Franzon, P.D.    8b.4

## G

Gadfort, P.    8b.4
Gadfort, P.    9b.2
Gamiz, F.    8a.3
Garant, J.    12.2
Ghibaudo, G.    6a.3
Ghibaudo, G.    6a.4
Ghibaudo, G.    7a.3
Ghoneim, M.T.    5.9
Ghosh, D.    5.1
Glisner, T.    11.1
Glowacki, F.    3a.2
Glisner, T.    11.1
Glowacki, F.    3a.2
Golz, J.    12.2
Gosset, G.    6b.5
Graves-Abe, T.    12.2
Grenouillet, L.    12.1
Grosse, P.    12.1
Grossmann, P.J.    3b.2
Guo, D.    3a.1
Guo, D.    3a.3

## H

Haddad, P.-A.    6b.5
Hairston, W.D.    12.3
Haond, M.    5.6
Haond, M.    6a.1
Haond, M.    6a.3
Hartmann, J.M.    4.1
Hartmann, J.-M.    3a.2
Hashimoto, M.    3b.1
He, H.    3a.4
Henning, A.    11.3
Herman, K.    12.2
Hiienkari, M.    5.1
Hiienkari, M.    3b.3
Hills, G.    4.5

Hioki, M.    6b.2
Hioki, M.    9b.5
Hiya, Y.    5.7
Hook, T.    3a.3
Hook, T.B.    3a.1
Horstmann, M.    2a.1
Hoshikawa, N.    5.7
Hu, V. P.-H.    7b.3
Hubbard, A.    12.2
Hughes, H.    9a.2
Hussain, M.M.    5.9

## I

Ionescu, A.M.    5.13
Ionescu, A.M.    8a.1
Ionica, I.    5.8
Ishibashi, K.    2b.2
Ishibashi, K.    2b.3
Ishikawa, O.    9a.1
Ishizuka, T.    9a.1
Iyer, S.    12.2

## J

Jacob, A.    3a.3
Jaud, M.A.    12.1
Joshi, A.    9a.4
Josse, E.    6a.4
Jovanović, N.    7b.4
Julián, P.    11.4

## K

Kaltiokallio, M.    3b.3
Kamohara, S.    2b.2
Karaulac, N.    5.2
Kelly, J.    12.2
Kenney, T.    10.2
Khakifirooz, A.    3a.3
Khakifirooz, A.    3a.4
Khare, M.    3a.3
Kilchytska, V.    5.6

Kilchytska, V.    9b.3
Kim, G.T.    6a.4
Kinoshita, K.    5.7
Kirihata, T.    12.2
Klinefelter, A.    3b.4
Kohen, D.A.    4.4
Koike, H.    6b.2
Koike, H.    9b.5
Koskinen, L.    5.10
Koskinen, L.    3b.3
Koyama, M.    3a.2
Kranti, A.    5.1

## L

Lacord, J.    6a.3
Lafond, D.    3a.2
Lang, J.H.    8b.2
Lattanzio, L.    8a.1
Le Royer, C.    8a.2
Le Royer, C.    12.1
Lee, K.H.    4.4
Lee, T.-Y.    9a.4
Levine, T.    12.2
Levisse, A.    7b.4
Li, J.    3a.4
Liao, Y.-B.    5.4
Lim, S.K.    4.2
Lin, W.    12.2
Lin, X.    6b.3
Linder, P.    11.1
Liu, F.    5.8
Liu, Z.H.    4.4
Loubet, N.    3a.3
Loubet, N.    3a.4
Lu, D.    3a.3
Lu, V.    4.1

# M

| | |
|---|---|
| Ma, C. | 9b.5 |
| Ma, J. | 8b.2 |
| MacDonald, E. | 8b.3 |
| Madolesi, P.S. | 11.4 |
| Maeda, H. | 11.2 |
| Maffini-Alvaro, V. | 3a.2 |
| Mäkipää, J. | 3b.3 |
| Makosiej, A. | 7a.4 |
| Marquez, C. | 8a.3 |
| Martinie, S. | 6a.3 |
| Martino, J. A. | 6a.5 |
| Mathieu, B. | 6a.3 |
| Mavilla, N. | 3a.1 |
| Mazuré, C. | 3a.2 |
| Mazurier, J. | 12.1 |
| McHerron, D. | 12.2 |
| Michallet, J.-E. | 4.1 |
| Mitra, S. | 4.5 |
| Mitsuishi, H. | 11.2 |
| Morin, P. | 3a.3 |
| Morin, P. | 3a.4 |
| Mouis, M. | 6a.4 |
| Mukhopadhyay, S. | 7b.2 |
| Mukhopadhyay, S. | 8b.1 |

# N

| | |
|---|---|
| Nagatomi, H. | 2b.2 |
| Nakagawa, T. | 6b.2 |
| Nakagawa, T. | 9b.5 |
| Nakano, M. | 9a.1 |
| Navarro, C. | 8a.2 |
| Navarro, C. | 8a.3 |
| Nemir, D. | 8b.3 |
| Nguyen, B.-Y. | 3a.2 |
| Nguyen, P. | 3a.2 |
| Nier, O. | 6a.2. |
| Noto, N. | 9a.1 |
| Nowak, E. | 3a.1 |

# O

| | |
|---|---|
| Ogasahara, Y. | 6b.2 |
| Ogasahara, Y. | 9b.5 |
| Okada, M. | 11.2 |
| Okamoto, K. | 11.2 |
| Oldiges, P. | 3a.3 |
| Omprakash, A.P. | 5.2 |
| Or-Bach, Z. | 11.3 |

# P

| | |
|---|---|
| Padilla, J.L. | 8a.1 |
| Palacios, T. | 4.4 |
| Palakurthi, P. | 8b.3 |
| Palestri, P. | 8a.1 |
| Palkuti, L. | 9a.2 |
| Panth, S. | 4.2 |
| Pasini, L. | 4.1 |
| Pasini, L. | 6a.3 |
| Pathak, D. | 5.12 |
| Patlolla, R. | 12.2 |
| Pavanello, M.A. | 5.3 |
| Pedram, M. | 6b.3 |
| Pedram, M. | 6b.4 |
| Peethala, B. | 12.2 |
| Peh, L.S. | 4.4 |
| Pelloux-Prayer, B. | 7.a2 |
| Pescovsky, A. | 9b.1 |
| Philip Wong, H.-S. | 4.5 |
| Planes, N. | 5.6 |
| Planes, N. | 7a.4 |
| Ponthenier, F. | 4.1 |
| Portal, J.-M. | 7b.4 |
| Previtali, B. | 4.1 |
| Previtali, B. | 6a.3 |
| Priyadarshini, D. | 12.2 |
| Proie, R. M. | 12.3 |

# Q

# R

| | |
|---|---|
| Ramadass, Y. | 8b.2 |
| Rambal, N. | 4.1 |
| Rambal, N. | 6a.3 |
| Ranica, R. | 7a.4 |
| Rantala, A. | 3b.3 |
| Raskin, J.-P. | 5.6 |
| Raskin, J.-P. | 6b.5 |
| Richard, O. | 9a.3 |
| Rideau, D. | 6a.2. |
| Rim, K. | 3a.3 |
| Rivallin, P. | 12.1 |
| Rivallin, P. | 6a.3 |
| Roche, P. | 7.a2 |
| Rodriguez, N. | 8a.3 |
| Rouchon, D. | 3a.2 |
| Roy, K. | 6b.1 |
| Rozeau, O. | 4.1 |
| Rozeau, O. | 12.1 |

# S

| | |
|---|---|
| Sabry, M.M. | 4.5 |
| Saha, P. | 5.2 |
| Saligane, M. | 7.a2 |
| Samal, S. | 4.2 |
| Samson, M.P. | 4.1 |
| Samson, M.-P. | 3a.2 |
| Sarhan, H. | 4.1 |
| Sasaki, K.R.A. | 6a.5 |
| Savidis, I. | 15.12 |
| Schabel, J. | 8b.4 |
| Schwarzenbach, W. | 3a.4 |
| Sekar, D.C. | 4.3 |
| Sekigawa, T. | 6b.2 |
| Sekigawa, T. | 9b.5 |
| Shafaei, A. | 6b.4 |
| Shi, C.-J. R. | 5.5 |
| Shi, M. | 6a.4 |
| Shin, M. | 6a.4 |

| | | | | | | | |
|---|---|---|---|---|---|---|---|
| Shkel, A. | 10.4 | | **V** | | | **Z** | |
| Shoudy, M. | 12.2 | | | | | | |
| Shulaker, M. M. | 4.5 | Vallet, M. | 9a.3 | | Zaslavsky, A. | 8a.2 | |
| Shuto, Y. | 9b.4 | Vianello, E. | 7b.4 | | Zhang B. | 5.11 | |
| Simoen, E. | 6a.5 | Villalon, A. | 8a.2 | | Zhao S. | 8b.2 | |
| Sklenard, B. | 4.1 | Vinet, M. | 4.1 | | Zhou X. | 4.4 | |
| Skordas, S. | 12.2 | Vinet, M. | 12.1 | | | | |
| Solaro, Y. | 8a.2 | Vinet, M. | 3a.2 | | | | |
| Sopanen, M. | 3b.3 | Vinet, M. | 6a.3 | | | | |
| Southwick, R. | 3a.1 | Vitale, S.A. | 3b.2 | | | | |
| Strane, J. | 3a.1 | Vizioz, C. | 3a.2 | | | | |
| Su, P. | 7b.3 | | | | | | |
| Sugahara, S. | 9b.4 | | **W** | | | | |
| Sugaya, I. | 11.2 | | | | | | |
| Sugii, N. | 2b.2 | Wagenleitner, T. | 11.1 | | | | |
| Sugii, N. | 2b.3 | Wan, J. | 8a.2 | | | | |
| Sun, X. | 3a.1 | Wang, A. | 1.2 | | | | |
| Sylvester, D. | 7.a2 | Wang, A. | 5.5 | | | | |
| | | Wang, Y. | 6b.3 | | | | |
| **T** | | Wang, Y. | 6b.4 | | | | |
| | | Wei, H. | 4.5 | | | | |
| Tabone, C. | 4.1 | Whitefield, D. | 9a.4 | | | | |
| Tabone, C. | 12.1 | Wimplinger, M. | 11.1 | | | | |
| Tabone, C. | 3a.2 | Winstel, K. | 12.2 | | | | |
| Taheri, B. | 10.1 | Wu, T.F. | 4.5 | | | | |
| Tan, C.S. | 4.4 | Wurman, Z. | 11.3 | | | | |
| Teittinen, J. | 3b.3 | Wyatt, P.W. | 3b.2 | | | | |
| Terkaly, B. | 1.1 | Wynne, J. | 12.2 | | | | |
| Thien, H.M. | 2b.3 | | | | | | |
| Thomas, O. | 7a.3 | | **X** | | | | |
| Thomas, O. | 7a.4 | | | | | | |
| Thomas, O. | 7b.4 | Xie Q. | 6b.3 | | | | |
| Thuries, S. | 4.1 | Xu W. | 8b.4 | | | | |
| Tomita, H. | 5.7 | | | | | | |
| Torres Sevilla, G.A. | 5.9 | | **Y** | | | | |
| | | | | | | | |
| **U** | | Yablonovitch E. | 7b.1 | | | | |
| | | Yamamoto S. | 9b.4 | | | | |
| Uhrmann, T. | 11.1 | Yokokawa I. | 9a.1 | | | | |
| | | Yoon S.F. | 4.4 | | | | |
| | | Yu Y.S. | 4.2 | | | | |